Physics of Wave Turbulence

A century ago, Lewis Fry Richardson introduced the concept of energy cascades in turbulence. Since this conceptual breakthrough, turbulence has been studied in diverse systems and our knowledge has increased considerably through theoretical, numerical, experimental, and observational advances. Eddy turbulence and wave turbulence are the two regimes we can find in nature. So far, most attention has been devoted to the former regime, eddy turbulence, which is often observed in water. However, physicists are often interested in systems for which wave turbulence is relevant. This textbook deals with wave turbulence and systems composed of a sea of weak waves interacting nonlinearly. After a general introduction which includes a brief history of the field, the theory of wave turbulence is introduced rigorously for surface waves. The theory is then applied to examples in hydrodynamics, plasma physics, astrophysics, and cosmology, giving the reader a modern and interdisciplinary view of the subject.

Sébastien Galtier is a professor at the University of Paris-Saclay. His research focuses on fundamental aspects of turbulence with applications to space plasmas and cosmology. He has published over 100 refereed papers and a graduate text, *Introduction to Modern Magnetohydrodynamics* (Cambridge University Press, 2016). He is a senior fellow of the prestigious Institut Universitaire de France.

Physics of Wave Turbulence

Sébastien Galtier
University of Paris-Saclay

Shaftesbury Road, Cambridge CB2 8EA, United Kingdom

One Liberty Plaza, 20th Floor, New York, NY 10006, USA

477 Williamstown Road, Port Melbourne, VIC 3207, Australia

314–321, 3rd Floor, Plot 3, Splendor Forum, Jasola District Centre, New Delhi – 110025, India

103 Penang Road, #05–06/07, Visioncrest Commercial, Singapore 238467

Cambridge University Press is part of Cambridge University Press & Assessment, a department of the University of Cambridge.

We share the University's mission to contribute to society through the pursuit of education, learning and research at the highest international levels of excellence.

www.cambridge.org
Information on this title: www.cambridge.org/9781009275897
DOI: 10.1017/9781009275880

© Sébastien Galtier 2023

This publication is in copyright. Subject to statutory exception and to the provisions of relevant collective licensing agreements, no reproduction of any part may take place without the written permission of Cambridge University Press & Assessment.

First published 2023

This is an adapted translation of *Physique de la turbulence* published in 2021 by CNRS Éditions.

A catalogue record for this publication is available from the British Library.

A Cataloging-in-Publication data record for this book is available from the Library of Congress.

ISBN 978-1-009-27589-7 Hardback

Cambridge University Press & Assessment has no responsibility for the persistence or accuracy of URLs for external or third-party internet websites referred to in this publication and does not guarantee that any content on such websites is, or will remain, accurate or appropriate.

To Cécile

Contents

Preface		*page* xi
1	**General Introduction**	1
	1.1 Brief History	2
	1.2 Chaos and Unpredictability	12
	1.3 Transition to Turbulence	14
	1.4 Statistical Tools and Symmetries	17
	References	20
	Part I Fundamentals of Turbulence	25
2	**Eddy Turbulence in Hydrodynamics**	27
	2.1 Navier–Stokes Equations	27
	2.2 Turbulence and Heating	27
	2.3 Kármán–Howarth Equation	33
	2.4 Locality and Cascade	35
	2.5 Kolmogorov's Exact Law	37
	2.6 Phenomenology of Eddy Turbulence	39
	2.7 Inertial Dissipation and Singularities	40
	2.8 Intermittency	46
	2.9 Compressible Turbulence	57
	References	65
3	**Spectral Theory in Hydrodynamics**	69
	3.1 Kinematics	69
	3.2 Detailed Energy Conservation	71
	3.3 Statistical Theory	74

	3.4	Two-Dimensional Eddy Turbulence	82
	3.5	Dual Cascade	93
	3.6	Nonlinear Diffusion Model	93
	References		96
Exercises I			99
	I.1	1D HD Turbulence: Burgers' Equation	99
	I.2	Structure Function and Spectrum	99
	I.3	2D HD Turbulence: Detailed Conservation	100
	References		100

Part II Wave Turbulence — 101

4 Introduction — 103
    4.1 Brief History — 104
    4.2 Multiple Scale Method — 111
    4.3 Weakly Nonlinear Model — 115
    References — 122

5 Theory for Capillary Wave Turbulence — 127
    5.1 Introduction — 127
    5.2 Phenomenology — 130
    5.3 Analytical Theory: Fundamental Equation — 132
    5.4 Analytical Theory: Statistical Approach — 136
    5.5 Detailed Energy Conservation — 140
    5.6 Exact Solutions and Zakharov's Transformation — 141
    5.7 Nature of the Exact Solutions — 145
    5.8 Comparison with Experiments — 147
    5.9 Direct Numerical Simulation — 149
    References — 152

6 Inertial Wave Turbulence — 155
    6.1 Introduction — 155
    6.2 What Do We Know About Rotating Turbulence? — 157
    6.3 Helical Inertial Waves — 161
    6.4 Phenomenological Predictions — 162
    6.5 Inertial Wave Turbulence Theory — 164
    6.6 Local Triadic Interactions — 167

	6.7 Perspectives	174
	References	175
7	**Alfvén Wave Turbulence**	179
	7.1 Incompressible MHD	180
	7.2 Strong Alfvén Wave Turbulence	181
	7.3 Phenomenology of Wave Turbulence	185
	7.4 Theory of Alfvén Wave Turbulence	189
	7.5 Direct Numerical Simulation	196
	7.6 Application: The Solar Corona	197
	7.7 Perspectives	200
	References	201
8	**Wave Turbulence in a Compressible Plasma**	205
	8.1 Multiscale Solar Wind	206
	8.2 Exact Law in Compressible Hall MHD	209
	8.3 Weakly Compressible Electron MHD	212
	8.4 Kinetic Alfvén Waves (KAW)	216
	8.5 Spectral Phenomenology	216
	8.6 Theory of Weak KAW Turbulence	219
	8.7 Inertial/Kinetic-Alfvén Wave Turbulence: A Twin Problem	223
	8.8 Perspectives	226
	References	227
9	**Gravitational Wave Turbulence**	231
	9.1 Primordial Universe	231
	9.2 Weak Gravitational Wave Turbulence	234
	9.3 Strong Turbulence and Inflation	241
	9.4 Perspectives	242
	References	242
Exercises II		245
	II.1 MHD Model of Nonlinear Diffusion	245
	II.2 Four-Wave Interactions	246
	II.3 Gravitational Wave Turbulence: Exact Solutions	246
	II.4 Inertial Wave Turbulence: Domain of Locality	247
	References	247
Appendix A Solutions to the Exercises		249

I.1 1D HD Turbulence: Burgers' Equation 249
I.2 Structure Function and Spectrum 253
I.3 2D HD Turbulence: Detailed Conservation 255
II.1 MHD Model of Nonlinear Diffusion 259
II.2 Four-Wave Interactions 261
II.3 Gravitational Wave Turbulence: Exact Solutions 262
II.4 Inertial Wave Turbulence: Domain of Locality 265
References 269

Appendix B Formulary 271

Index 275

Preface

Ink, this darkness from which a light comes out
Victor Hugo, Dernière gerbe

Anyone who has ever flown in an aircraft knows how to define a turbulence zone: it is characterized by unpredictable, sometimes violent, often unpleasant jolts, which can even cause some anxiety in the passenger. For the physicist, on the other hand, turbulence is a pleasant, fascinating, and mysterious subject. This book proposes a journey into the world of turbulence in which we will gradually unveil the main fundamental laws governing the physics of turbulence where waves are omnipresent. We will see that since Reynolds' first historical experiment on liquids in 1883, turbulence has been studied in a wide variety of systems: from surface waves on the sea to gravitational waves, turbulence is now ubiquitous in physics.

Eddy turbulence and *wave turbulence* are the two regimes that we may encounter in nature. The attention of fluid mechanics being mainly focused on incompressible hydrodynamics, it is usually the first regime that is treated in books on turbulence. However, physicists are interested in much more diverse systems where waves are often present and for which the second regime (the subject of this book) is relevant. Wave turbulence offers the possibility of developing an analytical theory. Beyond its mathematical beauty, this spectral theory allows a deep understanding of weakly nonlinear systems and to develop a physical intuition on strong wave turbulence. Weak and strong wave turbulence are not independent of each other. On the contrary, one can emerge from the other during the cascade process; the two regimes can also coexist and be in permanent interaction. Without being exhaustive, this book offers a relatively broad overview on wave turbulence which should enable beginning researchers to acquire fundamental knowledge on subjects which are sometimes under development.

The theoretical framework chosen in this book will be that of statistically homogeneous turbulence for which a universal behavior is expected. In Chapter 1, a general introduction to turbulence is given where we find a brief history of the evolution of ideas, and the emergence of the main concepts and results. This history is of particular importance today, a century after Richardson (1922) introduced the concept of energy cascade. The fundamentals of turbulence are outlined in the physical (Chapter 2) and spectral (Chapter 3) spaces, which constitutes Part I. This first part focuses on incompressible hydrodynamics and thus on eddy turbulence. With Part II, we enter into the core of the book. Wave turbulence is introduced in Chapter 4, with a brief history and a presentation of the multiple scale method for weakly nonlinear systems. In Chapter 5, the theory of weak wave turbulence is presented in great detail for capillary waves, which is one of the simplest systems (three-wave interactions, two-dimensional, Navier–Stokes equations). Various examples dealing with three-wave interactions are discussed in Chapters 6, 7, and 8. In Chapter 9, we conclude with a new topic – gravitational wave turbulence – which is far more complex and involves four-wave interactions.

This book is based on a course on turbulence that I have been giving for several years at École polytechnique to students of the Master's degree in plasma physics (from the University of Paris-Saclay, Institut Polytechnique de Paris, and Sorbonne University). It is thus, in part, the result of fruitful interactions with my students, whom I would like to thank. I would also like to thank all my colleagues with whom I share my passion on this subject and who have contributed, in their own way, to the writing of this book; I would like to thank in particular Nahuel Andrés, Supratik Banerjee, Amitava Battacharjee, Éric Buchlin, Pierre-Philippe Cortet, Vincent David, Éric Falcon, Stephan Fauve, Özgür Gürcan, Lina Hadid, Romain Meyrand, Frédéric Moisy, Sergey Nazarenko, Alan Newell, Hélène Politano, Fouad Sahraoui, and, of course, Annick Pouquet, who introduced me to turbulence.

General Introduction

Turbulence is often defined as the chaotic state of a fluid. The example that immediately comes to mind is that of water: turbulence in water takes the form of eddies whose size, location, and orientation are constantly changing. Such a flow is characterized by a very disordered behavior difficult to predict and by the existence of multiple spatial and temporal scales. There are many experiments of everyday life where the presence of turbulence can be verified: the agitated motions of a river downstream of an obstacle, those of smoke escaping from a chimney, or the turbulence zones that one sometimes crosses in an airplane.

Experiencing turbulence at our scale seems easy since it is not necessary to use powerful microscopes or telescopes. A detailed analytical understanding of turbulence remains, however, limited because of the intrinsic difficulty of nonlinear physics. As a result, we often read that turbulence is one of the last great unresolved problems of classical physics. This long-held message, found, for example, in Feynman et al. (1964), no longer corresponds to the modern vision. Indeed, even if turbulence remains a very active research topic, we have to date many theoretical, numerical, experimental, and observational results that allow us to understand in detail a part of the physics of turbulence.

This book deals mainly with wave turbulence. However, wave turbulence is not totally disconnected from eddy turbulence, from which the main concepts have been borrowed (e.g. inertial range, cascade, two-point correlation function, spectral approach). Moreover, very often, wave turbulence and eddy turbulence can coexist as in rotating hydrodynamics. This is why a broad introduction to eddy turbulence is given (Part I) before moving on to wave turbulence (Part II), giving this book, for the first time, a unified view on turbulence. We will see that many results have been obtained since the first steps taken by Richardson (1922), a century ago. The many examples discussed in this book reveal that the classical presentation of turbulence, based on the Navier–Stokes equations (Frisch, 1995; Pope, 2000), is somewhat too simplistic because turbulence is found in various environments, in various forms. If we restrict ourselves to the standard example of incompressible hydrodynamics, the simple introduction of a uniform rotation for describing geophysical fluids drastically changes the physics of turbulence by adding anisotropy. In astrophysics, 99 percent of the visible matter of the Universe is in the form of plasma, which is generally very turbulent, but plasma turbulence mixes waves and eddies. The regime of wave turbulence, described in Part II, can emerge from a vibrating steel plate; here, we are far from the classical image of eddies in water. Finally, recent studies reveal that the cosmological inflation that followed the Big Bang could have its origin in strong gravitational wave turbulence.

The objective of Part I, which follows this first chapter, is to present the fundamentals of turbulence. We will start with eddy turbulence, where the first concepts and laws have emerged. We will limit ourselves to the most important physical laws. The theoretical framework will be that of a statistically homogeneous turbulence for which a universal behavior is expected. The problems of inhomogeneity inherent to laboratory experiments will therefore not be dealt with. Through the examples discussed, we will gradually reveal the state of knowledge in turbulence. To help us in this task, we begin with a brief historical presentation.

1.1 Brief History

1.1.1 First Cognitive Advances

Leonardo da Vinci was probably the first to introduce the word *turbulence* (*turbulenza*) at the beginning of the sixteenth century to describe the tumultuous movements of water. However, the word was not commonly used by scientists until much later.[1]

The first notable scientific breakthrough in the field of turbulence can be attributed to Reynolds (1883): he showed experimentally that the transition between the laminar and turbulent regimes was linked to a dimensionless number – the Reynolds number.[2] The experiment, which can be easily reproduced in a laboratory, consists of introducing a colored stream of the same liquid as circulating in a straight transparent tube (see Figure 1.1). It can be shown that the transition to turbulence occurs when the Reynolds number becomes greater than a critical value. An important step in this discovery is the observation that the tendency to form eddies increases with the temperature of the water, and Reynolds knew that in this case the viscosity decreases. He also showed the important role played by the development of instabilities in this transition to turbulence.

World War I was a time of further important advances. The war efforts in Germany and, in particular, under the influence of Prandt in Göttingen, directed

---

[1] For example, the book of Boussinesq (1897) still bears the evocative title: "Theory of the Swirling and Tumultuous Flow of Liquids in Straight Beds with a Large Section."

[2] The Reynolds number measures the ratio between the inertial force and the viscous force. We will come back to this definition in Section 1.3.

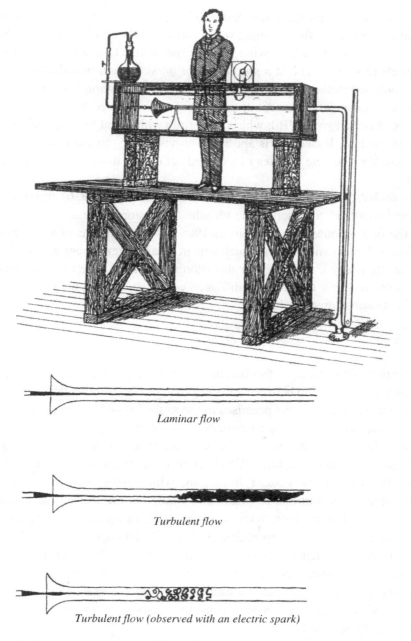

Figure 1.1 Historical experiment of Reynolds (1883) (top) and his observations (bottom). The original device is kept at the University of Manchester.

the research in the field of aerodynamics to the study of the fall of bombs in air or water. It is a question here of studying, for example, the drag of a sphere;

this work was then used for the design of airplanes. After the war, research in turbulence increased: for example, we can mention the results on the inhomogeneous effects due to walls in wind tunnel experiments (Burgers, 1925). But it is with Richardson (1922) that a second major breakthrough in turbulence arrives: in his book on weather predictions and numerical calculation.[3] Richardson introduced the fundamental concept of energy cascade. Inspired by the Irish writer J. Swift, Richardson wrote "Big whirls have little whirls that feed on their velocity. Little whirls have lesser whirls and so on to viscosity – in the molecular sense" (page 66). We find here the idea of a cascade of eddies from large to small spatial scales.

It is probably with this idea in mind that Richardson (1926) formulated the empirical 4/3 law[4] to describe the turbulent diffusion process. This law differs from the one proposed by Einstein in 1905 on the diffusion of small particles in a liquid (Brownian motion), which was in clear disagreement with turbulence experiments where a much higher diffusion was found.[5] The proposed new law is characterized by a nonconstant diffusion coefficient D_ℓ, which depends on the scale being considered, such that:

$$D_\ell \sim \ell^{4/3}. \tag{1.1}$$

This relationship reflects the fact that in a turbulent liquid the diffusivity increases with the mean separation between pairs of particles. This scaling law is fundamental because we find there the premises of the exact four-fifths law of Kolmogorov (1941a), with which it is in agreement dimensionally.

It was during this interwar period that the first works based on two-point correlations emerged (Taylor, 1935),[6] as well as works on the spectral analysis of fluctuations by Fourier transform, which have become the basis of modern research in turbulence (Motzfeld, 1938; Taylor, 1938). The correlation approach leads, in particular, to the Kármán–Howarth equation (von Kármán and Howarth, 1938) for an incompressible, statistically homogeneous, and isotropic[7] hydrodynamic turbulence. This equation describes the fluid dynamics through correlators – two-point measurements in physical space. As we will see in Chapter 2, this result is central for the establishment of the exact four-fifths law of Kolmogorov (1941a), which is not a dynamic equation but a statistical solution of Navier–Stokes equations.

---

[3] "Numerical calculation" here means calculation carried out by hand with a method essentially based on finite differences.
[4] This empirical law should not be confused with the exact four-thirds law which deals with structure functions (see Chapter 2).
[5] It is known that a cloud of milk dilutes more rapidly in tea if stirred with a spoon.
[6] It is the British Francis Galton (1822–1911) who seems to have been the first to correctly introduce the concept of correlation for statistical studies in biology.
[7] This is the strong isotropy that is considered here, which we will return to in Section 1.4.

1.1.2 Kolmogorov's Law and Intermittency

In the 1930s and under the leadership of the mathematician Kolmogorov, the Soviet school became very active in turbulence. At that time, Kolmogorov was working on stochastic processes and random functions. It was therefore natural that he turned his attention to turbulence, where a pool of data was available. Based on some of the work described in the Section 1.1.1, Kolmogorov and his student Obukhov set out to develop a theory for the standard case of incompressible, statistically homogeneous, and isotropic hydrodynamic turbulence. Based, in particular, on the Kármán–Howarth equation, Kolmogorov (1941a,b) established the first exact statistical law of turbulence – known as the four-fifths law – which relates a third-order structure function involving the difference of the component in direction ℓ of the velocity between two points separated by the vector $\boldsymbol{\ell}$, the distance ℓ, and the mean rate of dissipation of kinetic energy ε ($\langle\rangle$ means the ensemble average):[8]

$$-\frac{4}{5}\varepsilon\ell = \langle [u_\ell(\mathbf{x}+\boldsymbol{\ell}) - u_\ell(\mathbf{x})]^3 \rangle. \tag{1.2}$$

To establish this universal law, Kolmogorov assumes that fully developed turbulence becomes isotropic on a sufficiently small scale, regardless of the nature of the mean flow. He also assumes that ε becomes independent of viscosity within the limits of large Reynolds numbers (i.e. low viscosity); this is what is often referred to today as the zeroth law of turbulence. After several years of research, a first exact law was established for which it was possible to get rid of the nonlinear closure problem. The trick used to achieve this was to relate the cubic nonlinear term to the mean energy dissipation in the inertial range, that is, in a limited range of scales between the larger scales where inhomogeneous effects can be felt, and the smaller scales where viscosity efficiently damps the fluctuations. We will return at length to the law (1.2) in Chapter 2. Kolmogorov's law remained unnoticed for several years (outside the USSR). It was Batchelor (1946) who was the first to discover the existence of Kolmogorov's articles:[9] he immediately realized the importance of this work, which he shared with the scientific community at the Sixth International Congress of Applied Mathematics held in Paris in 1946 (Davidson et al., 2011).

For his part, independently of Kolmogorov but inspired by the ideas of Richardson (1922), Taylor (1938), and the work by Millionschikov (1939, 1941), who was another student of Kolmogorov, Obukhov (1941b) proposed a nonexact spectral theory of turbulence based on the relationship:

---

[8] Kolmogorov was probably the first to be interested in structure functions that are constructed from the differences and not from the products of a field (here the velocity field), as was the case with the Kármán–Howarth equation.

[9] The English version of the Russian papers had been received in the library of the Cambridge Philosophical Society.

$$\frac{\partial E}{\partial t} + D = T, \qquad (1.3)$$

with E the energy spectrum, D the viscous dissipation, and T the energy transfer (in Fourier space). The artificial closure proposed is based on an average over small scales. He obtained as a solution the energy spectrum:[10]

$$E(k) \sim k^{-5/3}, \qquad (1.4)$$

which is dimensionally compatible with Kolmogorov's exact law. In extending this study, Obukhov was then able to provide a theoretical justification for Richardson's (1926) empirical 4/3 law of diffusion. Later, Yaglom (1949) obtained a new exact law, applied this time to the passive scalar: this model describes how a scalar evolves, for example the temperature or the concentration of a product, in a turbulent fluid for which the velocity fluctuations are given.

For a short period of time Kolmogorov thought that the mean rate of energy dissipation was the key to establishing a more general exact law describing the statistics at any order in terms of a velocity structure function. This general law would have provided a complete statistical solution to the problem of hydrodynamic turbulence. But in 1944, Landau[11] pointed out the weakness of the demonstration (proposed by Kolmogorov during a seminar), which we will come back to in Chapter 2: it does not take into account the possible local fluctuations of ε, a property called intermittency. It took about 20 years for Kolmogorov (1962) and Oboukhov (1962) to propose, in response to Landau, a model (and not an exact law) of intermittency based on a log-normal statistics which incorporates the exact four-fifths law as a special case. Kolmogorov's answer was given (in French) at a conference held in Marseilles in 1961 to celebrate the opening of the Institut de Mécanique Statistique de la Turbulence. This conference became famous because it brought together for the first time all the major specialists (American, European, and Soviet) on the subject. It was also during this conference that the first energy spectrum in $k^{-5/3}$ measured at sea was announced (Grant et al., 1962).

Basically, the notion of intermittency is related to the concentration of dissipation in localized structures of vorticity. As mentioned by Kolmogorov, intermittency may slightly modify the $-5/3$ exponent of the energy spectrum, but its most important contribution is expected for statistical quantities of higher orders (the exact law is of course not affected). This new formulation is at the origin of work, in particular, on the concept of fractal dimension as a model of intermittency (Mandelbrot, 1974; Frisch et al., 1978) – see Chapter 2. It is interesting to note that we already find the concept of fractional dimension in Richardson's (1922) book, where the study of geographical boundaries is discussed.

---

[10] In general, this solution is called the Kolmogorov spectrum, but it would be more accurate to call it the Kolmogorov–Obukhov spectrum. This spectrum was also obtained independently by other researchers, such as Onsager (1945) and Heisenberg (1948).
[11] Landau's remark (Landau and Lifshitz, 1987) can be found in the original 1944 book (Davidson et al., 2011).

1.1.3 Spectral Theory and Closure

In this postwar period, the theoretical foundations of turbulence began to be established. The first book exclusively dedicated to this subject is that of Batchelor (1953), which still remains a standard reference on the subject: it deals with statistically homogeneous turbulence. From the 1950s, a major objective seemed to be within the reach of theorists: developing a theory for homogeneous and isotropic turbulence in order to rigorously obtain the energy spectrum. The work of Millionschikov (1941) (see also Chandrasekhar, 1955) based on the quasi-normal approximation (QN) had opened the way: this approximation – a closure – assumes that moments of order four and two are related as in the case of a normal (Gaussian) law without making this approximation for moments of order three (which would then be zero, making the problem trivial). Kraichnan (1957) was the first to point out that this closure was inconsistent because it violated some statistical inequalities (realizability conditions), and Ogura (1963) demonstrated numerically that this closure could lead to a negative energy spectrum for some wavenumbers.

In this quest, Kraichnan (1958, 1959) proposed a sophisticated theory which does not have the defects we have just mentioned: it is the direct interaction approximation (DIA), which is based on field theory methods, a domain in which Kraichnan was originally trained.[12] The fundamental idea of this approach is that a fluid perturbed over a wavenumber interval will have its perturbation spread over a large number of modes. Within the limit $L \to +\infty$, with L being the side of the cube in which the fluid is confined, this interval becomes infinite in size, which suggests that the mode coupling becomes infinitely weak. The response to the perturbation can then be treated in a systematic way. Under certain assumptions, two integro-differential equations are obtained for the correlation functions in two points of space and two of time, and the response function. The inferred prediction for the energy spectrum, in $k^{-3/2}$, is, however, not in dimensional agreement with Kolmogorov's theory, nor with the main spectral measurements. Improvements were then made (Lagrangian approach) to solve some problems (noninvariance by random Galilean transformation, Kolmogorov spectrum) (Kraichnan, 1966): this new theory can be seen as the most sophisticated closure model.[13] This work has led, in particular, to the development of the EDQNM (eddy-damped quasi-normal Markovian) closure model (Orszag, 1970), still widely used today, to which we will return in Chapter 3.

---

[12] Kraichnan became interested in turbulence in the early 1950s while he was Einstein's postdoctoral fellow. Together, they searched for nonlinear solutions to the unified field equations.

[13] In (strong) eddy turbulence, no exact spectral theory with an analytical closure has been found to date. This contrasts with the (weak) wave turbulence regime, for which an asymptotic closure is possible (see Chapter 4).

1.1.4 Inverse Cascade

Two-dimensional hydrodynamic (eddy) turbulence is the first example where an inverse cascade was suspected. The motivation for the study of such a system may seem on the face of it surprising, but several works showed that a two-dimensional approach could account for the atmospheric dynamics quite satisfactorily (Rossby and collaborators, 1939). We now know that the rotation, or stratification, of the Earth's atmosphere tends to confine its nonlinear dynamics to horizontal planes.[14] The first work on two-dimensional hydrodynamic turbulence dates back to the 1950s with, for example, Lee (1951), who demonstrated that a direct energy cascade would violate the conservation of enstrophy (proportional to vorticity squared), which is the second inviscid invariant (i.e. at zero viscosity) of the equations. Batchelor (1953) had also noted at the end of his book that the existence of this second invariant should contribute to the emergence, by aggregation, of larger and larger eddies. He concluded by asserting the very great difference between two- and three-dimensional turbulence. By using the two inviscid invariants, energy and enstrophy, Fjørtoft (1953) was able on his part to demonstrate, in particular with dimensional arguments, that the energy should cascade preferentially towards large scales.

It is in this context, clearly in favor of an inverse energy cascade, that Kraichnan became interested in two-dimensional turbulence. Using an analytical development of Navier–Stokes equations in Fourier space, the use of symmetries, and under certain hypotheses such as the scale invariance of triple moments, Kraichnan (1967) rigorously demonstrated the existence of a dual cascade – that is, in two different directions – of energy and enstrophy (see Chapter 3). This prediction is in agreement with previous analyses and the existence of a direct cascade of enstrophy and an inverse cascade of energy for which the proposed (nonexact) spectrum is in $k^{-5/3}$.

The existence in the same system of two different cascades was quite new in eddy turbulence. This prediction has since been accurately verified both experimentally and numerically (Leith, 1968; Pouquet et al., 1975; Paret and Tabeling, 1997; Chertkov et al., 2007). The second-best-known system where an inverse cascade exists is that of magnetohydrodynamics (MHD): using some arguments from Kraichnan (1967), Frisch et al. (1975) deduced in the three-dimensional case the possible existence of an inverse cascade of magnetic helicity, a quantity which plays a major role in the dynamo process in astrophysics (Galtier, 2016). To date, we know several examples of turbulent systems producing an inverse cascade (see, e.g., the review of Pouquet et al., 2019).

---

[14] Chapter 6 is devoted to inertial wave turbulence (i.e. incompressible hydrodynamic turbulence under a uniform and rapid rotation), for which it can be rigorously demonstrated that the cascade is essentially reduced to the direction transverse to the axis of rotation. However, it can be shown in this case that the energy cascade is direct.

Kraichnan's (1967) discovery was made at a period when the theory of wave turbulence, the regime that is the main subject of this book, was beginning to produce important results. The brief history presented in Chapter 4 allows us to appreciate the evolution of ideas on this subject, which finds a large part of its foundations in eddy turbulence (spectral approach, inertial range, cascade, closure problem). In this context, a problem that attracted a lot of attention was that of gravity wave turbulence (which is an example of surface waves). This problem deals with four-wave resonant interactions: in this case, there are two inviscid invariants, energy and wave action. The first is characterized by a direct cascade and the second by an inverse cascade. The study carried out[15] by Zakharov and Filonenko (1966) (see also Zakharov and Filonenko, 1967) focused only on the energy spectrum. The authors obtained the exact solution as a power law associated with energy, but curiously they did not focus on the second solution and therefore did not immediately realize that it corresponded to a new type of cascade. Starting from a similar study (involving four-wave resonant interactions) on Langmuir wave turbulence by Zakharov (1967), in which the energy spectrum had also been obtained, Kaner and Yakovenko (1970) found the second exact solution corresponding to an inverse cascade of wave action. It is thus in the field of plasmas that the existence of a dual cascade was finally demonstrated in wave turbulence.[16]

A major difference between the two turbulence regimes is that, unlike (strong) eddy turbulence, (weak) wave turbulence theory is analytical (see Chapter 4). In this case, one can develop a uniform asymptotic theory and obtain the dynamic equations of the system and then, if they exist, its exact spectral solutions. It is then possible to provide analytical proof of the type of cascade (direct or inverse). It is also possible to prove the local character of turbulence (by a study of the convergence of integrals) and thus be in agreement with one of Kolmogorov's fundamental hypotheses. For this reason, exact nontrivial solutions of wave turbulence are called Kolmogorov–Zakharov spectra. There are several examples in wave turbulence where there is an inverse cascade of wave action; in Chapter 9 we present the case of gravitational wave turbulence (Galtier and Nazarenko, 2017). It is less common to obtain an inverse cascade in the case of three-wave resonant interactions. An example is given by rotating magnetohydrodynamic turbulence: the energy cascades directly and the hybrid helicity (a modified magnetic helicity) cascades inversely (Galtier, 2014).

To conclude this section, let us note that Robert Kraichnan and Vladimir Zakharov received the Dirac medal in 2003 for their contributions to the theory of turbulence, particularly the exact results and the predictions of inverse

---

[15] Many other studies have been devoted to gravity wave turbulence. Chapter 4 discusses some of them.
[16] The second exact solution corresponding to an inverse cascade of wave action for gravity waves was published by Zaslavskii and Zakharov (1982).

cascade, and for identifying classes of turbulence problems for which in-depth understanding has been achieved.

1.1.5 Emergence of Direct Numerical Simulation

From the 1970s, a new method for analyzing turbulence emerged: direct numerical simulation (Patterson and Orszag, 1971; Fox and Lilly, 1972). By direct, we mean the simulation of the fluid equations themselves and not a model of these equations. We have already cited as a model the EDQNM approximation used in hydrodynamics (Orszag, 1970); there is also the case of magnetohydrodynamics with the study of the inverse cascade of magnetic helicity (Pouquet et al., 1976). There are other models such as nonlinear diffusion models (Leith, 1967) or shell models (Biferale, 2003) – which we will briefly discuss in Chapter 3.

Since its beginnings, direct numerical simulation has made steady progress. It currently represents a means of studying turbulence in great detail; it is also an indispensable complement to experimental studies. It is impossible to summarize in a few lines the numerous results obtained in the field of numerical simulation. Let us simply point out that the regular increase in spatial resolution makes it possible to increase the Reynolds number and to describe increasingly fine structures (see Figure 1.2). It is interesting to compare the current situation with the first direct numerical simulations of incompressible three-dimensional hydrodynamic turbulence. For example, Orszag and Patterson (1972) used a spatial resolution of 64^3 and, as explained by the authors, each time step then required a computation time of 30 seconds! It is also interesting to note that the diffusion of knowledge takes some time: for example, the first direct numerical simulation of incompressible three-dimensional magnetohydrodynamic turbulence was realized by Pouquet and Patterson (1978) with a spatial resolution of 32^3. Nowadays, a standard direct numerical simulation of turbulence is generally performed with a pseudospectral code, in a periodic box and with a spatial resolution of about 2048^3 – the highest to date being $16\,384^3$ (Iyer et al., 2019). For more information on the subject, the reader can consult the review article of Alexakis and Biferale (2018), where numerous examples of direct numerical simulation are presented in the context of various turbulence studies.

1.1.6 Turbulence Today

In the history of sciences on turbulence, the early 1970s were a turning point. Very schematically, we can consider that the theory of turbulence was built during the years 1922–1972, a period during which the main concepts were introduced, allowing the first exact results to be obtained.[17] The books of Monin and

---

[17] The year 1922 can be used as a reference since it is this year that Richardson introduced the fundamental concept of energy cascade.

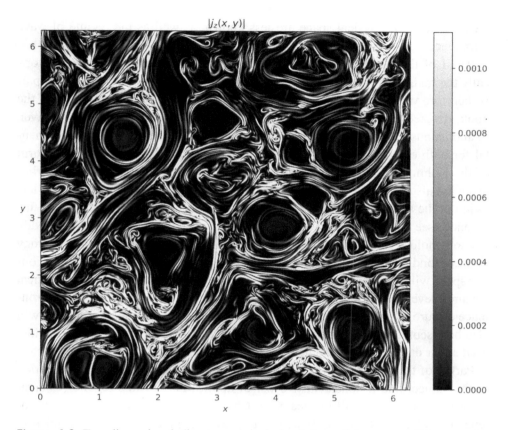

Figure 1.2 Two-dimensional direct numerical simulation of incompressible magneto-hydrodynamic turbulence (see Chapter 7). The image, with a spatial resolution of 2048×2048, shows the norm of the out-of-plane component of the electric current. The white regions correspond to the sites of energy dissipation.

Yaglom (1971, 1975) summarize the situation well. After this period, which can be described as exploration, the years 1972–2022 are rather a period of exploitation during which the results of incompressible hydrodynamics were generalized to other systems, often much more complex. However, it would be simplistic to limit this second period to a simple exploitation, because new concepts have also emerged and our knowledge has been considerably refined thanks, in particular, to numerous experiments and direct numerical simulations.

Today, the physics of turbulence appears in many fields (physics, geophysics, astrophysics, cosmology, aeronautics, biology) and it is impossible to draw up an exhaustive list of its applications. Given the difficulty of the subject, the use of simple – even simplistic – models of turbulence is quite common. The best-known result is probably the Kolmogorov energy spectrum. While there is no reason to think that this form of spectrum appears in other turbulence problems, it is often mentioned or even used. On the other hand, the exact laws based on two-point measurements in physical space are less known, as well as the regime of wave turbulence.

Part I of this book is a short introduction to eddy turbulence, where historically the main concepts and results of turbulence have been developed. This part is therefore very important to appreciate Part II, the second and main part of the book, devoted to wave turbulence. In Chapter 2, special attention will be given to anomalous dissipation and the zeroth law of turbulence, which are fundamental for both eddy and wave turbulence. It is then used to derive a modern form of the Kolmogorov exact law. We note in passing that the exact laws of turbulence are also valid for wave turbulence: for example, the Kolmogorov exact law (without the assumption of statistical isotropy) is also valid for inertial wave turbulence (see Chapter 6). In Chapter 2 we also introduce the Kolmogorov eddy phenomenology (which we will compare to the wave turbulence phenomenology in the second part) and intermittency models. The treatment of turbulence in Fourier space will be presented in Chapter 3. In particular, the discussion of statistical closures developed in the 1960s is relevant to the comparison with the wave turbulence closure presented in Chapter 4. The case of two-dimensional turbulence will also be discussed in great detail; we will show that the Zakharov transformation, used so far only for wave turbulence, can also be a powerful tool in this case.

Part II of the book is devoted to wave turbulence: after a general introduction to the subject and a nonexhaustive list of applications of this regime (Chapter 4), various examples will be treated in Chapters 5 to 9. Capillary wave turbulence is probably the simplest example to present the theory of wave turbulence. Therefore, in Chapter 5, we present this theory in great detail. This is an essential technical chapter to master the asymptotic development.

This book is an introduction to the physics of wave turbulence. The bias is to present fundamental results limited to the case of statistically homogeneous turbulence. Therefore, the problems of inhomogeneity that we encounter, especially in laboratory experiments, will not be discussed. Nevertheless, the results of laboratory experiments will be regularly presented, as well as those obtained from observations or numerical simulations.

1.2 Chaos and Unpredictability

Defining turbulence precisely requires the introduction of a number of notions that we will define in part in this chapter. Without going into detail, we can notice that the disordered – or chaotic – aspect seems to be the primary characteristic of turbulent flows. The chaotic nature of a system is of course related to nonlinearities. It is often said that a system is chaotic when two points initially very close to each other in phase space separate exponentially over time. This definition can be extended to the case of fluids.

1.2 Chaos and Unpredictability

The origin of the media success of chaos theory goes back to the early 1960s. It is indeed at this period that the meteorologist Lorenz from the Massachusetts Institute of Technology (MIT) decided to use his computer (a Royal McBee LGP-300, without screen, capable of performing 60 operations per second) to numerically integrate a system of nonlinear differential equations – the Lorenz system – which is a simplified version of the fluid equations of thermal convection and whose form is:

$$\frac{dX}{dt} = \sigma(Y - X), \quad (1.5a)$$

$$\frac{dY}{dt} = \rho X - Y - XZ, \quad (1.5b)$$

$$\frac{dZ}{dt} = XY - \beta Z, \quad (1.5c)$$

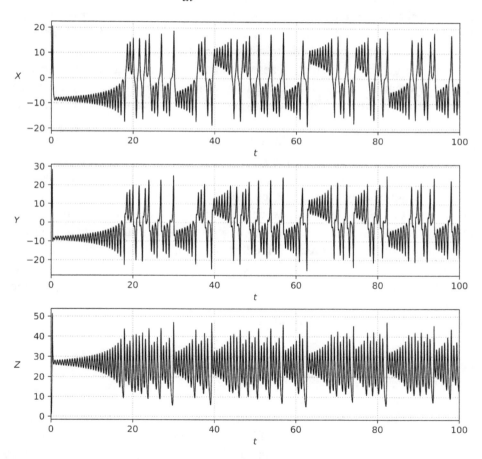

Figure 1.3 Numerical simulation of the Lorenz system (1.5) with $\sigma = 10$, $\rho = 28$ and $\beta = 8/3$. The three variables $X(t)$, $Y(t)$, and $Z(t)$ show a randomness, or unpredictability, in terms of variations or sign changes.

where σ, ρ, and β are real parameters. In Figure 1.3 we show the evolution over time of the three variables $X(t)$, $Y(t)$, and $Z(t)$.

By pure chance, Lorenz (1963) observed that two initial conditions very close to each other diverge quite rapidly.[18] Since linear functions imply results proportional to the initial uncertainties, the observed divergence could only be explained by the presence of nonlinear terms in the model equations. Lorenz then understood that even if some nonlinear phenomena are governed by rigorous and perfectly deterministic laws, precise predictions are impossible because of the sensitivity to initial conditions, which is, as we know, a major problem in meteorology. To make this result clear, Lorenz used an image that contributed to the media success of chaos theory: the famous butterfly effect. He explained that the laws of meteorology are so sensitive to initial conditions that the simple flapping of a butterfly's wings in Brazil can trigger a tornado in Texas. Lorenz had thus just demonstrated that the future is unpredictable. But what is unpredictable is not necessarily chaotic (i.e. disordered), as demonstrated, for example, by the existence of strange attractors (Hénon, 1976): we then speak of deterministic chaos. In phase space, this translates into trajectories irresistibly attracted by complex geometric figures. These systems wander randomly around these figures, without passing twice through the same point. In Figure 1.4, we show Lorenz's strange attractor: it appears when we plot the function $f(X, Y, Z)$ over time.

Turbulent flows are also unpredictable. Two initial conditions that are very close to each other diverge quite rapidly over time. Although the equations – such as those of Navier–Stokes – governing fluid motion are deterministic, it is not possible to predict exactly the state of the turbulent fluid at some distant future time. However, a distinction exists between turbulence and chaos: the word *chaos* is nowadays mainly used in mechanics to describe a deterministic dynamic system with a small number of degrees of freedom. In turbulence, flows have a very large number of degrees of freedom, which results, for example, in the nonlinear excitation of a wide range of spatial scales. As we will see in this book, turbulence is, on the other hand, predictable in the statistical sense, hence the importance of studying turbulence with statistical tools.

1.3 Transition to Turbulence

The observation of turbulence in fluid mechanics is often part of everyday life experiences. In fact, it is under this regime that most of the natural flows of the usual terrestrial fluids such as air and water occur. There is a very large variety of turbulent flows: for example, geophysical flows (atmospheric wind,

---

[18] Lorenz was not the first to wonder about unpredictability. Henri Poincaré addressed the question at the end of the nineteenth century in his study on the stability of the solar system (Poincaré, 1890). Later, Richardson (1922) also wondered about the effect of initial conditions on the predictability of atmospheric flows.

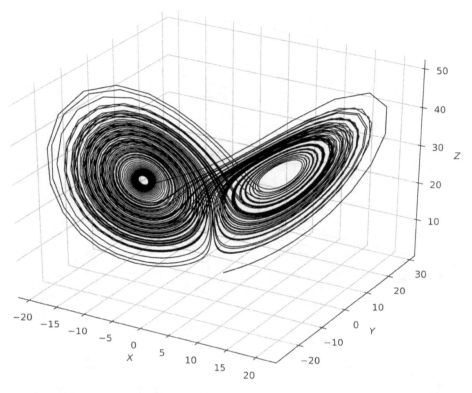

Figure 1.4 The Lorenz attractor appears when we plot the function $f(X, Y, Z)$ over time, with $X(t)$, $Y(t)$, and $Z(t)$ given by solving the Lorenz system.

river currents), astrophysical flows (sun, solar wind, interstellar cloud), biological (blood), quantum (superfluid), or industrial flows (aeronautical, hydraulic, chemical). Despite this diversity, these turbulent flows have a number of common properties.

Probably the most familiar example of turbulent flow is that of a river encountering an obstacle, such as a rock. Downstream, there is a random movement of water characterized by the presence of eddies of different sizes. As we will see in Chapter 2, the eddy is the central concept in the analysis of strong turbulence and, in particular, in the phenomenological description of the cascade of energy to spatial scales that are generally smaller. In Figure 1.5, one can see schematically how such a flow moves from the laminar regime with a low Reynolds number R_e, to the fully developed turbulence regime with a Reynolds number that exceeds 1000. In particular, during this transition a Kármán vortex street is formed for $R_e \sim 100$. Historically, it was Reynolds who was the first to study the transition between these two regimes in 1883 and who gave his name to the dimensionless parameter – the Reynolds number – measuring the degree of turbulence of a flow.

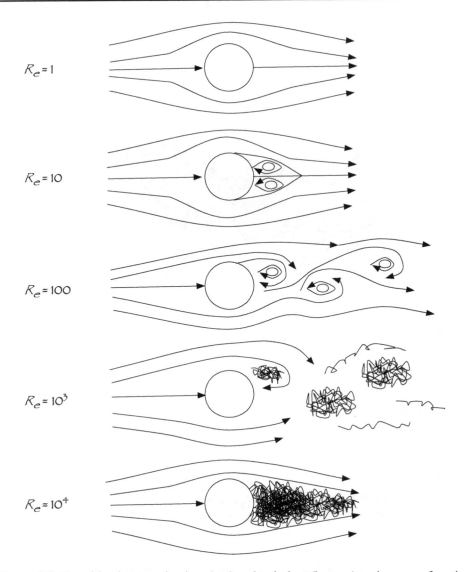

Figure 1.5 Transition between laminar (top) and turbulent (bottom) regimes as a function of the Reynolds number R_e for a flow coming from the left and encountering an obstacle (symbolized by a disc). Figure adapted from Feynman et al. (1964).

This number reflects the relative importance of nonlinear versus dissipative effects in the Navier–Stokes equations and is written as follows:

$$R_e = \frac{UL}{\nu}, \tag{1.6}$$

with U and L a velocity and a characteristic length of the flow respectively, while ν is the kinematic viscosity.

1.4 Statistical Tools and Symmetries

We have mentioned the importance of approaching the physics of turbulence with statistical tools in order to better understand its random nature. In this section, we recall some of these tools that are generally introduced in the course of statistical physics.

1.4.1 Ensemble Average

The ensemble average $\langle X \rangle$ of a quantity X is a statistical average performed on N independent realizations (with $N \to +\infty$) where we measure this quantity:

$$\langle X \rangle = \lim_{N \to +\infty} \frac{1}{N} \sum_{n=1}^{N} X_n. \quad (1.7)$$

If the averaged quantity is, for example, the velocity field, one has:

$$\langle \mathbf{u}(\mathbf{x}, t) \rangle = \lim_{N \to +\infty} \frac{1}{N} \sum_{n=1}^{N} \mathbf{u}_n(\mathbf{x}, t). \quad (1.8)$$

The average operation commutes with derivatives of different kinds, for example:

$$\left\langle \frac{\partial \mathbf{u}(\mathbf{x}, t)}{\partial \mathbf{x}} \right\rangle = \frac{\partial \langle \mathbf{u}(\mathbf{x}, t) \rangle}{\partial \mathbf{x}}. \quad (1.9)$$

The ensemble average operator is analogous to the one used in statistical thermodynamics. Generally, this is not equivalent to a spatial or temporal average, except under special conditions. For example, when turbulence is statistically homogeneous, the ergotic hypothesis can be used to calculate an ensemble average as a spatial average (Galanti and Tsinober, 2004). Note that, to date, no proof of the ergodic theorem is known for the Navier–Stokes equations.

1.4.2 Autocorrelation

To characterize the disorder in a signal $u(x, t)$, one uses the concept of correlation. The simplest correlation function is the autocorrelation:

$$R(x, t, T) = \langle u(x, t) u(x, t + T) \rangle, \quad (1.10)$$

which measures the resemblance of the function to itself, here at two different instants. The quantity $u(x, t)$ (for example, a velocity component) is a random function. To get statistical independence between $u(x, t)$ and $u(x, t + T)$, T cannot be too small, because the fundamental laws of turbulence lead us to expect a certain memory of the signal: T must therefore be larger than a value T_c, which is called the correlation time. A similar analysis can be done for two measurement points not in time, but in space. In this case, we arrive at the notion of correlation

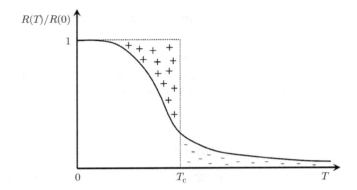

Figure 1.6 Illustration of the meaning of the correlation scale T_c from a given autocorrelation function: the surface $+$ is by definition equal to the surface $-$ (see relation (1.11)).

length L_c (also called the integral scale). Thus, a random flow is characterized by a spatiotemporal memory whose horizon is measured by T_c and L_c. The turbulence study consists in extracting information on the spatiotemporal memory of the flow which can thus be revealed only if we place ourselves on relatively small spatiotemporal correlation scales. Figure 1.6 illustrates the notion of correlation time; by definition we have:

$$T_c \equiv \frac{1}{R(0)} \int_0^{+\infty} R(T) dT, \qquad (1.11)$$

where the dependence in t has been forgotten under the assumption of statistical homogeneity.

1.4.3 Probability Distribution and PDF

Let us define $F_y(x)$ as the probability of finding a fluctuation of the random variable y in the interval $]-\infty, x]$: the function F_y is by definition a probability distribution. From this definition one has:

- $F_y(x)$ is an increasing function,
- $F_y(x)$ is a continuous function,
- $F_y(-\infty) = 0$ and $F_y(+\infty) = 1$.

If this function is differentiable then $F'_y(x)$ defines a probability density function (PDF), that is, $F'_y(x)$ is the probability of finding y in the interval $]x, x + dx]$. In the framework of intermittency (Chapter 2) we will see that the normal (or Gaussian) and Poisson PDFs play a central role.

1.4.4 Moments and Cumulants

The moments of a probability density function are the means of the powers:

$$M_n = \langle y^n \rangle = \int_{-\infty}^{+\infty} x^n F'_y(x) dx. \quad (1.12)$$

We may note that the moment of order one is the mean or expected value. The moments of $y - \langle y \rangle$ are said to be centered. The variance is the centered moment of order two:

$$\langle (y - \langle y \rangle)^2 \rangle, \quad (1.13)$$

while the mean quadratic deviation is the root of the variance.

Given any non-Gaussian random function of zero mean whose second-order moments are known, it is then possible to calculate the fictitious moments of order n that this function would have if it were a Gaussian function. The difference between the actual nth-order moment of the function and the corresponding Gaussian value is called the nth-order cumulant. Then, the odd cumulants are equal to the moments (since the odd moments of a Gaussian are zero) and by definition all the cumulants are zero for a Gaussian function. We will come back to moments, and cumulants in particular, in Chapter 4, when the asymptotic closure of wave turbulence is introduced.

1.4.5 Structure Functions

A structure function of order n of a quantity $f(\mathbf{x})$ is by definition:

$$S_n = \langle (f(\mathbf{x}_1) - f(\mathbf{x}_2))^n \rangle = \langle (\delta f)^n \rangle, \quad (1.14)$$

where $\mathbf{x}_1$ and $\mathbf{x}_2$ are two points of the space. We will see in Chapter 2 that the first rigorous law established in turbulence by Kolmogorov (1941a) involves the velocity structure function of order three.

1.4.6 Symmetries

In order to simplify the analytical study of turbulence, we often impose certain symmetries on the flow. Unless explicitly stated, the symmetries below are taken in the statistical sense.

- **Homogeneity:** This is the space translation invariance. It is the most classical assumption that is satisfied at the heart of turbulence, that is, far from the walls of an experiment. This assumption is essential in the theoretical treatment of turbulence insofar as it brings important simplifications both in physical space and in Fourier space. For a homogeneous turbulence the ergotic hypothesis allows one to calculate an ensemble average as a spatial average.

- **Stationarity:** This is the time translation invariance. It is a very classical hypothesis insofar as a system generally finds its balance between the external forces and the dissipation which occurs at a small scales by viscous friction. For stationarity turbulence the ergotic hypothesis allows one to calculate an ensemble average as a time average.
- **Isotropy:** This is an invariance under any arbitrary rotation. It is a classical assumption in hydrodynamics, which is less justified in presence of an external agent like, for example, rotation or stratification.
- **Mirror symmetry:** This is the invariance under any plane symmetry. It corresponds to an invariance when the sign of all vectors ($\mathbf{x} \to -\mathbf{x}$, $\mathbf{u} \to -\mathbf{u}$, etc.) is changed. It allows the removal of quantities such as the kinetic helicity. One speaks of strong isotropy when turbulence is both isotropic and mirror symmetric. Throughout the book we shall use the word *isotropy* in the weak sense to indicate invariance under rotations, but not necessarily under reflexions of the frame of reference.
- **Scale invariance:** This is the (nonstatistical) invariance by a transformation of the type $\mathbf{u}(\mathbf{x}, t) \to \lambda^h \mathbf{u}(\lambda \mathbf{x}, \lambda^{1-h} t)$. The solutions of the Navier–Stokes equations satisfy this symmetry if $h = -1$. If the viscosity is zero then h can be anything. In practice, this symmetry can be found in the turbulent regime if the scales considered are much greater than those at which the viscosity acts.

The statistical symmetries we have just defined can emerge in a fluid when the Reynolds number is large enough. A return to Figure 1.5 is instructive: comparison of the five images actually shows that the initial symmetries of the fluid disappear at an intermediate Reynolds number to reveal other symmetries at large Reynolds number.

References

Alexakis, A., and Biferale, L. 2018. Cascades and transitions in turbulent flows. *Phys. Rep.*, **767**, 1–101.

Batchelor, G. K. 1946. Double velocity correlation function in turbulent motion. *Nature*, **158**(4024), 883–884.

Batchelor, G. K. 1953. *The Theory of Homogeneous Turbulence*. Cambridge University Press.

Biferale, L. 2003. Shell models of energy cascade in turbulence. *Ann. Rev. Fluid Mech.*, **35**(35), 441–468.

Boussinesq, M. J. 1897. *Théorie de l'écoulement tourbillonnant et tumultueux des liquides dans les lits rectilignes à grande section*, vol. 1. Gauthier-Villars.

Burgers, J. M. 1925. The motion of a fluid in the boundary layer along a plane smooth surface. In C. B Biezeno and J. M. Burgers (eds.) *Proceedings of the First International Congress for Applied Mechanics, Delft 1924* (pages 113–128). J. Waltman, Jr.

Chandrasekhar, S. 1955. A theory of turbulence. *Proc. Roy. Soc. Lond. Series A*, **229**(1176), 1–19.

Chertkov, M., Connaughton, C., Kolokolov, I., and Lebedev, V. 2007. Dynamics of energy condensation in two-dimensional turbulence. *Phys. Rev. Lett.*, **99**(8), 084501.

Davidson, P. A., Kaneda, Y., Moffatt, K., and Sreenivasan, K. R. 2011. *A Voyage through Turbulence*. Cambridge University Press.

Feynman, R. P., Leighton, R. B., and Sands, M. 1964. *The Feynman Lectures on Physics*, vol. 2: *Mainly Electromagnetism and Matter*. Addison-Wesley.

Fjørtoft, R. 1953. On the changes in the spectral distribution of kinetic energy for two-dimensional, non-divergent flow. *Tellus*, **5**, 225–230.

Fox, D. G., and Lilly, D. K. 1972. Numerical simulation of turbulent flows. *Rev. Geophys. Space Phys.*, **10**, 51–72.

Frisch, U. 1995. *Turbulence*. Cambridge University Press.

Frisch, U., Pouquet, A., Leorat, J., and Mazure, A. 1975. Possibility of an inverse cascade of magnetic helicity in magnetohydrodynamic turbulence. *J. Fluid Mech.*, **68**, 769–778.

Frisch, U., Sulem, P. L., and Nelkin, M. 1978. A simple dynamical model of intermittent fully developed turbulence. *J. Fluid Mech.*, **87**, 719–736.

Galanti, B., and Tsinober, A. 2004. Is turbulence ergodic? *Phys. Lett. A*, **330**, 173–180.

Galtier, S. 2014. Weak turbulence theory for rotating magnetohydrodynamics and planetary flows. *J. Fluid Mech.*, **757**, 114–154.

Galtier, S. 2016. *Introduction to Modern Magnetohydrodynamics*. Cambridge University Press.

Galtier, S., and Nazarenko, S. V. 2017. Turbulence of weak gravitational waves in the early universe. *Phys. Rev. Lett.*, **119**(22), 221101.

Grant, H. L., Stewart, R. W., and Moilliet, A. 1962. Turbulence spectra from a tidal channel. *J. Fluid Mech.*, **12**(2), 241–268.

Heisenberg, W. 1948. Zur statistischen Theorie der Turbulenz. *Zeit. Physik*, **124**(7–12), 628–657.

Hénon, M. 1976. A two-dimensional mapping with a strange attractor. *Comm. Math. Physics*, **50**(1), 69–77.

Iyer, K. P., Sreenivasan, K. R., and Yeung, P. K. 2019. Circulation in high Reynolds number isotropic turbulence is a bifractal. *Phys. Rev. X*, **9**(4), 041006.

Kaner, É. A., and Yakovenko, V. M. 1970. Weak turbulence spectrum and second sound in a plasma. *J. Exp. Theor. Phys.*, **31**, 316–330.

Kolmogorov, A. N. 1941a. Dissipation of energy in locally isotropic turbulence. *Dokl. Akad. Nauk SSSR*, **32**, 16–18.

Kolmogorov, A. N. 1941b. The local structure of turbulence in incompressible viscous fluid for very large Reynolds number. *C. R. Acad. Sci. URSS*, **30**, 301–305.

Kolmogorov, A. N. 1962. A refinement of previous hypotheses concerning the local structure of turbulence in a viscous incompressible fluid at high Reynolds number. *J. Fluid Mech.*, **13**, 82–85.

Kraichnan, R. H. 1957. Relation of fourth-order to second-order moments in stationary isotropic turbulence. *Phys. Rev.*, **109**(5), 1407–1422.

Kraichnan, R. H. 1958. Irreversible statistical mechanics of incompressible hydromagnetic turbulence. *Phys. Rev.*, **111**(6), 1747 (1 page).

Kraichnan, R. H. 1959. The structure of isotropic turbulence at very high Reynolds numbers. *J. Fluid Mech.*, **5**, 497–543.

Kraichnan, R. H. 1966. Isotropic turbulence and inertial-range structure. *Phys. Fluids*, **9**(9), 1728–1752.

Kraichnan, R. H. 1967. Inertial ranges in two-dimensional turbulence. *Phys. Fluids*, **10**(7), 1417–1423.

Landau, L. D., and Lifshitz, E. M. 1987. *Fluid Mechanics*. Pergamon Press.

Lee, T. D. 1951. Difference between turbulence in a two-dimensional fluid and in a three-dimensional fluid. *J. Appl. Phys.*, **22**(4), 524 (1 page).

Leith, C. E. 1967. Diffusion approximation to inertial energy transfer in isotropic turbulence. *Phys. Fluids*, **10**(7), 1409–1416.
Leith, C. E. 1968. Diffusion approximation for two-dimensional turbulence. *Phys. Fluids*, **11**(3), 671–672.
Lorenz, E. N. 1963. Deterministic nonperiodic flow. *J. Atmos. Sciences*, **20**, 130–141.
Mandelbrot, B.B. 1974. Intermittent turbulence in self-similar cascades: Divergence of high moments and dimension of the carrier. *J. Fluid Mech.*, **62**, 331–358.
Millionschikov, M. D. 1939. Decay of homogeneous isotropic turbulence in a viscous incompressible fluid. *Dokl. Akad. Nauk SSSR*, **22**, 236–240.
Millionschikov, M. D. 1941. Theory of homogeneous isotropic turbulence. *Dokl. Akad. Nauk SSSR*, **22**, 241–242.
Monin, A. S., and Yaglom, A. M. 1971. *Statistical Fluid Mechanics*, Vol. I. MIT Press.
Monin, A. S., and Yaglom, A. M. 1975. *Statistical Fluid Mechanics*, Vol. 2. MIT Press.
Motzfeld, H. 1938. Frequenzanalyse turbulenter Schwankungen. *Z. Angew. Math. Mech.*, **18**(6), 362–365.
Obukhov, A. M. 1941. Spectral energy distribution in a turbulent flow. *Izv. Akad. Nauk SSSR Ser. Geogr. Geofiz.*, **5**, 453–466.
Obukhov, A. M. 1962. Some specific features of atmospheric turbulence. *J. Fluid Mech.*, **13**, 77–81.
Ogura, Y. 1963. A consequence of the zero-fourth-cumulant approximation in the decay of isotropic turbulence. *J. Fluid Mech.*, **16**, 33–40.
Onsager, L. 1945. The distribution of energy in turbulence. *Phys. Rev.*, **68**(11–12), 286 (1 page).
Orszag, S. A. 1970. Analytical theories of turbulence. *J. Fluid Mech.*, **41**, 363–386.
Orszag, S. A., and Patterson, G. S. 1972. Numerical simulation of three-dimensional homogeneous isotropic turbulence. *Phys. Rev. Lett.*, **28**(2), 76–79.
Paret, J., and Tabeling, P. 1997. Experimental observation of the two-dimensional inverse energy cascade. *Phys. Rev. Lett.*, **79**(21), 4162–4165.
Patterson, G. S., Jr., and Orszag, S. A. 1971. Spectral calculations of isotropic turbulence: Efficient removal of aliasing interactions. *Phys. Fluids*, **14**(11), 2538–2541.
Poincaré, H. 1890. *Sur le problème des trois corps et les équations de la dynamique Acta Mathematica*, **13**, 1–270.
Pope, S. B. 2000. *Turbulent Flows*. Cambridge University Press.
Pouquet, A., and Patterson, G. S. 1978. Numerical simulation of helical magnetohydrodynamic turbulence. *J. Fluid Mech.*, **85**, 305–323.
Pouquet, A., Lesieur, M., Andre, J. C., and Basdevant, C. 1975. Evolution of high Reynolds number two-dimensional turbulence. *J. Fluid Mech.*, **72**, 305–319.
Pouquet, A., Frisch, U., and Leorat, J. 1976. Strong MHD helical turbulence and the nonlinear dynamo effect. *J. Fluid Mech.*, **77**, 321–354.
Pouquet, A., Rosenberg, D., Stawarz, J. E., and Marino, R. 2019. Helicity dynamics, inverse, and bidirectional cascades in fluid and magnetohydrodynamic turbulence: A brief review. *Earth Space Science*, **6**(3), 351–369.
Reynolds, O. 1883. An experimental investigation of the circumstances which determine whether the motion of water shall be direct or sinuous and the law of resistance in parallel channels. *Phil. Trans. Roy. Soc.*, **174**, 935–982.
Richardson, L. F. 1922. *Weather Predictions by Numerical Process*. Cambridge University Press.
Richardson, L. F. 1926. Atmospheric diffusion shown on a distance-neighbour graph. *Proc. Roy. Soc. Lond. Series A*, **110**(756), 709–737.
Rossby, C. G., and collaborators. 1939. Relation between variations in the intensity of the zonal circulation of the atmosphere and the displacements of the semi-permanent centers of action. *J. Marine Res.*, **2**, 38–55.

References

Taylor, G. I. 1935. Statistical theory of turbulence. *Proc. Roy. Soc. Lond. Series A*, **151**(873), 421–444.

Taylor, G. I. 1938. The spectrum of turbulence. *Proc. Roy. Soc. Lond. Series A*, **164**(919), 476–490.

von Kármán, T., and Howarth, L. 1938. On the statistical theory of isotropic turbulence. *Proc. Roy. Soc. Lond. Series A*, **164**(917), 192–215.

Yaglom, A. M. 1949. Local structure of the temperature field in a turbulent flow. *Dokl. Akad. Nauk SSSR*, **69**, 743–746.

Zakharov, V. E. 1967. Weak-Turbulence spectrum in a plasma without a magnetic field. *J. Exp. Theor. Phys.*, **24**, 455–459.

Zakharov, V. E., and Filonenko, N. N. 1966. The energy spectrum for stochastic oscillations of a fluid surface. *Doclady Akad. Nauk. SSSR*, **170**, 1292–1295.

Zakharov, V. E., and Filonenko, N. N. 1967. Energy spectrum for stochastic oscillations of the surface of a liquid. *Soviet Phys. Dokl.*, **11**, 881–884.

Zaslavskii, M. M., and Zakharov, V. E. 1982. The kinetic equation and Kolmogorov spectra in the weak turbulence theory of wind waves. *Izv. Atmos. Ocean Phys.*, **18**, 747–753.

Part I

Fundamentals of Turbulence

2

Eddy Turbulence in Hydrodynamics

2.1 Navier–Stokes Equations

In this chapter, we shall focus on incompressible hydrodynamics and eddy turbulence. The associated equations are the Navier–Stokes equations, established by Claude Louis Marie Henri Navier and George Gabriel Stokes (see Figure 2.2):

$$\frac{\partial \mathbf{u}}{\partial t} + \mathbf{u} \cdot \nabla \mathbf{u} = -\nabla P/\rho_0 + \nu \Delta \mathbf{u}, \qquad (2.1a)$$

$$\nabla \cdot \mathbf{u} = 0, \qquad (2.1b)$$

where $\mathbf{u}$ is the flow velocity, P the pressure, ρ_0 the mass density which is supposed to be constant, and ν the kinematic viscosity. The second equation ensures the incompressibility of the fluid. We will assume that the reader knows the basics of fluid mechanics. Although this system is relatively simple, it already concentrates all the difficulties inherent to the treatment of turbulence with the presence of a nonlinear term. The relative importance of this term is usually measured by the Reynolds number (see Chapter 1):

$$R_e \equiv \frac{|\mathbf{u} \cdot \nabla \mathbf{u}|}{|\nu \Delta \mathbf{u}|} \sim \frac{UL}{\nu}, \qquad (2.2)$$

where U and L are the characteristic velocity and length of the flow, respectively. Subsequently, we will assume that the Reynolds number is very large: in practice, this means that the limit $R_e \to +\infty$ will be taken.

2.2 Turbulence and Heating

2.2.1 Joule's Experiment

Joule was a physicist known for his studies on energy conversion in the form of heating. We all know the Joule effect, which is the thermal manifestation of electrical resistance that occurs when an electric current passes through a conductive material. But he was also interested in mechanics. In a famous experiment, Joule

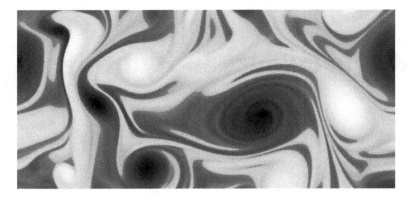

Figure 2.1 Eddy turbulence: vorticity produced by a direct numerical simulation of two-dimensional Navier–Stokes equations (see Chapter 3).

Figure 2.2 Claude Louis Marie Henri Navier (left) and George Gabriel Stokes (right) established the eponymous equations that describe viscous fluids (Navier, 1823; Stokes, 1845).

(1850) measured precisely the mechanical equivalent of heat. This experiment, which he did at home in his cellar, is relatively simple in its design (Young, 2015). A paddle wheel placed in a tank containing water is driven by a rope attached at the other end to a mass (see Figure 2.3). The fall of the mass from a given height is used to evaluate the mechanical work. The heating of the water set in motion by the blades can be deduced from the increase in temperature, measured with a thermometer placed in the tank. To obtain a measurable temperature increase (from 0.5 °F to 2 °F), the experiment, that is, the fall of the mass, would be repeated about

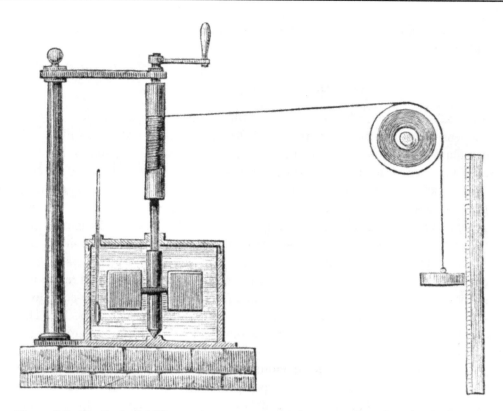

Figure 2.3 The Joule (1850) experiment to measure the conversion of mechanical energy into heat. The fall of the mass on the right causes the blades to move in a tank containing water (left). The water temperature is measured using a thermometer (vertical rod to the left of the tank). This experiment made it possible to demonstrate the equivalence between work and heat. From the point of view of turbulence, we can interpret this experiment as a manifestation of energy transfer from large scales (movement of the blades) to small viscous scales from which energy is dissipated and the fluid heated.

20 times in a row. Joule was thus able to demonstrate the mechanical equivalent of heat.

What is the link between Joule's experiment and turbulence? If we look at this experiment from a different angle, we can see here a proof of the transfer of the kinetic energy of the fluid from the large scales to the small viscous scales from which the energy is dissipated and the fluid heated.[1] The interpretation is then as follows: the setting in motion of the fluid by the blades is characterized by eddies whose size is comparable to that of the blades; then, by a process of direct cascade, the energy associated with these large eddies is finally found at the smallest scales in the form of small fluctuations in velocity, or, in the language of turbulence, in the form of small eddies, which disintegrate as a result of viscous friction. Without

---

[1] Of course, it is possible that part of the heating of the water is of a diffusive nature with a direct transmission between the heated objects (by friction with the rope for example) and water.

knowing it, Joule was probably the first to highlight a fundamental property of turbulence.

2.2.2 Mean Rate of Energy Dissipation

The Joule experiment highlights the heating process by viscous friction. What happens when the viscosity of the fluid is lower? In other words, in a fully developed turbulent regime, when the Reynolds number is increasing, and for identical experiment conditions, one can ask if the dissipation (or heating) through viscous friction decreases so that it becomes zero within the limit $R_e \to +\infty$ (since at $\nu = 0$ kinetic energy is a priori conserved). If this is the case, the smallest eddies, generated after a cascade, should accumulate since they are less dissipated. This thought experiment leads us to a surprising conclusion: within the limit of large Reynolds numbers, a fully developed turbulence excited at a large scale should be characterized by essentially small vortices and by the impossibility of reaching a statistically stationary regime, because the equilibrium between a large-scale forcing and the viscous dissipation at small scales would become impossible. This behavior is contrary to experimental measurements where the stationarity of the turbulence is observed.

It is difficult to know if Kolmogorov (1941b) followed the same reasoning; however, we can see that one of his hypotheses is that the mean rate of dissipation of kinetic energy ε becomes independent of viscosity within the large R_e. Note in passing that Taylor (1935) also reached this conclusion by proposing a semi-phenomenological demonstration to interpret some experimental data.

Let us see what this means in terms of energy conservation. From the Navier–Stokes equations, we get:

$$\frac{\rho_0}{2}\frac{\partial \mathbf{u}^2}{\partial t} = \rho_0 \mathbf{u} \cdot \frac{\partial \mathbf{u}}{\partial t} = \rho_0 \mathbf{u} \cdot (-\mathbf{u} \cdot \nabla \mathbf{u} - \nabla P/\rho_0 + \nu \Delta \mathbf{u}) \,. \tag{2.3}$$

The use of a vector identity (see Appendix B) allows us to write in the incompressible case:

$$\mathbf{u} \cdot \Delta \mathbf{u} = -\mathbf{u} \cdot (\nabla \times \mathbf{w}) = \nabla \cdot (\mathbf{u} \times \mathbf{w}) - \mathbf{w}^2 \,, \tag{2.4}$$

with by definition $\mathbf{w} \equiv \nabla \times \mathbf{u}$ the vorticity; then, one obtains:

$$\frac{\rho_0}{2}\frac{\partial \mathbf{u}^2}{\partial t} = -\rho_0 \mathbf{u} \cdot \nabla \left(\frac{\mathbf{u}^2}{2}\right) - \nabla \cdot (P\mathbf{u} - \rho_0 \nu \, \mathbf{u} \times \mathbf{w}) - \rho_0 \nu \, \mathbf{w}^2 \,. \tag{2.5}$$

After some manipulation, we finally come to the local form of the kinetic energy conservation ($E \equiv \rho_0 \mathbf{u}^2/2$):

$$\frac{\partial E}{\partial t} = -\nabla \cdot \left[\left(\frac{\rho_0 \mathbf{u}^2}{2} + P\right)\mathbf{u} - \rho_0 \nu \, \mathbf{u} \times \mathbf{w}\right] - \rho_0 \nu \, \mathbf{w}^2 \,. \tag{2.6}$$

The kinetic energy variation is therefore due to two types of term: a flux $\mathbf{F}$ and a source S. The source (last term on the right) is negative and depends on the viscosity. In other words, it is an irreversible source of energy dissipation for the turbulent system which is always zero in the inviscid case ($\nu = 0$). In the divergence operator, we find an energy flux and a nonlinear viscous contribution depending on the velocity. The integral form of the kinetic energy conservation is written:

$$\frac{d}{dt}\iiint E\, d\mathcal{V} = -\oiint \mathbf{F}\cdot\mathbf{n}\, d\mathcal{S} - \rho_0\nu \iiint \mathbf{w}^2\, d\mathcal{V}. \tag{2.7}$$

In the particular case of an integration on a volume large enough to contain all the fluid (so that on the surface of this volume $\mathbf{u} = \mathbf{0}$; one can also consider periodic boundary conditions), the flux is canceled out so that only a purely viscous dissipation contribution remains.

Subsequently, we will consider the ensemble average, which can be calculated via a spatial average when turbulence is statistically homogeneous (see Chapter 1). If we think of a direct numerical simulation carried out in a periodic box, this is equivalent to taking the spatial average of this box. In this case, and with the notations introduced previously, we obtain the exact relation:

$$\boxed{\frac{d\langle E\rangle}{dt} = -\rho_0\nu\langle \mathbf{w}^2\rangle \equiv -\varepsilon(\nu)}. \tag{2.8}$$

With this calculation, the mean rate of energy dissipation ε is now well defined.[2] Expression (2.8) allows the introduction of a quantity often used: the enstrophy,[3] $\Omega \equiv \mathbf{w}^2/2$.

2.2.3 Spontaneous Symmetry Breaking

In the previous expression, it is not obvious that ε becomes independent of viscosity within the limit of a large Reynolds number. This assertion assumes that there is, at the same time, an increase in vorticity such that $\langle \mathbf{w}^2\rangle \propto 1/\nu$. Physically, this behavior is not in contradiction with the idea of a direct cascade since small eddies are associated with small and rapid fluctuations in velocity. Since vorticity involves a spatial derivative, we can be convinced that its modulus increases when moving towards small scales. This property of ε cannot be rigorously demonstrated; however, it can be highlighted numerically. In Figure 2.4, we report the results of four freely decaying numerical simulations, that is to say, without external forcing. The initial condition consists in exciting the flow on a large scale by depositing energy only on a few (small) Fourier modes. The parameters of the

---
[2] Usually this is the mean rate of energy dissipation per unit of mass, ε/ρ_0, which is used. The comparison with the compressible case (see Section 2.9) shows that our definition is actually more universal.
[3] The word "enstrophy" comes from the Greek στροφή, which means rotation.

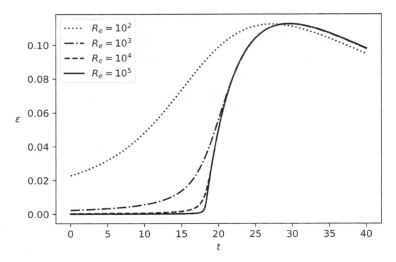

Figure 2.4 Temporal variation of the mean rate of energy dissipation ε for different Reynolds numbers: $R_e = 10^2$ (dot), 10^3 (dash-dot), 10^4 (dash), and 10^5 (solid). Freely decaying numerical simulations.

calculation are the same, only the viscosity changes, and thus the Reynolds number, which varies by a power of 10 between $R_e = 10^2$ and $R_e = 10^5$. We see that the curves tend to overlap when R_e increases, which shows that indeed $\varepsilon(t)$ tends to become independent of viscosity. The asymptotic shape of this curve reflects the fact that turbulence first passes through a phase of development characterized by essentially a zero dissipation rate. This is the period during which turbulence develops: the direct cascade generates smaller and smaller eddies; when the smallest modes are reached, viscous dissipation becomes significant and ε increases sharply. The last phase of decay is related to the fact that the system is not forced: consequently, the dissipated energy is not renewed and the dissipation rate can only decrease and tend towards zero after an infinite time. In the presence of a stationary external force, the behavior is different: the turbulent system adjusts itself so that the energy injection and dissipation rate compensate each other on average; in this case ε also becomes independent of time. The fact that the mean rate of energy dissipation tends towards a finite value at large R_e is observed numerically (Kaneda et al., 2003) and also experimentally with a Reynolds number that can approach 10^8 using superfluids (Sreenivasan, 1984; Pearson et al., 2002; Ravelet et al., 2008; Saint-Michel et al., 2014).

Suppose that ε becomes asymptotically independent of viscosity at high R_e. Moreover, we know that the Euler equations, that is, the Navier–Stokes equations at zero viscosity, conserve energy. Therefore, with expression (2.8) we obtain the property:

$$\boxed{\lim_{\nu \to 0} \frac{d\langle E \rangle}{dt} = -\varepsilon \neq 0}. \tag{2.9}$$

This property is a fundamental law of turbulence – often called the zeroth law. Behind this result hides a remarkable physical property. Viscosity is the parameter that breaks the symmetry of the system by evacuating energy irreversibly by heating or, in other words, by producing entropy. Viscosity thus induces a symmetry breaking of the time reversal in the Navier–Stokes equations. Now, relation (2.9) tells us that when the parameter ν, at the source of this symmetry breaking, tends towards zero, its effect (the symmetry breaking) remains nonzero: we have here an example of spontaneous symmetry breaking.[4] In turbulence, we call this anomalous dissipation.

2.3 Kármán–Howarth Equation

The first exact law in turbulence was obtained by Kolmogorov (1941a): it is the four-fifths law whose name is simply due to the value of the constant that appears in the expression. The path taken by Kolmogorov requires a delicate tensor analysis which – although elegant and historical – tends to make the formalism more cumbersome. We choose to follow, here, a more modern path which allows us to obtain, with less computation, an exact law with a slightly different form;[5] it is the four-thirds law obtained by Antonia et al. (1997). These Kolmogorov laws express how third-order structure functions in velocity are related to the distance between the two measurement points in the case of statistically homogeneous and isotropic turbulence. To achieve this, we have to go through the Kármán–Howarth equation (von Kármá and Howarth 1938).

Let us write the Navier–Stokes equations at points $\mathbf{x}$ and $\mathbf{x}'$ for the components i and j, respectively. Using Einstein's notations, we get:

$$\partial_t u_i + \partial_k(u_k u_i) = -\partial_i P/\rho_0 + \nu \partial^2_{kk} u_i, \qquad (2.10a)$$

$$\partial_t u'_j + \partial'_k(u'_k u'_j) = -\partial'_j P'/\rho_0 + \nu \partial'^2_{kk} u'_j, \qquad (2.10b)$$

where the incompressibility condition writes $\partial_k u_k = 0$. To simplify the writing, we define $\mathbf{u}(\mathbf{x}) \equiv \mathbf{u}$, $\mathbf{u}(\mathbf{x}') \equiv \mathbf{u}'$, $\partial/\partial x' \equiv \partial'$ and we will consider that the velocity fluctuations are of zero mean, that is to say, $\langle \mathbf{u} \rangle = 0$. We multiply the first equation by u'_j and the second by u_i. The addition of the two equations gives us, taking the ensemble average, a dynamic equation for the second-order correlation tensor:

$$\partial_t \langle u_i u'_j \rangle + \langle \partial_k(u_k u_i u'_j) \rangle + \partial'_k(u'_k u'_j u_i) \rangle = \qquad (2.11)$$
$$-\langle \partial_i(Pu'_j/\rho_0) + \partial'_j(P'u_i/\rho_0) \rangle + \nu \langle \partial^2_{kk}(u_i u'_j) + \partial'^2_{kk}(u'_j u_i) \rangle .$$

To get this expression, we used the relation: $\partial_k u'_j = \partial'_k u_i = 0$. Subsequently, we will make extensive use of properties that are valid in statistically homogeneous

---

[4] Turbulence was the first example in physics where such a breaking was reported. The second would come later in quantum electrodynamics (Schwinger, 1951; Davidson et al., 2011).
[5] This is also the approach used in MHD for electrically conducting fluids (see Chapter 8).

turbulence (see Batchelor, 1953). Since turbulence is statistically homogeneous, the two-point correlation tensors depend only on the relative distance ℓ, with $\mathbf{x}' = \mathbf{x} + \ell$, and not on the absolute positions $\mathbf{x}$ and $\mathbf{x}'$ (see Figure 2.5). In particular, we have:

$$\langle \partial_k(u_k u_i u_j') \rangle = -\partial_{\ell_k} \langle u_k u_i u_j' \rangle, \qquad (2.12a)$$

$$\langle \partial_k'(u_k' u_j' u_i) \rangle = \partial_{\ell_k} \langle u_k' u_j' u_i \rangle, \qquad (2.12b)$$

hence the expression:

$$\partial_t \langle u_i u_j' \rangle + \partial_{\ell_k} \langle u_k' u_j' u_i - u_k u_i u_j' \rangle = \qquad (2.13)$$
$$+\partial_{\ell_i} \langle P u_j'/\rho_0 \rangle - \partial_{\ell_j} \langle P' u_i/\rho_0 \rangle + 2\nu \partial^2_{\ell_k \ell_k} \langle u_i u_j' \rangle.$$

We will now restrict ourselves to the trace of the second-order correlation tensor. In this case, the equation simplifies for two reasons: on the one hand, the contribution of the pressure disappears by incompressibility:

$$\partial_{\ell_i} \langle P u_i'/\rho_0 \rangle = \langle \partial_i'(P u_i'/\rho_0) \rangle = \langle (P/\rho_0) \partial_i' u_i' \rangle = 0, \qquad (2.14)$$

and on the other hand, we have by homogeneity:

$$\partial_{\ell_k} \langle u_k u_i u_i' \rangle(\ell) \equiv \partial_{\ell_k} \langle u_k(\mathbf{x}) u_i(\mathbf{x}) u_i(\mathbf{x}+\ell) \rangle$$
$$= \partial_{\ell_k} \langle u_k(\mathbf{x}-\ell) u_i(\mathbf{x}-\ell) u_i(\mathbf{x}) \rangle \equiv \partial_{\ell_k} \langle u_k' u_i' u_i \rangle(-\ell)$$
$$= -\partial_{\ell_k} \langle u_k' u_i' u_i \rangle(\ell). \qquad (2.15)$$

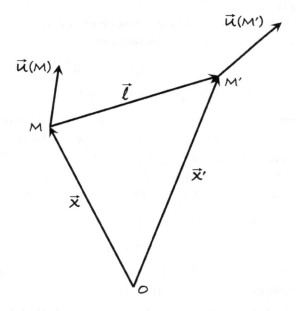

Figure 2.5 In statistically homogeneous turbulence, only the relative position of the two points M and M' is relevant.

Hence the expression:

$$\partial_t \langle u_i u_i' \rangle = -2\partial_{\ell_k} \langle u_k' u_i' u_i \rangle + 2\nu \partial^2_{\ell_k \ell_k} \langle u_i u_i' \rangle$$
$$= -2\nabla_\ell \cdot \langle u_i u_i' \mathbf{u}' \rangle + 2\nu \partial^2_{\ell_k \ell_k} \langle u_i u_i' \rangle. \quad (2.16)$$

The introduction of the third-order structure function gives (with the assumption of homogeneity):

$$\langle (\delta \mathbf{u} \cdot \delta \mathbf{u}) \delta \mathbf{u} \rangle \equiv \langle (u_i' - u_i)^2 (\mathbf{u}' - \mathbf{u}) \rangle$$
$$= -2\langle u_i u_i' \mathbf{u}' \rangle + 2\langle u_i u_i' \mathbf{u} \rangle + \langle u_i^2 \mathbf{u}' \rangle - \langle u_i'^2 \mathbf{u} \rangle. \quad (2.17)$$

By incompressibility, then homogeneity, we find:

$$\nabla_\ell \cdot \langle (\delta \mathbf{u} \cdot \delta \mathbf{u}) \delta \mathbf{u} \rangle = -2\nabla_\ell \cdot \langle u_i u_i' \mathbf{u}' \rangle + 2\nabla_\ell \cdot \langle u_i u_i' \mathbf{u} \rangle$$
$$= -4\nabla_\ell \cdot \langle u_i u_i' \mathbf{u}' \rangle. \quad (2.18)$$

Finally, we obtain the dynamic equation for a homogeneous turbulence:

$$\boxed{\partial_t \left\langle \frac{u_i u_i'}{2} \right\rangle = \frac{1}{4} \nabla_\ell \cdot \langle (\delta \mathbf{u} \cdot \delta \mathbf{u}) \delta \mathbf{u} \rangle + 2\nu \partial^2_{\ell_k \ell_k} \left\langle \frac{u_i u_i'}{2} \right\rangle.} \quad (2.19)$$

This is the Kármán–Howarth equation obtained by the authors[6] in 1938, which reflects the dynamic evolution of the velocity correlation at two points.

2.4 Locality and Cascade

To go further in the analysis and obtain the exact Kolmogorov law, it is necessary to introduce a locality hypothesis. This hypothesis asserts that there is an interval of scales – called the inertial range – where the physics of turbulence is insensitive (i) to large-scale flow motions, whose dynamics are correlated in particular to the action of a nonuniversal external force, and (ii) to small-scale motions whose dynamics are governed by viscous dissipation. In practice, this means that the inertial range[7] lies on the interval:

$$\ell_{diss} \ll \ell \ll \ell_0, \quad (2.20)$$

with ℓ_0 the typical scale where an external force applies and ℓ_{diss} the typical scale where the viscous dissipation becomes important. Therefore, the inertial range is the domain where nonlinear physics dominates; it is here that we can hope to find some universality in the flow behavior.

A fundamental hypothesis proposed by Kolmogorov (1941b) is that, within this scale interval, turbulence can be considered statistically homogeneous and isotropic. Note that the assumption of statistical homogeneity has already been used

---

[6] In fact, its original form was slightly different because it involved the longitudinal and transverse correlations (von Kármán and Howarth, 1938).
[7] The name *inertial* comes of course from the inertial term, $\mathbf{u} \cdot \nabla \mathbf{u}$, from the Navier–Stokes equations.

to obtain the Kármán–Howarth equation: the correlation scale ℓ in this equation is therefore assumed to be sufficiently small. In Figure 2.6, we schematically represent the physics of turbulence in the inertial range: it is characterized by a cascade of eddies from large to small scales. This classical image of the cascade process has, above all, a pedagogical dimension, because the reality is a bit different. An illustration of the vortex structures actually present in hydrodynamic turbulence is visible in Figure 2.7: direct numerical simulations show instead that we have nested vorticity tubes (like Russian dolls) which are thinning over time.

A more realistic phenomenological interpretation of the cascade was proposed by Taylor (1937). His reasoning is based on the topological conservation of a vorticity tube in the inviscid case. We know that in turbulence two points tend statistically to move away from each other by turbulent diffusion. From this property, we can be convinced that a vorticity tube will tend to stretch over time. Since the volume of the tube is preserved by incompressibility, a stretching of this tube implies at the same time a thinning. On the other hand, we know that the vorticity flow in the tube is conserved (Kelvin's theorem), therefore, a thinning of the tube implies an increase in the vorticity norm. (It can be shown that this norm increases in proportion to the elongation of the tube.) This process stops when the radius of the tube reaches the viscous dissipation scale. The vorticity tube is then destroyed. It is interesting to relate the increase of vorticity at small scales to relation (2.8): when $\nu \to 0$, the system will produce thinner and thinner vorticity tubes as $\mathbf{w}^2 \to +\infty$. We find here a beginning of explanation of the anomalous dissipation. We will come back to this in Section 2.7.1.

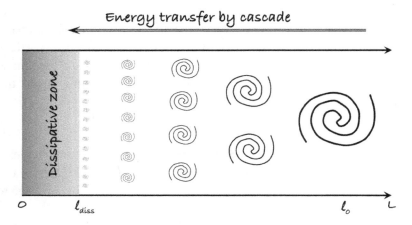

Figure 2.6 Three-dimensional hydrodynamic turbulence is characterized by a cascade of eddies from large to small scales. This energy transfer process is carried out from scale to scale from the energy injection scale ℓ_0 via an external forcing, up to its dissipation by viscous friction from a typical scale ℓ_{diss}. The inertial range is located on the interval $\ell_{diss} \ll \ell \ll \ell_0$.

2.5 Kolmogorov's Exact Law

Figure 2.7 Direct numerical simulation of Navier–Stokes equations with a spatial resolution 1024 × 1024 × 1024. Two iso-surfaces of vorticity are drawn in a localized region of the simulation box. Results from the database JHU Turbulence Database Cluster; http://turbulence.pha.jhu.edu. Reprinted with permission (Galtier, 2016).

2.5 Kolmogorov's Exact Law

The Kármán–Howarth equation was obtained without introducing an external force; this is also the situation considered by Kolmogorov (1941a). However, the calculation assumes in its final resolution the stationarity of the turbulence; it is therefore more consistent to introduce an external force to maintain a fully developed turbulence. In practice, this corresponds to adding to the Navier–Stokes equations (2.1a) a term $\mathbf{f}$ in the right-hand side. This force is assumed to be random, homogeneous, and stationary in nature. The Kármán–Howarth equation is then modified as follows:

$$\partial_t \left\langle \frac{u_i u_i'}{2} \right\rangle = \frac{1}{4} \nabla_\ell \cdot \langle (\delta u_i)^2 \delta \mathbf{u} \rangle + 2\nu \partial^2_{\ell_k \ell_k} \left\langle \frac{u_i u_i'}{2} \right\rangle + \mathcal{F}(\ell), \qquad (2.21)$$

with the correlator $\mathcal{F}(\ell) = \langle f_i u_i' + f_i' u_i \rangle / 2$. To appreciate its shape, we have to come back to the energy conservation equation. In the presence of $\mathbf{f}$, we get:

$$\frac{d\langle E \rangle}{dt} = \rho_0 \langle \mathbf{f} \cdot \mathbf{u} \rangle - \varepsilon = \rho_0 \mathcal{F}(0) - \varepsilon. \qquad (2.22)$$

Therefore, the stationary regime corresponds to a statistical equilibrium between forcing and dissipation.

We will place ourselves in the inertial range in order to highlight the universal behavior of turbulence. We will assume that the external force acts only at the largest scales of the system. In this case, we can evaluate its contribution by a simple Taylor development:

$$\mathcal{F}(\ell) \simeq \mathcal{F}(0) + \ell \cdot \nabla_\ell \mathcal{F}(\ell), \qquad (2.23)$$

with $\rho_0 \mathcal{F}(0) = +\varepsilon$. Consequently, under the stationarity hypothesis we obtain:

$$0 = \frac{1}{4} \nabla_\ell \cdot \langle (\delta u_i)^2 \delta \mathbf{u} \rangle + 2\nu \partial_{\ell_k \ell_k} \left\langle \frac{u_i u'_i}{2} \right\rangle + \varepsilon/\rho_0. \qquad (2.24)$$

The relation (2.24) is valid for homogeneous, stationary turbulence such that $\ell \ll \ell_0$ (which is not equivalent to the limit $\ell \to 0$). It becomes simpler if we assume $\ell_{diss} \ll \ell$ to be at the heart of the inertial range (which means that we consider the limit $\nu \to 0$). In this case, the dissipation term of equation (2.24) is negligible and the zeroth law of turbulence can be used. A primitive form of Kolmogorov's law is obtained:[8]

$$\boxed{-4\bar{\varepsilon} = \nabla_\ell \cdot \langle (\delta u_i)^2 \delta \mathbf{u} \rangle}, \qquad (2.25)$$

with $\bar{\varepsilon} \equiv \varepsilon/\rho_0$ the mean rate of energy dissipation per unit of mass. This expression is valid for both isotropic and nonisotropic turbulence.[9]

Kolmogorov's exact law, in its classical form, can be obtained by finally assuming that turbulence is statistically isotropic. In this case, all that remains to be done is to integrate equation (2.25) on a full sphere of radius ℓ; one obtains the four-thirds law valid when $\ell_{diss} \ll \ell \ll \ell_0$:

$$\boxed{-\frac{4}{3} \bar{\varepsilon} \ell = \langle (\delta u_i)^2 \delta u_\ell \rangle}, \qquad (2.26)$$

where u_ℓ is the longitudinal component of the velocity, that is, that along the direction of separation ℓ between the two measuring points. As already mentioned, the law originally obtained by Kolmogorov is a bit different: it is the exact four-fifths law, which is written:

$$\boxed{-\frac{4}{5} \bar{\varepsilon} \ell = \langle (\delta u_\ell)^3 \rangle}, \qquad (2.27)$$

with the same notations. Note that this law is experimentally easier to use because it only requires knowing one component of velocity. Contrary to appearances, the mathematical link between the four-thirds law and the four-fifths law is not trivial (e.g. one must be careful to project the components of velocity before calculating their statistical value).

Kolmogorov's law (2.27) is the first exact result in turbulence. It is a universal law invariant by Galilean transformation, which describes (in the inertial range) the statistically homogeneous, stationary, and isotropic hydrodynamic turbulence.

---

[8] Note that this primitive form is not unique: for example, it is possible to obtain a law that involves the vorticity but not the divergence operator (Banerjee and Galtier, 2017).

[9] For its derivation we did not make any assumption of the type of regime, so this law is also valid for wave turbulence such as inertial wave turbulence, that is, fast-rotating hydrodynamic turbulence which is anisotropic with a cascade mainly in the direction transverse to the axis of rotation (see Chapter 6).

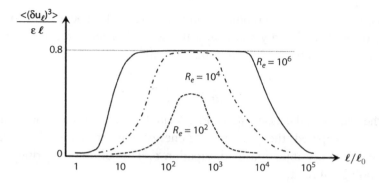

Figure 2.8 Schematic reproduction of experimental measurements in a wind tunnel (see, e.g., Antonia and Burattini, 2006): the higher the Reynolds number, the better the Kolmogorov four-fifths law holds.

By exact, we mean a law obtained rigorously after an analytical calculation. Of course, hypotheses have been made, but they are all mathematically well defined. Kolmogorov's laws are well verified[10] numerically and experimentally (Moisy et al., 1999; Antonia and Burattini, 2006): in Figure 2.8, we report schematically a series of experimental measurements of the four-fifths law. Since then, other exact laws have been obtained: this is essentially an extension of the four-thirds law to other incompressible fluids. Note that the first compressible exact law was obtained only 70 years after Kolmogorov: it describes the compressible hydrodynamic turbulence in the isothermal case (Galtier and Banerjee, 2011). The most interesting applications of this law concern turbulence in interstellar clouds where stars are formed. Indeed, observations reveal that the turbulent Mach number in these clouds approaches 100 (see Section 2.9).

2.6 Phenomenology of Eddy Turbulence

To obtain the exact four-thirds law, we have done a rigorous analysis which is, therefore, quite long. This law can in fact be found more quickly by a phenomenological approach – therefore less rigorous. However, this approach is very useful because it highlights the main quantities required to describe the physics of turbulence.

We will rely on Figure 2.6, focusing on the energy transfer in the inertial range at a ℓ scale such that $\ell_{diss} \ll \ell \ll \ell_0$. We can associate with this scale an eddy of size ℓ. The velocity of this eddy is noted u_ℓ: we sometimes speak of the eddy turnover velocity. We can also define this velocity by the relation $\sqrt{\langle(\delta u_\ell)^2\rangle}$. From these two quantities, we can construct a (nonlinear) eddy turnover time $\tau_{NL} \sim \ell/u_\ell$. Note that this is also the time that emerges dimensionally from the Navier–Stokes

---

[10] Note that it is easier to highlight the dimensional (nonexact) law $\langle(\delta u_\ell)^2\rangle \propto \ell^{2/3}$ for the second-order structure function because it is a positive definite quantity (Antonia and Burattini, 2006; Ishihara et al., 2020).

equations. In this phenomenological approach, we will associate τ_{NL} to the characteristic time of the energy cascade at the scale ℓ. Therefore, the mean rate of energy transfer from the scale ℓ to a smaller scale can be written ($\rho_0 = 1$):

$$\bar{\varepsilon}_\ell \sim \frac{dE_\ell}{dt} \sim \frac{u_\ell^2}{\tau_{NL}} \sim \frac{u_\ell^3}{\ell}, \qquad (2.28)$$

with E_ℓ the energy of the associated eddy. As in the inertial range, energy is neither injected nor dissipated; the mean rate of energy transfer ε_ℓ must be equal to the mean rate of energy injection or dissipation ε. This ultimately brings us to the phenomenological relation:

$$\bar{\varepsilon}\ell \sim u_\ell^3, \qquad (2.29)$$

which is the dimensional analogue of Kolmogorov's (four-thirds or four-fifths) law. To obtain expression (2.29), we used the locality of the cascade, as well as the isotropy hypothesis. We will see later how this last hypothesis can be relaxed to allow us to predict the right energy spectrum in anisotropic turbulence (see Chapter 6).

2.7 Inertial Dissipation and Singularities

2.7.1 Onsager's Conjecture

In this section, we will discuss the consequences of the anomalous dissipation (2.9), which asserts that the mean rate of energy dissipation tends to a positive nonzero value when $\nu \to 0^+$. We always place ourselves in the three-dimensional case. We will introduce the inertial dissipation, which is linked to the existence of singularities, a concept that belongs more to mathematics than to physics. However, it is interesting to see what we can learn from mathematics to better understand physics.

Since we have by definition $\varepsilon = \rho_0 \nu \langle \mathbf{w}^2 \rangle$, the anomalous dissipation necessarily implies that the vorticity (or enstrophy) tends to become infinite in a region of the flow within the $\nu \to 0^+$ limit. Let us note first of all that with the phenomenology of Taylor (1937) presented in Section 2.4, we obtain a vorticity that tends towards infinity in the inviscid limit. This behavior is also supported by a phenomenological analysis of the vorticity equation, which can be written as follows:

$$\frac{D\mathbf{w}}{Dt} \equiv \frac{\partial \omega}{\partial t} + \mathbf{u} \cdot \nabla \mathbf{w} = \mathbf{w} \cdot \nabla \mathbf{u}, \qquad (2.30)$$

with D/Dt the Lagrangian derivative. Since dimensionally $\nabla \mathbf{u} \sim \mathbf{w}$, we get:

$$\frac{D\mathbf{w}}{Dt} \sim \mathbf{w}^2, \qquad (2.31)$$

whose solution is of the form $|\mathbf{w}| \sim 1/(t_* - t)$. This solution implies an explosion of vorticity in a finite time t_*.[11] The works dedicated to the appearance of singularities in fluid mechanics are of diverse nature. For example, there are those based on a spectral technique that allows us to follow a complex singularity over time, and to measure its proximity to the real axis. If it touches the real axis, it becomes a true singularity. In practice, only a trend can be identified because, in the inviscid case, direct numerical simulations become unstable in a finite time, whereas in the presence of viscosity the motion of the singularity is modified by the latter in a finite time (Brachet et al., 1983; Sulem et al., 1983; Fournier and Galtier, 2001). Let us note that a new physical scenario has recently been proposed: the appearance of singularities in Euler's equations could be linked to the collision of vorticity tubes and their transformation into layers (Brenner et al., 2016). But to date, the proof of the appearance of singularity in a finite time has not yet been obtained. This is also a major research topic for mathematicians.[12]

A singular behavior of vorticity may seem difficult to accept physically: it means in particular that the velocity becomes an irregular field within the $\nu \to 0^+$ limit (see Figure 2.9 for an illustration). Derivative operations are then no longer applicable, and expression (2.25) becomes inappropriate. This situation was considered by the mathematician Leray (1934), who introduced the notion of weak solution for Navier–Stokes equations (behind this lies the concept of distribution

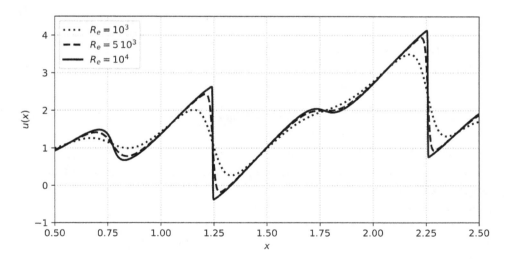

Figure 2.9 Velocity fluctuations for three Reynolds numbers R_e. With the increase of R_e, the irregularity of the field emerges as discontinuities.

---

[11] For two-dimensional turbulence, the situation is different since the vorticity equation (which is then a scalar function) becomes a simple transport equation. In this particular case, theorems on the existence and uniqueness of the solutions can be established (Bardos, 1972).

[12] The question of the underlying regularity of the velocity field, at any instant, is one of the seven Millennium mathematical problems proposed by the Clay Institute. A million-dollar reward is promised for a significant breakthrough on this question.

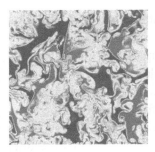 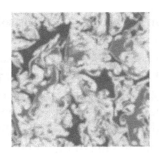 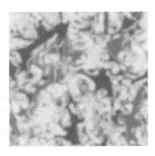

Figure 2.10 The convolution operation (2.32) applied to the image on the left gives new images that are increasingly blurry as ℓ becomes larger (images on the right). Images made by V. David.

later formalized by Laurent Schwartz). The underlying idea of this approach is to introduce a local average of the field values in an increasingly tightened volume around the point of study. In practice, a field mean weighted by a φ^ℓ function is computed on a full sphere of radius ℓ such that:

$$\mathbf{u}^\ell(\mathbf{x}, t) \equiv \int_{\mathbb{R}^3} \varphi^\ell(\xi) \mathbf{u}(\mathbf{x} + \xi, t) d\xi, \qquad (2.32)$$

with:

$$\varphi^\ell(\xi) \equiv \frac{\varphi(\xi/\ell)}{\ell^3}. \qquad (2.33)$$

By definition, $\varphi \in \mathbf{C}^\infty$, and is compactly supported on $\mathbb{R}^3$, even, positive, and such that $\int_{\mathbb{R}^3} \varphi(\xi) d\xi = 1$. These properties assure that φ^ℓ tends towards the distribution $\delta(\xi)$ within the limit $\ell \to 0$ and that therefore:

$$\mathbf{u}^\ell(\mathbf{x}, t) \xrightarrow[\ell \to 0]{} \mathbf{u}(\mathbf{x}, t). \qquad (2.34)$$

With this mathematical tool, it is possible to obtain a small-scale smoothed version of the Navier–Stokes equations while keeping intact information coming from the largest scales: in mathematical language, we speak of a weak (or regularized) formulation of the equations. Figure 2.10 shows the consequences of the smoothing operation on an image.

As we will see in the Section 2.7.2, it is possible to demonstrate with this tool that an irregular velocity field can potentially lead to local energy dissipation. This result has been proven by French mathematicians Duchon and Robert (2000), but it was actually Onsager (1949) (see also Eyink and Sreenivasan, 2006) who was the first to speculate that energy could be dissipated without the assistance of viscosity, by the simple fact of irregularities in the velocity field. This potential dissipation of energy is called inertial dissipation, as opposed to viscous dissipation. The importance of this conjecture lies in the fact that the anomalous dissipation in turbulence could have its origin in the velocity field irregularities of the Euler equations.[13]

2.7.2 Weak Formulation

In this section, we will establish the expression for the inertial dissipation $\mathcal{D}_I$. To do this, let us apply the smoothing operator to the Navier–Stokes equations:

$$\partial_t \mathbf{u}^\ell + \partial_j (u_j \mathbf{u})^\ell = -\nabla P^\ell / \rho_0 + \nu \Delta \mathbf{u}^\ell . \tag{2.35}$$

After multiplying by $\mathbf{u}$ (a weak solution) and some manipulations, we get:

$$\partial_t (\mathbf{u} \cdot \mathbf{u}^\ell) + \mathbf{u} \cdot \partial_j (u_j \mathbf{u})^\ell + \mathbf{u}^\ell \cdot \partial_j (u_j \mathbf{u}) = \\ -\mathbf{u} \cdot \nabla P^\ell / \rho_0 - \mathbf{u}^\ell \cdot \nabla P / \rho_0 + \nu \mathbf{u} \cdot \Delta \mathbf{u}^\ell + \nu \mathbf{u}^\ell \cdot \Delta \mathbf{u} . \tag{2.36}$$

With the relation:

$$\nabla \cdot \left[(\mathbf{u}^\ell \cdot \mathbf{u}) \mathbf{u} \right] = u_i u_j \partial_j u_i^\ell + u_i^\ell \partial_j (u_i u_j) , \tag{2.37}$$

and the condition of zero divergence on the velocity, we can rewrite the previous expression in the form:

$$\partial_t (\mathbf{u} \cdot \mathbf{u}^\ell) + \nabla \cdot \left[(\mathbf{u} \cdot \mathbf{u}^\ell) \mathbf{u} + (P^\ell / \rho_0) \mathbf{u} + (P / \rho_0) \mathbf{u}^\ell \right] = \\ -u_i \partial_j (u_i u_j)^\ell + u_i u_j \partial_j u_i^\ell + \nu \mathbf{u} \cdot \Delta \mathbf{u}^\ell + \nu \mathbf{u}^\ell \cdot \Delta \mathbf{u} . \tag{2.38}$$

We define the inertial dissipation (per unit mass) from the third-order structure function:

$$\boxed{\mathcal{D}_I^\ell \equiv \frac{1}{4} \int_{\mathbb{R}^3} \nabla \varphi^\ell(\xi) \cdot \left[(\delta \mathbf{u})^2 \delta \mathbf{u} \right] d\xi} . \tag{2.39}$$

After an integration by part and the development of the structure function, we obtain:

$$4 \mathcal{D}_I^\ell = -\partial_j (u^2 u_j)^\ell + 2 u_i \partial_j (u_i u_j)^\ell + u_j \partial_j (u^2)^\ell - 2 u_i u_j \partial_j u_i^\ell . \tag{2.40}$$

Introducing this expression into relation (2.38) gives:

$$\partial_t \left(\frac{\mathbf{u} \cdot \mathbf{u}^\ell}{2} \right) + \nabla \cdot \left[\left(\frac{\mathbf{u} \cdot \mathbf{u}^\ell}{2} \right) \mathbf{u} + \frac{P^\ell \mathbf{u} + P \mathbf{u}^\ell}{2 \rho_0} \right] = \\ \frac{1}{4} \partial_j \left[u_j (u^2)^\ell - (u^2 u_j)^\ell \right] - \mathcal{D}_I^\ell + \frac{\nu}{2} u_i \Delta u_i^\ell + \frac{\nu}{2} u_i^\ell \Delta u_i . \tag{2.41}$$

---

[13] The physicist may, however, wonder whether this mathematical origin of the anomalous dissipation is physically relevant, because the irregularities in question involve a physical description on an infinitely small scale, for which the Euler equations are in theory no longer valid.

After using vector identities (see Appendix B), we find the exact equation for energy conservation (per unit mass):

$$\partial_t \left(\frac{\mathbf{u} \cdot \mathbf{u}^\ell}{2} \right) + \nabla \cdot \mathbf{J}^\ell = -\mathcal{D}_I^\ell - \mathcal{D}_\nu^\ell, \qquad (2.42)$$

where by definition the energy flux (per unit mass) is:

$$\mathbf{J}^\ell \equiv \frac{\mathbf{u} \cdot \mathbf{u}^\ell}{2} \mathbf{u} + \frac{P^\ell \mathbf{u} + P \mathbf{u}^\ell}{2\rho_0} - \frac{\nu \mathbf{u}^\ell \times \mathbf{w} + \nu \mathbf{u} \times \mathbf{w}^\ell}{2} + \frac{(u^2 \mathbf{u})^\ell - \mathbf{u}(u^2)^\ell}{4}, \qquad (2.43)$$

and the viscous dissipation (per unit mass):

$$\mathcal{D}_\nu^\ell \equiv \nu \, \mathbf{w} \cdot \mathbf{w}^\ell. \qquad (2.44)$$

Expression (2.42) can be seen as a weak formulation of the Kármán–Howarth equation (2.19). Unlike that famous equation, the weak formulation does not use the ensemble average: it is therefore a local form of energy conservation. The correlation is made between a field at position $\mathbf{x}$ and another field averaged over a full sphere of radius ℓ centered at the point $\mathbf{x}$; this correlation can be rewritten in the form:

$$\mathbf{u} \cdot \mathbf{u}^\ell = \int_{\mathbb{R}^3} \varphi^\ell(\xi) \mathbf{u}(\mathbf{x}) \mathbf{u}(\mathbf{x} + \xi) d\xi. \qquad (2.45)$$

With this new expression, we see that it is a correlation similar to the Kármán–Howarth equation, but locally averaged. The two right-hand terms of expression (2.42) also have an equivalent: they can be associated with the inertial (after integration by part) and dissipative (after some vectorial manipulations) terms. On the other hand, the energy flux on the left has no equivalent in the Kármán–Howarth equation: it is a purely local contribution that disappears on average (Dubrulle, 2019).

We can now take the limit $\ell \to 0$ (in the sense of distribution) of expression (2.42). In this case, the last two terms of the energy flux compensate each other and we obtain the equation (which is a theorem) of Duchon and Robert (2000):

$$\partial_t \left(\frac{\mathbf{u}^2}{2} \right) + \nabla \cdot \left[\frac{u^2}{2} \mathbf{u} + \frac{P}{\rho_0} \mathbf{u} - \nu \mathbf{u} \times \mathbf{w} \right] = -\mathcal{D}_I - \mathcal{D}_\nu, \qquad (2.46)$$

with by definition:

$$\mathcal{D}_I \equiv \lim_{\ell \to 0} \mathcal{D}_I^\ell \quad \text{and} \quad \mathcal{D}_\nu \equiv \lim_{\ell \to 0} \mathcal{D}_\nu^\ell = \nu \mathbf{w}^2. \qquad (2.47)$$

Equation (2.46) takes all its importance when $\nu = 0$: in the case of Euler's equations, the inertial dissipation is the only term on the right. The evaluation of this dissipation does not pose a problem, since the derivative is applied (before taking the limit) on the function φ^ℓ and the possible discontinuities of the velocity field are absorbed when calculating the integral. This leads to a surprising conclusion for the physicist: an irregular field can potentially contribute to dissipating energy. The anomalous dissipation could therefore have its origin in the nonregularity of the velocity field. Laboratory experiments are compatible with this idea: indeed, measurements show that the mean rate of energy dissipation tends towards a finite value within the limit of a large Reynolds number ($R_e \sim 10^8$) (Ravelet et al., 2008; Saint-Michel et al., 2014). Note that extreme events potentially related to inertial dissipation have been identified experimentally (Saw et al., 2016; Debue et al., 2021).

From a mathematical point of view, the absence of anomalous dissipation requires checking the so-called Hölder's condition.[14] Let us introduce $\delta u_\ell \equiv \sup_{\xi < \ell} |\mathbf{u}(\mathbf{x} + \xi) - \mathbf{u}(\mathbf{x})|$ and assume that this increment satisfies the following relation for small enough ℓ:

$$\delta u_\ell < C\ell^h, \tag{2.48}$$

with C a constant and h the Hölder exponents. From expression (2.39), we can show that:[15]

$$\mathcal{D}_I^\ell = \mathcal{O}(\ell^{3h-1}). \tag{2.49}$$

Therefore, for $h > 1/3$, $\mathcal{D}_I^\ell \to 0$ when $\ell \to 0$ and energy is conserved in the inviscid case. On the other hand, for $h \leq 1/3$, we are in presence of an anomalous dissipation. We can see that Kolmogorov's law is dimensionally compatible with $h = 1/3$, and thus with the existence of an anomalous dissipation. Under certain mathematical conditions concerning the velocity field, we can also show that $\mathcal{D}_I \geq 0$ (Duchon and Robert, 2000; Eyink, 2008). The Euler and Navier–Stokes three-dimensional equations have been the subject of many mathematical studies. For example, in the first case, it has been shown that if a nonregular solution exists then the vorticity is not bounded (Beale et al., 1984); in the second case, it has been shown that the regularity of the velocity is ensured if it remains bounded, and that the singularity associated with the nonregularity is necessarily punctual in time and space (Caffarelli et al., 1982; Constantin, 2008). For the physicist, the result on Euler is intuitive; on the other hand, the result on Navier–Stokes with the appearance of furtive singularities seems not very physical, because the speed of light is an impassable limit.

2.7.3 1D Shocks

To conclude this section, we first note that the anomalous dissipation can be calculated exactly in the particular case of Burgers' equation, which is a one-dimensional hydrodynamic model of shocks (see Exercise I.1). In this case, we can show analytically that $\langle \mathcal{D}_I \rangle = \varepsilon$. This result gives us an indication – but not a

---

[14] See also the work of mathematicians Constantin et al. (1994) carried out in the Besov space.
[15] For a regular velocity field, using a Taylor expansion gives us $h = 1$. In that case, $\mathcal{D}_I^\ell \to 0$ when $\ell \to 0$.

proof – that for three-dimensional hydrodynamics the situation could be the same. However, the task required to conduct such an analysis, for example with data, seems impractical because it requires to study locally all the events in order to sum up all the contributions to inertial dissipation.

We also note that the weak formulation introduced in this section can be used in other systems. For example, the theorem of Duchon and Robert (2000) has been generalized to incompressible magnetohydrodynamics (Galtier, 2018) and, as suggested by the author, the inertial dissipation can be a relevant proxy to measure the local heating in space plasmas (David et al., 2022). Using the one-dimensional Burgers–MHD model for shocks, it is also possible to generalize the proof of the zeroth law of turbulence (i.e. $\langle \mathcal{D}_I \rangle = \varepsilon$) and to use it to evaluate, for example, the heating produced by collisionless shocks near Jupiter (David and Galtier, 2021).

2.8 Intermittency

Intermittency is a still poorly understood property of turbulence. This is often the reason why it is claimed that turbulence is an unresolved problem. This assertion is, however, contradicted by the immense knowledge acquired on the subject over more than 50 years. For example, we have at our disposal intermittency models that can correctly reproduce experimental data for structure functions of the order $p \leq 15$. Even if our knowledge is still limited, we are able to understand an important part of the physics of intermittency. The objective of this section is to present, first of all, a simple and pedagogical model based on the notion of fractals. Then, we will present the two best-known models of intermittency: the log-normal and log-Poisson models.

More information on intermittency is available in the eponymous chapter of Frisch's (1995) book.

2.8.1 What Is Intermittency?

From the exact Kolmogorov four-fifths law (2.27), one may wonder whether it would be possible to extend the analytical development to higher-order structure functions.[16] Until now all attempts have failed, and the closure problem seems insurmountable (see Chapter 1). We can, however, introduce a simple self-similarity hypothesis to extend the Kolmogorov law dimensionally to higher orders. In this case, we get:

$$S_p(\ell) \equiv \langle (\delta u_\ell)^p \rangle = C_p (\varepsilon \ell)^{\zeta_p}, \qquad (2.50)$$

[16] In general, a random variable follows a probability law whose form is all the better known, as one can calculate its moments of high order. In our case, it is the longitudinal velocity increment δu_ℓ that serves as basic random variable.

2.8 Intermittency

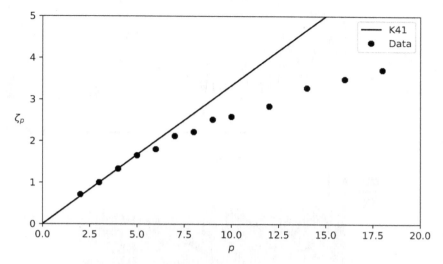

Figure 2.11 Reproduction of experimental measurements (black octagons) of the velocity structure function exponents ζ_p (after Anselmet et al., 1984). The self-similar Kolmogorov law (K41) is plotted for comparison. The deviation from the self-similar law is called intermittency.

with:

$$\zeta_p = p/3, \qquad (2.51)$$

to satisfy the third-order structure function law. It is now widely recognized that the self-similar model of Kolmogorov is not satisfactory, because the structure functions of higher order ($p > 3$) show unambiguously a large deviation of the scaling exponents. Figure 2.11 illustrates schematically this property: the higher the order p, the greater the discrepancy between the measures and the Kolmogorov self-similar law. It is this discrepancy that is called intermittency. In addition, we speak of anomalous exponents for ζ_p, because the latter cannot be predicted by simple dimensional arguments. Note that intermittency is also detected in natural environments such as stratocumulus clouds (Siebert et al., 2010).

A relatively simple way to visualize intermittency is to plot the spatial variation of the velocity as well as its derivative. In Figure 2.12, we see schematically that the intermittent character is amplified on the derivative with the presence of intermittent bursts, that is, the sudden appearance of large-amplitude fluctuations. A Gaussian random function has, however, the same behavior as its derivative. If we now plot the probability density function of the velocity increment δu_ℓ, we see that non-Gaussian wings appear. The smaller the separation between the two points, the larger the wings. In other words, turbulent events of large amplitude are more likely than they would be if the velocity at a point are the result of the sum of independent random events (which follow a Gaussian law). Physically, this means that in a turbulent flow the velocity fluctuations at a point are the result of

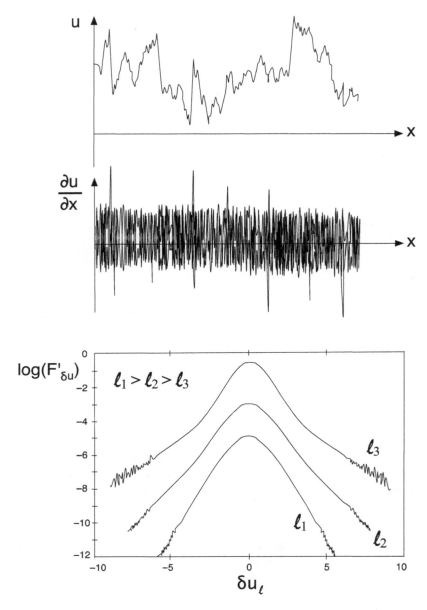

Figure 2.12 Schematic representation of the spatial variation of a turbulent velocity $u(x)$ (top), its derivative (middle), and its probability density function $F'_{\delta u}$ (bottom) for different separations ℓ (see, e.g., Anselmet et al., 1984). The greatest non-Gaussian wings correspond to the smallest separations. Reprinted with permission (Galtier, 2016).

the superposition of the influence of a large number of eddies that operate at different scales. These eddies are not completely independent of each other but have a spatiotemporal memory whose origin lies mainly in the cascade process.

2.8.2 Fractal Model

Among the existing intermittency models, one of the simplest is undoubtedly the fractal model, known as the β–model, introduced by Frisch et al. (1978) (see also Mandelbrot, 1974). As we will see, this model is based on the idea of a fractal cascade; it is therefore by nature a self-similar model. However, since the exponents of the structure functions are not those predicted by the self-similar Kolmogorov law (2.50), we speak of intermittency and anomalous exponents.

The idea underlying the β–fractal model is Richardson's cascade (see Figure 2.6): at each stage of the cascade, the number of eddies is chosen so that the volume (or the surface in the two-dimensional case) occupied by these eddies decreases by a factor β (with $0 < \beta < 1$) compared with the volume (or surface) of the parent eddy. The β factor is a parameter of the model less than one to reflect the fact that the filling factor varies according to the scale considered: the smallest eddies occupy less space than the largest.

We define by ℓ_n the discrete scales of our system: the fractal cascade is characterized by jumps from the scale ℓ_n to the scale ℓ_{n+1}. We show an example of a fractal cascade in Figure 2.13: at each stage of the cascade, the elementary scale is divided by two.[17] So we have:

$$\ell_n = \frac{\ell_0}{2^n}, \qquad (2.52)$$

with ℓ_0 the integral scale, that is, the largest scale of the inertial range. Let p_n be the probability of finding an "active" region after n steps (this is our filling factor), hence:

$$p_n = \beta^n. \qquad (2.53)$$

We suppose that initially $p_0 = 1$; in other words, that in the initial state (at the integral scale) we have a single eddy of the size of the system. We can show that:

$$p_n = \beta^{\frac{\ln(\ell_n/\ell_0)}{\ln(1/2)}}. \qquad (2.54)$$

Now we seek an expression of the form:

$$p_n = \left(\frac{\ell_n}{\ell_0}\right)^{d-D}, \qquad (2.55)$$

where d is the dimension of the system and D its fractal dimension. In our case, another way to introduce the fractal dimension is to define the following relationship between the number of children N that each parent produces and D:

$$N = 2^d \beta \equiv 2^D. \qquad (2.56)$$

---

[17] Since an inverse cascade occurs in two-dimensional hydrodynamic turbulence, Figure 2.13 should be seen as a simple illustration of the concept of fractal cascades. In Figure 2.15, it is a three-dimensional fractal model that is considered.

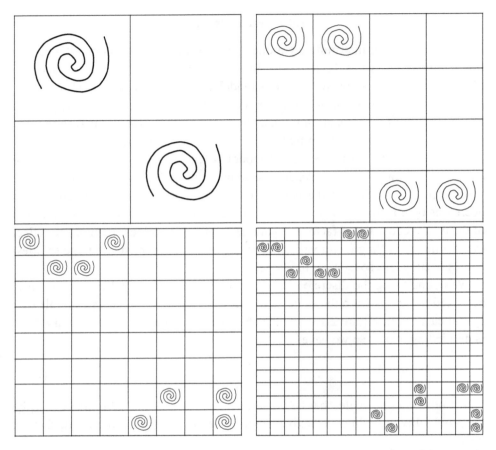

Figure 2.13 A fractal cascade in two dimensions for $\beta = 1/2$. At each stage of the cascade the elementary scale is divided by two and the children eddies occupy only a fraction β of the surface of the parent eddy. At the integral scale (not shown) a single eddy occupies the entire available surface. The fractal dimension of this cascade is $D = 1$.

This finally gives us the general relationship:

$$D = d + \frac{\ln \beta}{\ln 2}. \tag{2.57}$$

The example in Figure 2.13 therefore corresponds to a fractal dimension $D = 1$, whereas that of Figure 2.14 corresponds to $D = \ln 3 / \ln 2 \simeq 1.585$.

From this fractal model, it is possible to predict the scaling laws for the energy spectrum and more generally for the structure functions of order p. Eddies of size ℓ_n fill only a fraction p_n of the considered volume (we only study the three-dimensional case and take $d = 3$). Therefore, one can assume that the energy per unit mass associated with motions at the scale ℓ_n is such that:

$$E_n = u_{\ell_n}^2 p_n = u_{\ell_n}^2 \left(\frac{\ell_n}{\ell_0}\right)^{3-D}. \tag{2.58}$$

2.8 Intermittency

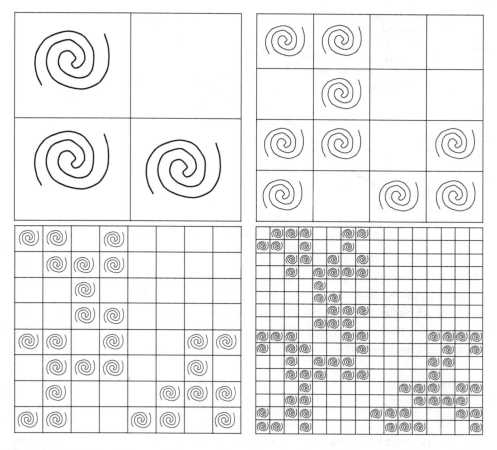

Figure 2.14 A fractal cascade in two dimensions for $\beta = 3/4$ (see the comments in the caption of Figure 2.13). The fractal dimension of this cascade is $D \simeq 1.585$.

Following the usual Kolmogorov phenomenology, one gets:

$$\varepsilon \sim \frac{u_{\ell_n}^3}{\ell_n} p_n. \tag{2.59}$$

We have in particular:

$$\varepsilon \sim \frac{u_{\ell_0}^3}{\ell_0}, \tag{2.60}$$

hence:

$$u_{\ell_n} \sim u_{\ell_0} \left(\frac{\ell_n}{\ell_0}\right)^{(D-2)/3}. \tag{2.61}$$

Finally, we arrive at the following spectral prediction:

$$E_n \sim E(k)k \sim u_{\ell_0}^2 \left(\frac{\ell_n}{\ell_0}\right)^{(5-D)/3}, \tag{2.62}$$

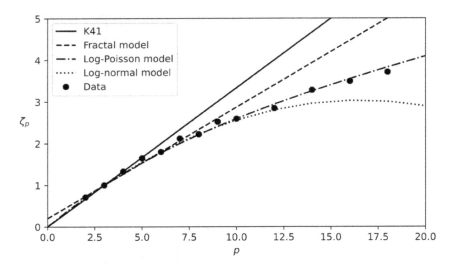

Figure 2.15 Reproduction of experimental measurements (black octagons) of the velocity structure function exponents ζ_p (after Anselmet et al., 1984). Four theoretical models of reference are plotted for comparison: the self-similar Kolmogorov model (K41), the three-dimensional β fractal model with $D = 2.8$ (dashes), the log-Poisson model with $\beta = 2/3$ (dash-dotted line), and the log-normal model with $\mu = 0.2$ (dots).

or, in other words:

$$E(k) \sim k^{-5/3-(3-D)/3}. \tag{2.63}$$

The energy spectrum associated with the fractal cascade is therefore steeper (since $D < 3$) than that of Kolmogorov.

We can generalize this result to the case of structure functions of order p; it yields:

$$S_p(\ell_n) = \langle (\delta u_{\ell_n})^p \rangle \sim p_n u_{\ell_n}^p \sim u_{\ell_0}^p \left(\frac{\ell_n}{\ell_0}\right)^{\zeta_p}, \tag{2.64}$$

with the fractal law:

$$\zeta_p = p/3 + (3-D)(1-p/3). \tag{2.65}$$

As expected, this fractal – self-similar – model gives us a linear relationship between the scaling exponents. This correction is zero for $p = 3$ so that we recover the Kolmogorov exact four-fifths law. For $p = 2$, we find a result compatible with the energy spectrum that we have calculated just before. Furthermore, we note that the Kolmogorov law is obtained for $D = 3$: a fractal dimension identical to the space dimension simply means that all space is filled with eddies. Otherwise (for $D < 3$), the correction is negative and corresponds quantitatively to the measurements made where, for example, $D \simeq 2.8$ has been found for $p < 8$ (see Figure

2.15). This model has subsequently been generalized to bifractal, then multifractal, cases to better describe the curvature of the ζ_p function. In the most trivial case, the bifractal model corresponds to the combination of the Kolmogorov law for $p \leq 3$ and the fractal model for $p > 3$, a situation that we encounter in Burgers' turbulence (see Exercise I.2).

2.8.3 The Log-Normal Model

We have just seen that the β fractal model gives a different law from that of Kolmogorov in introducing the fundamental idea that intermittency is rooted in the nonuniform spatial distribution of turbulent structures. This property is observed very well in direct numerical simulations, with the appearance of clusters of vorticity filaments (Kaneda and Ishihara, 2006). As we see in Figure 2.15, however, the β fractal model becomes less relevant for high values of p, where the data show clearly a curvature of the $\zeta_p(p)$ function. The log-normal model that we present in this section brings this improvement by predicting a nonlinear law in p. In fact, it was historically the first intermittency model: it was introduced by Kolmogorov (1962) and Oboukhov (1962) in order to respond to a criticism made by Landau regarding the potential problems of the self-similar Kolmogorov theory (see Chapter 1).

The new idea introduced in the log-normal model is to rewrite the relationship (2.50) in the form:

$$S_p(\ell) \equiv \langle (\delta u_\ell)^p \rangle = C_p \langle \varepsilon_\ell^{p/3} \rangle \ell^{p/3} , \qquad (2.66)$$

with, by definition:

$$\varepsilon_\ell \equiv \frac{1}{4/3\pi \ell^3} \int_{|\xi| \leq \ell} \varepsilon(\mathbf{x} + \xi) d\xi , \qquad (2.67)$$

which is the (local) dissipation of energy averaged in a full sphere of radius ℓ centered at $\mathbf{x}$. Note in passing that we have the trivial relation $\langle \varepsilon_\ell \rangle = \varepsilon$. In this way, we will take into consideration Landau's remark about the possible fluctuations of the dissipation: for example, $\langle \varepsilon_\ell \rangle^p$ will be even more different from $\langle \varepsilon_\ell^p \rangle$ the stronger the fluctuations of ε_ℓ, but these fluctuations tend to increase as the volume decreases, since it is at small scales that dissipative structures are concentrated.

In this model, it is assumed that the probability density function of ε_ℓ follows a log-normal behavior, that is, the probability density function of $\ln(\varepsilon_\ell)$ is a Gaussian of variance σ_ℓ centered on m_ℓ:

$$F'_{\ln(\varepsilon_\ell)} \equiv \frac{1}{\sqrt{2\pi \sigma_\ell^2}} \exp\left[-\frac{(\ln(\varepsilon_\ell/\varepsilon) - m_\ell)^2}{2\sigma_\ell^2} \right] . \qquad (2.68)$$

The log-normal hypothesis originated in a study by Oboukhov on the pulverization of ore (Davidson et al., 2011). In this process, the pieces of ore get smaller and smaller. After n steps, the size of the ore is ε_n, a product of the initial size ε and the n random fragmentation factors f_i: $\varepsilon_n = \varepsilon f_1 f_2 \ldots f_n$, with $f_i < 1$. If these factors are assumed to be independent, then $\log \varepsilon_n$ is a sum of independent random numbers. When n increases, the statistics of this sum tends to a Gaussian (or normal) law with a variance proportional to n. The analogy between the size of the fragments and that of the eddies justifies this hypothesis because the dissipation rate is fundamentally related to filamentary vorticity structures (see Figure 2.7).

By a change of variables,[18] we obtain:

$$F'_{\varepsilon_\ell} \equiv \frac{1}{\varepsilon_\ell} \frac{1}{\sqrt{2\pi \sigma_\ell^2}} \exp\left[-\frac{(\ln(\varepsilon_\ell/\varepsilon) + \sigma_\ell^2/2)^2}{2\sigma_\ell^2}\right], \qquad (2.69)$$

where we have used the relation $m_\ell = -\sigma_\ell^2/2$, which allows us to verify $\langle \varepsilon_\ell \rangle = \varepsilon$. We deduce:[19]

$$\langle \varepsilon_\ell^n \rangle = \epsilon^n \exp\left(\frac{n\sigma_\ell^2}{2}(n-1)\right). \qquad (2.70)$$

Insofar as we are looking for power-law solutions, it is convenient to introduce the following form for the variance:

$$\sigma_\ell^2 \equiv \mu \ln\left(\frac{\ell_0}{\ell}\right), \qquad (2.71)$$

where μ is a constant assumed to be universal. We introduce (2.70) into (2.66), which finally gives:

$$S_p(\ell) = C_p(\varepsilon \ell_0)^{p/3} \left(\frac{\ell}{\ell_0}\right)^{\zeta_p}, \qquad (2.72)$$

with the log-normal law:

$$\boxed{\zeta_p = \frac{p}{3} - \frac{\mu}{18} p(p-3)}. \qquad (2.73)$$

The first intermittency measurements gave, with a relatively good accuracy, the anomalous exponents ζ_p for $p < 10$, and allow us to conclude that $\mu \simeq 0.2$. More recent measurements have, however, shown that for $p > 10$ a clear divergence appears (see Figure 2.15), invalidating the log-normal model. Despite these limitations, it is interesting to see what happens for $p = 2$; we obtain $S_2(\ell) = C_2 \varepsilon^{2/3} \ell^{2/3} (\ell/\ell_0)^{\mu/9}$, and hence the spectrum:

$$\boxed{E(k) \sim \varepsilon^{2/3} k^{-5/3} (k\ell_0)^{-\mu/9}}. \qquad (2.74)$$

---

[18] We recall the following mathematical property: the probability of a differential surface remains invariant under a change of variables, that is, $|F'_{\ln(\varepsilon_\ell)} d\ln(\varepsilon_\ell)| = |F'_{\varepsilon_\ell} d\varepsilon_\ell|$.

[19] We recall that: $\int_{-\infty}^{+\infty} \exp(-\alpha x^2 - \beta x) dx = \sqrt{\pi/\alpha} \exp(\beta^2/4\alpha)$, with $\alpha > 0$.

We see that the Kolmogorov energy spectrum undergoes a slight correction of approximately -0.02 (with $\mu \simeq 0.2$) and thus becomes steeper. This is actually a trend observed in the experiments and direct numerical simulations.

In conclusion, we can say that as a first approximation the log-normal law is relatively clearly observed (Anselmet et al., 2001). However, the model shows signs of weakness for $p > 10$: this finding motivated new work which culminated in the log-Poisson model.

2.8.4 The Log-Poisson Model

The log-Poisson model of intermittency was proposed by She and Leveque (1994). It is currently a very used model insofar as it reproduces very well the data up to values of $p \simeq 16$. It is based on the following three assumptions:

- Existence of a scaling law for structure functions:

$$S_p(\ell) \equiv \langle (\delta u_\ell)^p \rangle = C_p \langle \varepsilon_\ell^{p/3} \rangle \ell^{p/3}. \tag{2.75}$$

- Existence of a recurrence relation between the moments of the distribution of the local energy dissipation ε_ℓ:

$$\frac{\langle \varepsilon_\ell^{p+1} \rangle}{\varepsilon_\ell^\infty \langle \varepsilon_\ell^p \rangle} = A_p \left(\frac{\langle \varepsilon_\ell^p \rangle}{\varepsilon_\ell^\infty \langle \varepsilon_\ell^{p-1} \rangle} \right)^\beta, \quad 0 < \beta < 1, \tag{2.76}$$

where A_p are some constants and $\varepsilon_\ell^\infty \equiv \lim_{p \to \infty} \langle \varepsilon_\ell^{p+1} \rangle / \langle \varepsilon_\ell^p \rangle$ is a quantity that is mainly sensitive to the tail of the distribution of ε_ℓ. (We always have $\langle \varepsilon_\ell \rangle = \varepsilon$.)

- Existence of divergence scale dependence associated with the most intermittent dissipative structures:

$$\lim_{\ell \to 0} \varepsilon_\ell^\infty \sim \ell^{-2/3}. \tag{2.77}$$

Upon introducing the scaling relation $\langle \varepsilon_\ell^p \rangle \sim \ell^{\tau_p}$, expression (2.76) leads to the following recurrence relation:

$$\tau_{p+1} - (1 + \beta)\tau_p + \beta \tau_{p-1} + \frac{2}{3}(1 - \beta) = 0. \tag{2.78}$$

By defining $\tau_p = -2p/3 + 2 + f_p$, we get a simple form of linear recursive progression of order two:

$$f_{p+1} - (1 + \beta)f_p + \beta f_{p-1} = 0, \tag{2.79}$$

which we can solve easily; we find $f_p = \lambda + \mu \beta^p$. The coefficients λ and μ are determined from the initial conditions $\tau_0 = \tau_1 = 0$, hence:

$$\tau_p = \frac{2}{3}\left(\frac{1 - \beta^p}{1 - \beta} - p\right). \tag{2.80}$$

Finally, we obtain the following log-Poisson law (with by definition, $S_p(\ell) \sim \ell^{\zeta_p}$):

$$\boxed{\zeta_p = \frac{p}{3} + \frac{2}{3}\left(\frac{1-\beta^{p/3}}{1-\beta} - \frac{p}{3}\right)}. \qquad (2.81)$$

We see that $\lim_{\beta \to 1} \zeta_p = p/3$: the parameter β measures the degree of intermittency in the sense that the smaller the value, the stronger the intermittency. The best agreement with the data is obtained for $\beta = 2/3$ (see Figure 2.15). The log-Poisson law correctly predicts the anomalous exponents up to approximately $p = 16$. The latest measurements seem to show that beyond 16 a divergence appears, perhaps revealing the limits of the model. Note, however, that precautions have to be taken in the statistical analysis because the uncertainties increase with the order p since the statistical sample is finite. Probably 16 is already too high for a serious comparison with a model.

The log-Poisson model owes its name to the fact that the recurrence (2.76) can be interpreted as an underlying statistical property of the distribution ε_ℓ (Dubrulle, 1994). Let us rewrite this recurrence for the variable, $\pi_\ell = \varepsilon_\ell/\varepsilon_\ell^\infty$; we get by a simple substitution:

$$\langle \pi_\ell^{p+1} \rangle = A_p \frac{\langle \pi_\ell^p \rangle^{\beta+1}}{\langle \pi_\ell^{p-1} \rangle^\beta}. \qquad (2.82)$$

We show easily that:

$$\langle \pi_\ell^p \rangle = A_{p-1}\left(A_{p-2}\frac{\langle \pi_\ell^{p-2} \rangle^{\beta+1}}{\langle \pi_\ell^{p-3} \rangle^\beta}\right)^{\beta+1} \frac{1}{\langle \pi_\ell^{p-2} \rangle^\beta} = A_{p-1}A_{p-2}^{\beta+1}\frac{\langle \pi_\ell^{p-2} \rangle^{\beta^2+\beta+1}}{\langle \pi_\ell^{p-3} \rangle^{\beta(\beta+1)}}$$

$$= \ldots = A_{p-1}A_{p-2}^{\beta+1}A_{p-3}^{\beta^2+\beta+1}\ldots A_0^{\beta^{p-1}+\ldots+1}\langle \pi_\ell \rangle^{\beta^{p-1}+\ldots+1}$$

$$= B_p \langle \pi_\ell \rangle^{\frac{1-\beta^p}{1-\beta}}, \qquad (2.83)$$

where the coefficients B_p depend on A_q (with $0 \le q < p$). The distribution F'_{π_ℓ} corresponding to the moments (2.83) is a generalized Poisson distribution for the variable $Y = \ln \pi_\ell / \ln \beta$, namely:

$$F'_Y = \frac{e^{-\lambda}\lambda^Y}{Y!}, \qquad (2.84)$$

where λ is the variance. With the change of variables, we get:

$$\langle \pi_\ell^p \rangle = \int \pi_\ell^p \frac{e^{-\lambda}\lambda^Y}{Y!}e^{-Y\ln\beta}d\pi_\ell = \int e^{pY\ln\beta}\frac{e^{-\lambda}\lambda^Y}{Y!}dY = e^{-\lambda}\int \frac{(\lambda\beta^p)^Y}{Y!}dY$$

$$= e^{-\lambda(1-\beta^p)}. \qquad (2.85)$$

The condition $p = 0$ fixes the variance:

$$\lambda = \frac{1}{\beta - 1}\ln\langle \pi_\ell \rangle. \qquad (2.86)$$

Hence, finally:
$$\langle \pi_\ell^p \rangle = e^{\frac{1-\beta^p}{1-\beta} \ln\langle \pi_\ell \rangle} = \langle \pi_\ell \rangle^{\frac{1-\beta^p}{1-\beta}}. \tag{2.87}$$

We can even show that the relationship (2.83) is still satisfied if the distribution F'_Y is the convolution of a Poisson distribution with any other distribution. In this case, a coefficient appears in the calculation and $B_p \neq 1$.

2.8.5 Exact Constraints

To conclude on this subject of intermittency, let us point out that there are exact constraints that must be satisfied by the exponents ζ_p (Constantin and Fefferman, 1994; Frisch, 1995). The first is obviously $\zeta_3 = 1$. A second is the convexity constraint:
$$\boxed{(p_3 - p_1)\zeta_{2p_2} \geq (p_3 - p_2)\zeta_{2p_1} + (p_2 - p_1)\zeta_{2p_3}}, \tag{2.88}$$
for exponent orders such as $p_1 \leq p_2 \leq p_3$. A third states that if:
$$\boxed{\zeta_{2p} > \zeta_{2p+2}}, \tag{2.89}$$
then the velocity fluctuations become unbounded. It is therefore necessary to consider exponents which do not decrease strictly to ensure that these fluctuations remain subsonic. We note, in particular, that the log-normal model does not satisfy this third constraint, while it is actually the case for the log-Poisson model.

2.9 Compressible Turbulence

Most of the fundamental results in strong turbulence are limited to incompressible fluids. We can find at least three reasons for this: (i) the study of turbulence is more difficult in the compressible case and the basic concepts are still under debate, (ii) in laboratory experiments where the compressibility is strong, turbulence is generally statistically inhomogeneous, which limits the universality of the behavior, (iii) the supersonic homogeneous turbulence regime mainly concerns astrophysics, a field where important advances in turbulence have been made only recently. In Section 2.9.1 present in a few lines the main properties of interstellar turbulence, where the turbulent Mach numbers are often approaching 100. In this medium, it is believed that turbulence plays a key role by regulating the rate of star formation. Then, we will show how Kolmogorov's exact law in hydrodynamics can be generalized to compressible fluids.

2.9.1 Supersonic Turbulence in Astrophysics

Understanding star formation is one of the major challenges of modern astrophysics (Mac Low and Klessen, 2004; McKee and Ostriker, 2007). Although it is well established that stars form in molecular clouds by collapse, many fundamental questions remain controversial, because the star formation is governed by the interstellar cycle, which involves a wide range of spatial and temporal scales, and a great diversity of coupled physical phenomena such as turbulence, magnetic field, gravity, radiation, and cosmic rays. A major problem is the low rate of star formation observed. To understand this problem, we can evaluate the free-fall time τ_{ff}, that is, the time it takes for a spherical cloud without pressure to collapse to a point due to its own gravity. For our galaxy, we get (Spitzer, 1978):

$$\tau_{ff} = \sqrt{\frac{3\pi}{32G\rho}} \simeq 4 \times 10^6 \text{ years}. \qquad (2.90)$$

The Milky Way contains a mass of molecular clouds of about $M_c \simeq 2 \times 10^9 M_\odot$ (with $M_\odot$ the mass of the sun) (Solomon et al., 1987), which gives us a rate of formation:

$$\chi_* \sim \frac{M_c}{\tau_{ff}} \simeq 500 \, M_\odot/\text{year}. \qquad (2.91)$$

The problem is that the rate χ_* measured in our galaxy is about $2 \, M_\odot/\text{year}$ (Chomiuk and Povich, 2011), which is more than two orders of magnitude lower than the previous estimate. Although relatively simple, this evaluation highlights a major problem whose origin could well be turbulence: turbulent fluctuations may act as an additional pressure which slows down the collapse of clouds.

The first observational signatures of interstellar turbulence date back to the 1950s, when spectral line widening, such as that of the ionized oxygen atom OIII, was detected (Münch, 1958). The widening of spectral lines coming from cold clouds proves to be much larger than one would expect for low-temperature gases. This anomaly is attributed to turbulence whose velocity exceeds by far that of sound ($c_s \sim 0.5$ km/s), which means that the turbulence is supersonic. Molecular clouds were discovered in the 1970s with the detection of an intense radiation line emitted by the Orion nebula. This line was attributed to a transition of carbon monoxide (CO) molecules (Wilson et al., 1970; Penzias et al., 1972). Since then, there has been a lot of discussion about molecular clouds. CO molecules are considered to be one of the best tracers for measuring interstellar dynamics: for example, we know by spectroscopic analysis that molecular clouds are cold ($T < 10$ K) and dense ($n > 10^9$ m^{-3}). Nowadays, their observations in the galaxy cover a wide range of length scales, from a few 100 pc to around a millisparsec (1pc $\simeq 3 \times 10^{16}$ m). New instruments are now available to probe interstellar matter. Perhaps the most important is the ALMA (Atacama Large Millimeter Array) antenna array in Chile, which considerably increases the available spatial resolution with the best submillimeter and millimeter interferometers. In this way it is possible to map cold and dark regions that cannot be observed in optics, where new stars and their associated planetary systems form. In particular, the observations give

2.9 Compressible Turbulence

Figure 2.16 Interstellar filaments at a distance of 500 pc from the Earth observed in the infrared domain (70, 250, and 500 μm), using the European space telescope Herschel. Stars are born along these filaments. Credits: ESA/Herschel/SPIRE/PACS/D. Arzoumanian (CEA Saclay) for the "Gould Belt survey" Key Programme Consortium.

access to fine structures at scales close to the dissipative scale of turbulence in molecular clouds (Baudry et al., 2016).

The systematic observation of molecular clouds in the Milky Way has made it possible to highlight the scaling law (Heyer and Brunt, 2004):

$$\langle \delta v \rangle \sim \ell^{+0.56}, \tag{2.92}$$

with δv the velocity increment between two points separated by the distance ℓ. This law is observed on more than three orders of magnitude, between 0.04 pc and 40 pc. Dimensionally, the law (2.92) is compatible with the spectrum of velocity fluctuations:

$$E^u(k) \sim k^{-2.1}, \tag{2.93}$$

which is significantly steeper than the Kolmogorov spectrum. It is interesting to note that this scaling law is close to the Burgers' spectrum in k^{-2} where the dynamics are governed by shocks (see Exercise 1).

The imaging allows us to characterize the interstellar medium and to show that this medium is in the form of gigantic networks of filaments in which stars are born (see Figure 2.16). Recent measurements with the Herschel/ESA space telescope (Arzoumanian et al., 2011) or ALMA (Fukui et al., 2019) suggest that these filaments, which extend over tens of parsecs, have approximately the same thickness ($\sim$ 1pc) regardless of the density or length of these filaments (see, however, Panopoulou et al., 2022). This characteristic scale may correspond to the sonic scale, that is, the scale under which interstellar turbulence passes from a supersonic to a subsonic regime. This property has been observed in numerical simulations where, furthermore, it is found that the velocity dispersion inside the filaments is subsonic and supersonic outside (Federrath, 2016). Compressible turbulence thus appears as a key ingredient to understand the dynamics of interstellar clouds and in particular the rate of star formation (Elmegreen and Scalo, 2004; Chabrier and Hennebelle, 2011).

2.9.2 Generalized Exact Law

As mentioned at the start of Section 2.9, the study of compressible turbulence is more difficult than in the incompressible case, and the number of results is more limited. Note that compressible turbulence can be strong or weak. In the latter case, the regime – called acoustic wave turbulence – is characterized by relatively small mass density fluctuations with a turbulent Mach number M_t much smaller than one (subsonic turbulence); this regime will be discussed in Chapter 4. Intermediate turbulent Mach numbers ($0 < M_t < 1$) is a domain widely studied both experimentally and theoretically whose description would require much more than a chapter (Sagaut and Cambon, 2008). In this domain, we note that recent progress has been made in finding a universal behavior (Donzis and John, 2020). In the present section, the density variation is, however, not limited and the turbulence can possibly be supersonic ($M_t > 1$) and even hypersonic ($M_t \gg 1$), with no possibility to find a small parameter.

For the derivation of the Kolmogorov exact law, we shall consider the simplest case: that of isothermal compressible hydrodynamics for which the equations are:

$$\frac{\partial \rho}{\partial t} + \nabla \cdot (\rho \mathbf{u}) = 0, \quad (2.94a)$$

$$\frac{\partial \rho \mathbf{u}}{\partial t} + \nabla \cdot (\rho \mathbf{u}\mathbf{u}) = -\nabla P + \mu \Delta \mathbf{u} + \frac{\mu}{3}\nabla \theta, \quad (2.94b)$$

with ρ the mass density, μ the dynamic viscosity, and $\theta \equiv \nabla \cdot \mathbf{u}$ the dilatation. The isothermal closure leads to the pressure-density relationship:

$$P = c_s^2 \rho, \quad (2.95)$$

with c_s the sound speed, which is constant in this approximation. This system of equations is often used to study interstellar supersonic turbulence (Kritsuk et al., 2007; Federrath et al., 2008). The implicit assumption made with this model is that the thermal processes (which include necessarily a cooling function like radiation), are much faster than the turbulence dynamics, so that on the turbulence timescale the environment appears isothermal. We will see that the filamentary structures produced in supersonic isothermal turbulence are very similar to observations of molecular clouds (see Figure 2.16). Equations (2.94) verify the conservation of free energy E (also called total energy), namely:[20]

$$\frac{\partial \langle E \rangle}{\partial t} = -\mu \langle \mathbf{w}^2 \rangle - \frac{4}{3}\mu \langle \theta^2 \rangle, \qquad (2.96)$$

with $E = \rho \mathbf{u}^2/2 + \rho U$ and the internal energy $U = c_s^2 \ln(\rho/\rho_0)$ (ρ_0 is the mean mass density). Among the questions currently being debated there is that concerning the anomalous dissipation. While it seems likely that the first term in the right-hand side of equation (2.96) will lead to a nonzero contribution in the small viscosity μ limit (as observed in incompressible turbulence), it is not clear whether the second term will also lead to an anomalous contribution. Numerical studies in this direction exist but are limited to relatively small Reynolds numbers and, more importantly, to mainly subsonic turbulence (Jagannathan and Donzis, 2016).

The formalism presented previously in this chapter will be applied to equations (2.94) in order to obtain an exact law for homogeneous isothermal compressible turbulence. This law, which was obtained for the first time by Galtier and Banerjee (2011), brings up a new type of term – a source S – in addition to the classical flux vector $\mathbf{F}$. Within the stationary limit and a large Reynolds number, the exact law is written (no proof is given here since the technique of derivation is similar to incompressible turbulence):

$$\boxed{-4\varepsilon = \nabla_\ell \cdot \mathbf{F} + S}, \qquad (2.97)$$

with ε the mean rate of free energy dissipation; the expressions for the flux and source are respectively:

$$F \equiv \langle \bar{\delta}\rho (\delta \mathbf{u})^2 \delta \mathbf{u} \rangle, \qquad (2.98a)$$

$$S \equiv -\frac{1}{2} \langle (\rho\theta' + \rho'\theta)(\delta \mathbf{u})^2 \rangle, \qquad (2.98b)$$

where by definition $\bar{\delta}\rho \equiv (\rho + \rho')/2$. Note that we show here the modern form of the law (Ferrand et al., 2020). The isothermal compressible hydrodynamic law is reminiscent of the classical Kolmogorov law, which is found when the density is constant: in the incompressible limit $\theta \to 0$ the flux is identified with the expression obtained previously in Section 2.5 (it is necessary to introduce $\bar{\delta}\rho = \rho_0$ and use $\bar{\varepsilon} = \varepsilon/\rho_0$). We note in passing that expressions (2.98) are Galilean invariant.

The relative simplicity of expression (2.98b) allows us to understand easily its role in compressible turbulence (see Figure 2.17). We can see that in a compression phase, $\theta < 0$ and thus $S > 0$, whereas in a dilatation phase $\theta > 0$ and $S < 0$.

---

[20] By free energy we mean the kinetic energy plus the work part of the internal energy. As usual in turbulence (compressible or not), the heat term related to entropy is not included (the isothermal closure used can be seen as isentropic).

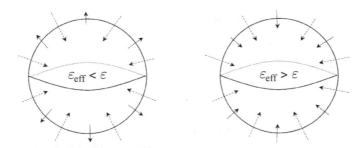

Figure 2.17 Dilatation (left) and compression (right) phases in space correlation for isotropic turbulence. In a direct cascade scenario, the flux vectors (dashed arrows) are oriented towards the center of the sphere. Dilatation and compression (solid arrows) are additional effects which act, respectively, in the opposite or in the same direction as the flux vectors.

Therefore, in a compression phase S will modify the rate ε so that the effective rate $\varepsilon_{\text{eff}} \equiv \varepsilon + S/4$ will be bigger than ε, while ε_{eff} will be smaller than ε in a dilatation phase. The source can therefore be interpreted as a term whose overall effect is to modify the apparent cascade rate, increasing or decreasing it as follows if the turbulence compresses or expands the fluid. Note that the primary form of the exact compressible law has been studied in great detail using three-dimensional supersonic numerical simulations (Kritsuk et al., 2013): an extended inertial range was found where the theoretical predictions have been verified.

The derivation of the isothermal law (2.97) relies on the existence of a range of scales where the physics is immune from direct effects of viscosity and large-scale forcing. Such a property has been studied by Aluie (2011, 2013) under a rigorous framework based on coarse-graining. The technique used is similar to that introduced previously in this chapter in the context of the inertial dissipation. The same author also shows that under mild assumptions on the structure functions (for velocity and density), the mean kinetic and internal energy budgets decouple statistically beyond a transitional conversion range and behave therefore like two invariants, and that the cascade is dominated by local interactions. Several comments can be made. First, the decoupling in question is found numerically at subsonic scales (Aluie et al., 2012), which means that supersonic turbulence is probably more complex to describe. Second, the locality of the interactions prevents the direct transfer of kinetic energy from a large scale to small (dissipative) scales, such as into shocks, as is commonly believed.[21] Third, the fact that the kinetic energy behaves as an invariant, while it is not, does not mean that subsonic turbulence is incompressible (Ferrand et al., 2020). Overall, one can say that these results found with a coarse-graining approach provide strong arguments in favor of the universal law (2.97).

---

[21] This belief is partially based on our knowledge of Burgers' equation (see Exercise 1).

2.9.3 Phenomenology

Compressible turbulence is physically expected to be subsonic at small scales, that is, where density fluctuations and the source term are relatively small. The question is then to know from which scale the source becomes comparable, or even superior, to the first/flux term in the right-hand side of expression (2.97). This transition scale, ℓ_s, can be roughly evaluated from the following ratio:

$$\frac{\nabla_\ell \cdot \mathbf{F}}{S} \sim \frac{\langle \bar{\delta}\rho(\delta u)^3 \rangle}{\ell \langle \rho\theta(\delta u)^2 \rangle}. \quad (2.99)$$

When this ratio is of the order of unity, one obtains: $\ell_s \sim \langle \delta u \rangle / \langle \theta \rangle$. Comparison with numerical simulations shows that ℓ_s coincides well with the sonic scale.

Based on the exact law (2.97), we can propose a phenomenology for compressible turbulence. The law suggests a (isotropic) spectrum of the velocity having the shape:

$$E(k) \sim \varepsilon_{\text{eff}}^{2/3} k^{-5/3}, \quad (2.100)$$

with an effective energy flux ε_{eff} which depends on the scale considered. At supersonic scales ($k < k_s$), this flux varies according to a scaling law deduced from the source: a possible estimate deduced from numerical simulation (Ferrand et al., 2020) suggests a variation of the source in $k^{-1/2}$, which leads us to a velocity spectrum in k^{-2}. At subsonic scales ($k > k_s$) this effective flux is approximately constant because the source is negligible. The associated spectrum could therefore be that of Kolmogorov in $k^{-5/3}$. A schematic illustration of this compressible turbulence regime is given in Figure 2.18. Note that we can also expect to observe the acoustic wave turbulence regime at subsonic scales if the conditions are satisfied (see Chapter 4). In this case, we may find a spectrum in $k^{-3/2}$ for the dilatation (i.e. compressible) component of the velocity (Zakharov and Sagdeev, 1970).

We show in Figure 2.19 the result of a numerical simulation of the isothermal compressible hydrodynamic equations at very high spatial resolution ($10\,048^3$ points). A mixed compressible–solenoidal external force is applied at a large scale and the turbulent Mach number is around 4. A visualization of θ in a slice of the simulation cube reveals a structured environment with large regions where θ is mainly positive and others in the form of filaments where θ is negative. These compression filaments resemble the interstellar filaments (see Figure 2.16) in which stars are formed. As mentioned in Section 2.9.1, interstellar filaments might be characterized by a thickness of the order of the sonic scale (Arzoumanian et al., 2011), a property also found in some numerical simulations (Federrath, 2016). In the light of numerical simulations, it appears that supersonic turbulence produces planar shocks whose collisions lead to the formation of filaments. Their thickness

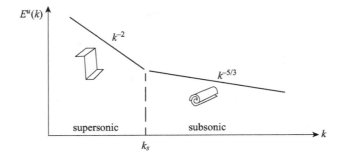

Figure 2.18 A compressible phenomenology based on the exact law suggests a velocity spectrum $E^u(k)$ characterized by an effective flux of energy nonconstant at supersonic scales ($k < k_s$) and approximately constant at subsonic scales ($k > k_s$). These spectra correspond to a physics dominated by shocks ($E^u(k) \sim k^{-2}$) and vortices ($E^u(k) \sim k^{-5/3}$), respectively. We can also envisage a physics dominated by waves at subsonic scales ($E^u(k) \sim k^{-3/2}$).

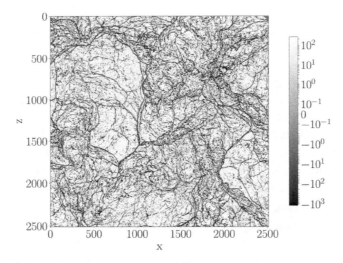

Figure 2.19 Dilatation θ in a x-z plane of a cube of simulation of supersonic isothermal compressible hydrodynamic turbulence with spatial resolution $10\,048^3$ and a turbulent Mach number of 4 (Ferrand et al., 2020; Federrath et al., 2021). We can distinguish large white regions where θ is mainly positive (dilatation zone) and narrow dark filamentary regions where θ is negative (compression zone). Image produced by R. Ferrand.

is limited at the sonic scale because the physics underneath is governed by eddies. If gravity is also present, numerical simulations show that at even smaller scales it can override the turbulent dynamics to lead to a gravitational fragmentation and star formation (Federrath, 2016).

References

Aluie, H. 2011. Compressible turbulence: The cascade and its locality. *Phys. Rev. Lett.*, **106**(17), 174502.

Aluie, H. 2013. Scale decomposition in compressible turbulence. *Physica D Nonlinear Phenomena*, **247**(1), 54–65.

Aluie, H., Li, S., and Li, H. 2012. Conservative cascade of kinetic energy in compressible turbulence. *Astrophys. J. Lett.*, **751**(2), L29 (6 pages).

Anselmet, F., Gagne, Y., Hopfinger, E. J., and Antonia, R. A. 1984. High-order velocity structure functions in turbulent shear flows. *J. Fluid Mech.*, **140**, 63–89.

Anselmet, F., Antonia, R. A., and Danaila, L. 2001. Turbulent flows and intermittency in laboratory experiments. *Plan. Space Sci.*, **49**(12), 1177–1191.

Antonia, R. A., and Burattini, P. 2006. Approach to the 4/5 law in homogeneous isotropic turbulence. *J. Fluid Mech.*, **550**, 175–184.

Antonia, R. A., Ould-Rouis, M., Anselmet, F., and Zhu, Y. 1997. Analogy between predictions of Kolmogorov and Yaglom. *J. Fluid Mech.*, **332**, 395–409.

Arzoumanian, D., André, P., Didelon, P. et al. 2011. Characterizing interstellar filaments with Herschel in IC 5146. *Astron. Astrophys.*, **529**, L6.

Banerjee, S., and Galtier, S. 2017. An alternative formulation for exact scaling relations in hydrodynamic and magnetohydrodynamic turbulence. *J. Phys. A: Math. Theor.*, **50**(1), 015501.

Bardos, C. 1972. Existence et unicité de la solution de l'équation d'Euler en dimension deux. *J. Math. Anal. Appl.*, **40**(3), 769–790.

Batchelor, G. K. 1953. *The Theory of Homogeneous Turbulence*. Cambridge University Press.

Baudry, A., Brouillet, N., and Despois, D. 2016. Star formation and chemical complexity in the Orion nebula: A new view with the IRAM and ALMA interferometers. *Comptes Rendus Physique*, **17**, 976–984.

Beale, J. T., Kato, T., and Majda, A. 1984. Remarks on the breakdown of smooth solutions for the 3-D Euler equations. *Comm. Math. Physics*, **94**(1), 61–66.

Brachet, M. E., Meiron, D. I., Orszag, S. A. et al. 1983. Small-scale structure of the Taylor–Green vortex. *J. Fluid Mech.*, **130**, 411–452.

Brenner, M. P., Hormoz, S., and Pumir, A. 2016. Potential singularity mechanism for the Euler equations. *Phys. Rev. Fluids*, **1**(8), 084503.

Caffarelli, L., Kohn, R., and Nirenberg, L. 1982. Partial regularity of suitable weak solutions of the Navier–Stokes equations. *Comm. Pure Appl. Math.*, **35**(6), 771–831.

Chabrier, G., and Hennebelle, P. 2011. Dimensional argument for the impact of turbulent support on the stellar initial mass function. *Astron. Astrophys.*, **534**, A106 (4 pages).

Chomiuk, L., and Povich, M. S. 2011. Toward a unification of star formation rate determinations in the Milky Way and other galaxies. *Astronom. J.*, **142**, 197 (16 pages).

Constantin, P. 2008. Euler and Navier–Stokes equations. *Pub. Math.*, **52**(2), 235–265.

Constantin, P., and Fefferman, C. 1994. Scaling exponents in fluid turbulence: Some analytical results. *Nonlinearity*, **7**, 41–57.

Constantin, P., Weinan, E., and Titi, S. 1994. Onsager's conjecture on the energy conservation for solutions of Euler's equation. *Commun. Math. Phys.*, **165**, 207–209.

David, V., and Galtier, S. 2021. Proof of the zeroth law of turbulence in one-dimensional compressible MHD and shock heating. *Phys. Rev. E*, **103**(6), 063217.

David, V., Galtier, S., Sahraoui, F., and Hadid, L. Z. 2022. Energy transfer, discontinuities and heating in the inner solar wind measured with a weak and local formulation of the Politano-Pouquet law. *Astrophys. J.*, arXiv:2201.02377.

Davidson, P. A., Kaneda, Y., Moffatt, K., and Sreenivasan, K. R. 2011. *A Voyage Through Turbulence*. Cambridge University Press.

Debue, P., Valori, V., Cuvier, C. et al. 2021. Three-dimensional analysis of precursors to non-viscous dissipation in an experimental turbulent flow. *J. Fluid Mech.*, **914**, A9 (20 pages).

Donzis, D. A., and John, J. P. 2020. Universality and scaling in homogeneous compressible turbulence. *Phys. Rev. Fluids*, **5**(8), 084609.

Dubrulle, B. 1994. Intermittency in fully developed turbulence: Log-Poisson statistics and generalized scale covariance. *Phys. Rev. Lett.*, **73**, 959–962.

Dubrulle, B. 2019. Beyond Kolmogorov cascades. *J. Fluid Mech.*, **867**, P1 (63 pages).

Duchon, J., and Robert, R. 2000. Inertial energy dissipation for weak solutions of incompressible Euler and Navier–Stokes equations. *Nonlinearity*, **13**, 249–255.

Elmegreen, B. G., and Scalo, J. 2004. Interstellar turbulence I: Observations and processes. *Ann. Rev. Astron. Astrophys.*, **42**, 211–273.

Eyink, G. L. 2008. Dissipative anomalies in singular Euler flows. *Physica D*, **237**(14–17), 1956–1968.

Eyink, G. L., and Sreenivasan, K. R. 2006. Onsager and the theory of hydrodynamic turbulence. *Rev. Mod. Phys.*, **78**(1), 87–135.

Federrath, C. 2016. On the universality of interstellar filaments: Theory meets simulations and observations. *MNRAS*, **457**(1), 375–388.

Federrath, C., Klessen, R. S., and Schmidt, W. 2008. The density probability distribution in compressible isothermal turbulence: Solenoidal versus compressive forcing. *Astrophys. J. Lett.*, **688**, L79–L82.

Federrath, C., Klessen, R. S., Iapichino, L., and Beattie, J. R. 2021. The sonic scale of interstellar turbulence. *Nat. Astron.*, **5**, 365–371.

Ferrand, R., Galtier, S., Saharoui, F., and Federrath, C. 2020. Compressible turbulence in the interstellar medium: New insights from a high-resolution supersonic turbulence simulation. *Astrophys. J.*, **904**, 160 (7 pages).

Fournier, J.-D., and Galtier, S. 2001. Meromorphy and topology of localized solutions in the Thomas–MHD model. *J. Plasma Phys.*, **65**(5), 365–406.

Frisch, U. 1995. *Turbulence*. Cambridge University Press.

Frisch, U., Sulem, P.-L., and Nelkin, M. 1978. A simple dynamical model of intermittent fully developed turbulence. *J. Fluid Mech.*, **87**, 719–736.

Fukui, Y., Tokuda, K., Saigo, K. et al. 2019. An ALMA view of molecular filaments in the Large Magellanic Cloud. I. The formation of high-mass stars and pillars in the N159E-Papillon Nebula triggered by a cloud-cloud collision. *Astrophys. J.*, **886**(1), 14.

Galtier, S. 2016. *Introduction to Modern Magnetohydrodynamics*. Cambridge University Press.

Galtier, S. 2018. On the origin of the energy dissipation anomaly in (Hall) magnetohydrodynamics. *J. Physics A: Math. Theo.*, **51**(20), 205501.

Galtier, S., and Banerjee, S. 2011. Exact relation for correlation functions in compressible isothermal turbulence. *Phys. Rev. Lett.*, **107**(13), 134501.

Heyer, M. H., and Brunt, C. M. 2004. The universality of turbulence in galactic molecular clouds. *Astrophys. J.*, **615**, L45–L48.

Ishihara, T., Kaneda, Y., Morishita, K., Yokokawa, M., and Uno, A. 2020. Second-order velocity structure functions in direct numerical simulations of turbulence with R_λ up to 2250. *Phys. Rev. Fluids*, **5**(10), 104608.

Jagannathan, S., and Donzis, D. A. 2016. Reynolds and Mach number scaling in solenoidally-forced compressible turbulence using high-resolution direct numerical simulations. *J. Fluid Mech.*, **789**, 669–707.

Joule, J. P. 1850. On the mechanical equivalent of heat. *Phil. Trans. R. Soc. Lond.*, **140**, 61–82.

Kaneda, Y., and Ishihara, T. 2006. High-resolution direct numerical simulation of turbulence. *J. Turbulence*, **7**(20), N20.

Kaneda, Y., Ishihara, T., Yokokawa, M., Itakura, K., and Uno, A. 2003. Energy dissipation rate and energy spectrum in high resolution direct numerical simulations of turbulence in a periodic box. *Phys. Fluids*, **15**(2), L21–L24.

Kolmogorov, A. N. 1941a. Dissipation of energy in locally isotropic turbulence. *Dokl. Akad. Nauk SSSR*, **32**, 16–18.

Kolmogorov, A. N. 1941b. The local structure of turbulence in incompressible viscous fluid for very large Reynolds number. *C. R. Acad. Sci. URSS*, **30**, 301–305.

Kolmogorov, A. N. 1962. A refinement of previous hypotheses concerning the local structure of turbulence in a viscous incompressible fluid at high Reynolds number. *J. Fluid Mech.*, **13**, 82–85.

Kritsuk, A. G., Norman, M. L., Padoan, P., and Wagner, R. 2007. The statistics of supersonic isothermal turbulence. *Astrophys. J.*, **665**, 416–431.

Kritsuk, A. G., Wagner, R., and Norman, M. L. 2013. Energy cascade and scaling in supersonic isothermal turbulence. *J. Fluid Mech.*, **729**(127), 140–142.

Leray, J. 1934. Sur le mouvement d'un liquide visqueux emplissant l'espace. *Acta Math.*, **63**, 192–248.

Mac Low, M.-M., and Klessen, R. S. 2004. Control of star formation by supersonic turbulence. *Rev. Modern Phys.*, **76**, 125–194.

Mandelbrot, B. B. 1974. Intermittent turbulence in self-similar cascades: Divergence of high moments and dimension of the carrier. *J. Fluid Mech.*, **62**, 331–358.

McKee, C. F., and Ostriker, E. C. 2007. Theory of star formation. *Annu. Rev. Astron. Astrophys.*, **45**, 565–687.

Moisy, F., Tabeling, P., and Willaime, H. 1999. Kolmogorov equation in a fully developed turbulence experiment. *Phys. Rev. Lett.*, **82**(20), 3994–3997.

Münch, G. 1958. Internal motions in the Orion Nebula. *Rev. Mod. Phys.*, **30**, 1035–1041.

Navier, C. L. M. H. 1823. Mémoire sur les lois du mouvement des fluides. *Mém. Acad. Roy. Sci.*, **6**, 389–416.

Oboukhov, A. M. 1962. Some specific features of atmospheric turbulence. *J. Fluid Mech.*, **13**, 77–81.

Onsager, L. 1949. Statistical hydrodynamics. *Il Nuovo Cimento*, **6**, 279–287.

Panopoulou, G. V., Clark, S. E., Hacar, A. et al. 2022. The width of Herschel filaments varies with distance. *Astron. Astrophys.*, **657**, L13.

Pearson, B. R., Krogstad, P.-A., and van de Water, W. 2002. Measurements of the turbulent energy dissipation rate. *Phys. Fluids*, **14**(3), 1288–1290.

Penzias, A. A., Solomon, P. M., Jefferts, K. B., and Wilson, R. W. 1972. Carbon monoxide observations of dense interstellar clouds. *Astrophys. J. Lett.*, **174**, L43–L48.

Ravelet, F., Chiffaudel, A., and Daviaud, F. 2008. Supercritical transition to turbulence in an inertially driven von Karman closed flow. *J. Fluid Mech.*, **601**, 339–364.

Sagaut, P., and Cambon, C. 2008. *Homogeneous Turbulence Dynamics*. Cambridge University Press.

Saint-Michel, B., Herbert, E., Salort, J. et al. 2014. Probing quantum and classical turbulence analogy in von Karman liquid helium, nitrogen, and water experiments. *Phys. Fluids*, **26**(12), 125109.

Saw, E.-W., Kuzzay, D., Faranda, D. et al. 2016. Experimental characterization of extreme events of inertial dissipation in a turbulent swirling flow. *Nature Comm.*, **7**, 12466.

Schwinger, J. 1951. On gauge invariance and vacuum polarization. *Phys. Review*, **82**(5), 664–679.

She, Z.-S., and Leveque, E. 1994. Universal scaling laws in fully developed turbulence. *Phys. Rev. Lett.*, **72**, 336–339.

Siebert, H., Shaw, R. A., and Warhaft, Z. 2010. Statistics of small-scale velocity fluctuations and internal intermittency in marine stratocumulus clouds. *J. Atm. Sciences*, **67**(1), 262–273.

Solomon, P. M., Rivolo, A. R., Barrett, J., and Yahil, A. 1987. Mass, luminosity, and line width relations of galactic molecular clouds. *Astrophys. J.*, **319**, 730–741.

Spitzer, L. 1978. *Physical Processes in the Interstellar Medium*. Wiley-Interscience.

Sreenivasan, K. R. 1984. On the scaling of the turbulence energy dissipation rate. *Phys. Fluids*, **27**(5), 1048–1051.

Stokes, G. G. 1845. On the theory of the internal friction of fluids in motion and the equilibrium and motion of elastic solids. *Trans. Cam. Phil. Soc*, **8**, 287–319.

Sulem, C., Sulem, P. L., and Frisch, H. 1983. Tracing complex singularities with spectral methods. *J. Comp. Physics*, **50**(1), 138–161.

Taylor, G. I. 1935. Statistical theory of turbulence. *Proc. Roy. Soc. Lond. Series A*, **151**(873), 421–444.

Taylor, G. I. 1937. The statistical theory of isotropic turbulence. *J. Aero. Sci.*, **4**(8), 311–315.

von Kármán, T., and Howarth, L. 1938. On the statistical theory of isotropic turbulence. *Proc. Roy. Soc. Lond. Series A*, **164**(917), 192–215.

Wilson, R. W., Jefferts, K. B., and Penzias, A. A. 1970. Carbon monoxide in the Orion Nebula. *Astrophys. J. Lett.*, **161**, L43–L44.

Young, J. 2015. Heat, work and subtle fluids: A commentary on Joule (1850) "On the mechanical equivalent of heat." *Phil. Trans. R. Soc. Lond. A*, **373**(2039), 20140348.

Zakharov, V. E., and Sagdeev, R. Z. 1970. Spectrum of acoustic turbulence. *Sov. Phys. Dok.*, **15**, 439–440.

3

Spectral Theory in Hydrodynamics

The spectral approach in eddy turbulence offers an indispensable complementary way of study to the one based on correlations in physical space (see Chapter 2). The use of spectral space can be justified in particular from (i) a theoretical point of view with the possibility of analyzing the mechanisms of interaction and energy exchange (to mention only this invariant) between wavenumbers, (ii) modeling with the introduction of statistical closure hypotheses (e.g. EDQNM, DIA), and (iii) direct numerical simulation using spectral type codes. As we will see in Part II, wave turbulence is mainly studied in spectral space, therefore the three points mentioned are also relevant for this regime.

We will first define some indispensable statistical tools, and then discuss the most used spectral quantity: the energy spectrum. We will derive several exact results of the spectral theory in incompressible hydrodynamics, and we will consider in particular the two-dimensional case, for which we can prove the presence of a dual cascade. For this, we will show that the Zakharov transformation, usually used only in wave turbulence, can be an elegant tool for the demonstration. Finally, the best-known closure models will be presented as well as the nonlinear diffusion model.

3.1 Kinematics

3.1.1 Spectral Tensor

The three-dimensional Fourier transformation of an integrable function $f(\mathbf{x})$ is by definition:

$$\hat{f}(\mathbf{k}) \equiv \hat{f}_{\mathbf{k}} \equiv \frac{1}{(2\pi)^3} \int_{\mathbb{R}^3} f(\mathbf{x}) e^{-i\mathbf{k}\cdot\mathbf{x}} d\mathbf{x}, \qquad (3.1)$$

with $\mathbf{k}$ a wavevector (its norm will be written k); then the inverse Fourier transformation is:

$$f(\mathbf{x}) \equiv \int_{\mathbb{R}^3} \hat{f}_\mathbf{k}\, e^{i\mathbf{k}\cdot\mathbf{x}} d\mathbf{k}. \tag{3.2}$$

From these definitions, we define the correlator:

$$\langle \hat{f}(\mathbf{k})\hat{f}(\mathbf{k}')\rangle = \frac{1}{(2\pi)^6} \int_{\mathbb{R}^6} \langle f(\mathbf{x})f(\mathbf{x}')\rangle\, e^{-i(\mathbf{k}\cdot\mathbf{x}+\mathbf{k}'\cdot\mathbf{x}')} d\mathbf{x}d\mathbf{x}', \tag{3.3}$$

with $\langle\rangle$ the ensemble average (see Chapter 1).

In the statistical homogeneous case, we can define:

$$Q(\boldsymbol{\ell}) \equiv \langle f(\mathbf{x})f(\mathbf{x}')\rangle = \langle f(\mathbf{x})f(\mathbf{x}+\boldsymbol{\ell})\rangle, \tag{3.4}$$

which leads to:

$$\begin{aligned}
\langle \hat{f}(\mathbf{k})\hat{f}(\mathbf{k}')\rangle &= \frac{1}{(2\pi)^6} \int_{\mathbb{R}^6} Q(\boldsymbol{\ell})\, e^{-i(\mathbf{k}+\mathbf{k}')\cdot\mathbf{x}}\, e^{-i\mathbf{k}'\cdot\boldsymbol{\ell}} d\mathbf{x}d\boldsymbol{\ell} \\
&= \frac{1}{(2\pi)^3} \int_{\mathbb{R}^3} Q(\boldsymbol{\ell})\, e^{-i\mathbf{k}'\cdot\boldsymbol{\ell}}\delta(\mathbf{k}+\mathbf{k}')d\boldsymbol{\ell} \\
&= \delta(\mathbf{k}+\mathbf{k}')\hat{Q}(\mathbf{k}'),
\end{aligned} \tag{3.5}$$

with $\hat{Q}(\mathbf{k}')$ a two-point spectral correlator.

In hydrodynamic turbulence, we are mainly interested in the velocity field. The three-dimensional spectral tensor of the velocity for a homogeneous turbulence is defined as

$$\Phi_{ij}(\mathbf{k}) \equiv \frac{1}{(2\pi)^3} \int_{\mathbb{R}^3} R_{ij}(\boldsymbol{\ell})\, e^{-i\mathbf{k}\cdot\boldsymbol{\ell}} d\boldsymbol{\ell}, \tag{3.6}$$

with $R_{ij}(\boldsymbol{\ell}) \equiv \langle u_i(\mathbf{x})u_j(\mathbf{x}+\boldsymbol{\ell})\rangle$ the second-order velocity correlator tensor. This quantity is well defined mathematically (it is an integrable function) to the extent that the correlation between two points tends to zero when the distance between these points increases to infinity. We have the Hermitian property:

$$\Phi_{ji}(\mathbf{k}) = \Phi_{ij}(-\mathbf{k}) = \Phi_{ij}^*(\mathbf{k}), \tag{3.7}$$

where the symbol * means the complex conjugate (we will use c.c. for bigger terms). This property is the result of the homogeneity relationship in the physical space $R_{ij}(\boldsymbol{\ell}) = R_{ji}(-\boldsymbol{\ell})$. Since turbulence is assumed incompressible, we also have the relation:

$$k_i \Phi_{ij}(\mathbf{k}) = 0. \tag{3.8}$$

3.1.2 Energy Spectrum

We will now define one of the best-studied turbulence quantities, namely the kinetic energy spectrum per unit mass (we assume $\rho_0 = 1$). We have:

$$\langle E \rangle \equiv \frac{1}{2}\langle \mathbf{u}^2 \rangle = \frac{1}{2} R_{ii}(\mathbf{0}) = \frac{1}{2} \int_{\mathbb{R}^3} \Phi_{ii}(\mathbf{k}) d\mathbf{k}. \tag{3.9}$$

This relation allows us to define the energy spectrum:

$$E(\mathbf{k}) \equiv \frac{1}{2}\Phi_{ii}(\mathbf{k}),\qquad(3.10)$$

which reflects the distribution of energy by spectral band $[k, k+dk]$.

In the particular case of isotropic turbulence, the spectral tensor depends only on the modulus of $\mathbf{k}$; also the definition of the spectrum is reduced to:

$$E(k) \equiv E_k = \oiint E(\mathbf{k})dS(\mathbf{k}) = 2\pi k^2 \Phi_{ii}(k),\qquad(3.11)$$

where $S(\mathbf{k})$ is a sphere of radius $\mathbf{k}$ in the Fourier space. The mean kinetic energy of the system is found by a simple integration over k:

$$\langle E \rangle = \int_0^{+\infty} E_k dk.\qquad(3.12)$$

3.2 Detailed Energy Conservation

In this section, we will study a fundamental property of turbulence: the detailed conservation of invariants. To illustrate our point, we will consider the three-dimensional hydrodynamic Navier–Stokes equations (2.1) and show that the kinetic energy is conserved by triadic interaction. For that, let us apply a Fourier transform to these equations (we use Einstein's notations and we do not write explicitly the time variable):[1]

$$\partial_t \hat{u}_i(\mathbf{k}) + \widehat{u_m \partial_m u_i}(\mathbf{k}) = -ik_i \hat{P}(\mathbf{k})/\rho_0 - \nu k^2 \hat{u}_i(\mathbf{k}),\qquad(3.13a)$$

$$k_i \hat{u}_i(\mathbf{k}) = 0.\qquad(3.13b)$$

We can rewrite the nonlinear term by showing a convolution product of Fourier transforms:

$$\widehat{u_m \partial_m u_i}(\mathbf{k}) = \hat{u}_m \star \widehat{\partial_m u_i}(\mathbf{k})\qquad(3.14)$$

$$= i \int_{\mathbb{R}^6} \hat{u}_m(\mathbf{p}) q_m \hat{u}_i(\mathbf{q}) \delta(\mathbf{k} - \mathbf{p} - \mathbf{q}) d\mathbf{p}d\mathbf{q}.$$

In applying the divergence operator to equation (3.13a), one obtains:

$$ik_i \widehat{u_m \partial_m u_i}(\mathbf{k}) = k^2 \hat{P}(\mathbf{k})/\rho_0,\qquad(3.15)$$

hence, the expression for the pressure:

$$\hat{P}(\mathbf{k})/\rho_0 = -\frac{k_n}{k^2} \int_{\mathbb{R}^6} \hat{u}_m(\mathbf{p}) q_m \hat{u}_n(\mathbf{q}) \delta(\mathbf{k} - \mathbf{p} - \mathbf{q}) d\mathbf{p}d\mathbf{q}.\qquad(3.16)$$

---

[1] The application of a Fourier transform to the Navier–Stokes equations supposes that it is mathematically possible to define the Fourier transform of the velocity field, that is, that this field decreases sufficiently quickly to infinity.

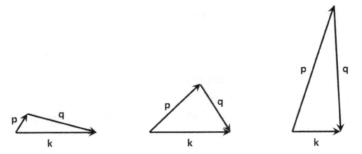

Figure 3.1 Examples of local (middle) and nonlocal (left and right) triadic interactions.

The incompressibility condition allows us to affirm that for a triadic interaction $q_m \hat{u}_m(\mathbf{p}) = k_m \hat{u}_m(\mathbf{p})$. Then, one obtains:

$$\boxed{\partial_t \hat{u}_i(\mathbf{k}) = -ik_m P_{in} \int_{\mathbb{R}^6} \hat{u}_m(\mathbf{p})\hat{u}_n(\mathbf{q})\delta(\mathbf{k}-\mathbf{p}-\mathbf{q})d\mathbf{p}d\mathbf{q} - \nu k^2 \hat{u}_i(\mathbf{k})}, \quad (3.17)$$

with $P_{in} = \delta_{in} - k_i k_n/k^2$ the projection operator which ensures the incompressibility of the fluid. Expression (3.17) is often the starting point of the spectral analysis: it is the spectral equations of Navier–Stokes. They involve triadic interactions and therefore a geometric aspect of the equations (see Figure 3.1).

To show the detailed energy conservation, we will write an evolution equation for the modulus of velocity, which is nothing other than the kinetic energy (per unit mass and) per mode:[2]

$$\partial_t |\hat{u}(\mathbf{k})|^2 + 2\nu k^2 |\hat{u}(\mathbf{k})|^2 =$$
$$- ik_m P_{in} \int \hat{u}_i^*(\mathbf{k})\hat{u}_m(\mathbf{p})\hat{u}_n(\mathbf{q})\delta(\mathbf{k}-\mathbf{p}-\mathbf{q})d\mathbf{p}d\mathbf{q}$$
$$+ ik_m P_{in} \int \hat{u}_i(\mathbf{k})\hat{u}_m^*(\mathbf{p})\hat{u}_n^*(\mathbf{q})\delta(\mathbf{k}-\mathbf{p}-\mathbf{q})d\mathbf{p}d\mathbf{q}. \quad (3.18)$$

After development, we have in vector writing:

$$\partial_t |\hat{\mathbf{u}}(\mathbf{k})|^2 + 2\nu k^2 |\hat{\mathbf{u}}(\mathbf{k})|^2 = -i \int [(\mathbf{k}\cdot\hat{\mathbf{u}}(\mathbf{p}))(\hat{\mathbf{u}}^*(\mathbf{k})\cdot\hat{\mathbf{u}}(\mathbf{q})) \quad (3.19)$$
$$- (\mathbf{k}\cdot\hat{\mathbf{u}}^*(\mathbf{p}))(\hat{\mathbf{u}}(\mathbf{k})\cdot\hat{\mathbf{u}}^*(\mathbf{q}))]\delta(\mathbf{k}-\mathbf{p}-\mathbf{q})d\mathbf{p}d\mathbf{q}.$$

This equation simplifies by using the relation $\hat{\mathbf{u}}(-\mathbf{k}) = \hat{\mathbf{u}}^*(\mathbf{k})$ and by playing with the mute variables $\mathbf{p}$ and $\mathbf{q}$; one obtains:

$$\partial_t |\hat{\mathbf{u}}(\mathbf{k})|^2 + 2\nu k^2 |\hat{\mathbf{u}}(\mathbf{k})|^2 = -i \int [(\mathbf{k}\cdot\hat{\mathbf{u}}^*(\mathbf{p}))(\hat{\mathbf{u}}^*(\mathbf{k})\cdot\hat{\mathbf{u}}^*(\mathbf{q})) \quad (3.20)$$
$$- (\mathbf{k}\cdot\hat{\mathbf{u}}(\mathbf{p}))(\hat{\mathbf{u}}(\mathbf{k})\cdot\hat{\mathbf{u}}(\mathbf{q}))]\,\delta(\mathbf{k}+\mathbf{p}+\mathbf{q})d\mathbf{p}d\mathbf{q}$$
$$= i \int (\mathbf{k}\cdot\hat{\mathbf{u}}(\mathbf{p}))(\hat{\mathbf{u}}(\mathbf{k})\cdot\hat{\mathbf{u}}(\mathbf{q}))\delta(\mathbf{k}+\mathbf{p}+\mathbf{q})d\mathbf{p}d\mathbf{q} + c.c.,$$

[2] From now, the multiple integral will be denoted simply $\int$.

3.2 Detailed Energy Conservation

with *c.c.* the complex conjugate. By symmetrizing the equation with respect to the variables **p** and **q**, we can finally write:

$$\partial_t |\hat{\mathbf{u}}(\mathbf{k})|^2 + 2\nu k^2 |\hat{\mathbf{u}}(\mathbf{k})|^2 = \int S(\mathbf{k},\mathbf{p},\mathbf{q})\delta(\mathbf{k}+\mathbf{p}+\mathbf{q})d\mathbf{p}d\mathbf{q}, \quad (3.21)$$

with by definition:

$$S(\mathbf{k},\mathbf{p},\mathbf{q}) \equiv -\Im[(\mathbf{k}\cdot\hat{\mathbf{u}}(\mathbf{p}))(\hat{\mathbf{u}}(\mathbf{k})\cdot\hat{\mathbf{u}}(\mathbf{q})) + (\mathbf{k}\cdot\hat{\mathbf{u}}(\mathbf{q}))(\hat{\mathbf{u}}(\mathbf{k})\cdot\hat{\mathbf{u}}(\mathbf{p}))], \quad (3.22)$$

and $\Im$ the imaginary part. The detailed conservation of energy is obtained by using the symmetry properties of the operator $S(\mathbf{k},\mathbf{p},\mathbf{q})$. After some manipulations and use of the incompressibility condition, we arrive at the exact relation by triad of interaction:

$$\boxed{S(\mathbf{k},\mathbf{p},\mathbf{q}) + S(\mathbf{p},\mathbf{q},\mathbf{k}) + S(\mathbf{q},\mathbf{k},\mathbf{p}) = 0}. \quad (3.23)$$

The conservation of kinetic energy (per unit mass) can be written:

$$\partial_t \int \frac{1}{2}|\hat{\mathbf{u}}(\mathbf{k})|^2 d\mathbf{k} + \nu k^2 \int |\hat{\mathbf{u}}(\mathbf{k})|^2 d\mathbf{k} = \frac{1}{6}\int [S(\mathbf{k},\mathbf{p},\mathbf{q}) + S(\mathbf{p},\mathbf{q},\mathbf{k}) \quad (3.24)$$
$$+ S(\mathbf{q},\mathbf{k},\mathbf{p})]\delta(\mathbf{k}+\mathbf{p}+\mathbf{q})d\mathbf{k}d\mathbf{p}d\mathbf{q} = 0.$$

When $\nu = 0$, there is, as expected, conservation of kinetic energy because the integral on the right (second line) is zero, but in fact this conservation is carried out locally (before integration) by triadic interaction: energy is exchanged by triadic interaction without loss or gain. This remarkable property was discovered by Kraichnan (1959) in his work on the direct interaction approximation (DIA).[3] This exact result is important because it informs us about the way in which energy is redistributed between modes.

The detailed conservation is a property that is verified for all invariants of a system (without dissipation), regardless of the regime under consideration. For example, in wave turbulence (see Chapter 4) this detailed conservation is performed for resonant interactions (triadic, quartic, etc.). This property has its roots outside of statistical physics since the ensemble average is not applied to obtain expression (3.24); it must therefore also be verified as a statistical average. In Figure 3.1 we show examples of triadic interactions: we can distinguish the (two) nonlocal interactions from the local one. Energy is exchanged within each triad: the direct cascade suggests that, on average, the smallest modes yield energy to the largest modes.

---

[3] Kraichnan was an original researcher whose work profoundly influenced our understanding of turbulence. Its originality is due in part to the fact that he obtained his PhD thesis at the age of 21, then left the academic world in 1962 to become an independent researcher, regularly funded by American agencies.

3.3 Statistical Theory

3.3.1 Flux and Transfer

Let us apply the ensemble average to equation (3.21); one gets:

$$\partial_t \langle |\hat{\mathbf{u}}(\mathbf{k})|^2 \rangle + 2\nu k^2 \langle |\hat{\mathbf{u}}(\mathbf{k})|^2 \rangle = \int \langle S(\mathbf{k}, \mathbf{p}, \mathbf{q}) \rangle \delta(\mathbf{k} + \mathbf{p} + \mathbf{q}) d\mathbf{p} d\mathbf{q}. \tag{3.25}$$

It is easy to demonstrate that this equation conserves energy in detail. By introducing the energy spectrum, we get an equation whose form was suggested by Batchelor (1953) (see also Obukhov, 1941b):

$$\boxed{\partial_t E(\mathbf{k}) + 2\nu k^2 E(\mathbf{k}) = T(\mathbf{k})}, \tag{3.26}$$

with by definition the transfer function:

$$T(\mathbf{k}) \equiv \int \frac{1}{2} \langle S(\mathbf{k}, \mathbf{p}, \mathbf{q}) \rangle \delta(\mathbf{k} + \mathbf{p} + \mathbf{q}) d\mathbf{p} d\mathbf{q}. \tag{3.27}$$

Equation (3.26) is exact: it expresses the evolution over time of the three-dimensional energy spectrum. This equation should not make people believe that we have solved the difficult problem of closure: in fact, the whole problem now is to find an analytical form for the transfer function $T(\mathbf{k})$.

Let us simplify the problem and assume that turbulence is statistically isotropic. We then arrive at the equation of temporal evolution of the one-dimensional energy spectrum:

$$\partial_t E(k) + 2\nu k^2 E(k) = T(k), \tag{3.28}$$

with by definition:

$$T(k) \equiv \oiint 4\pi k^2 T(\mathbf{k}) dS(\mathbf{k}), \tag{3.29}$$

where the integral is performed on a sphere S of radius k. The transfer function can be expressed as an energy flux term $\Pi(k)$ with the following definition (Kraichnan, 1959):

$$T(k) \equiv -\frac{\partial \Pi(k)}{\partial k}. \tag{3.30}$$

The detailed conservation of energy (3.23) allows us to affirm that:

$$\boxed{\int_0^{+\infty} T(k) dk = 0}, \tag{3.31}$$

therefore, we have the flux-transfer relations:

$$\Pi(k) = \int_k^{+\infty} T(k') dk' = -\int_0^k T(k') dk', \tag{3.32}$$

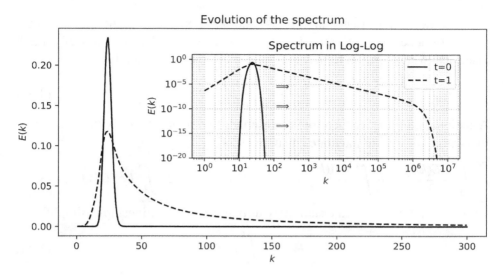

Figure 3.2 Temporal evolution of the energy spectrum in linear scales. In general, we prefer to show this spectrum in logarithmic scales (insert), where it is easier to distinguish the inertial range.

with the condition $\Pi(0) = \Pi(+\infty) = 0$. This condition is satisfied insofar as the energy tends towards zero within the limit of small and large wavenumbers. In Figure 3.2 we show schematically the evolution of the energy spectrum between the instants $t = 0$ and $t = 1$. The energy, localized initially over a narrow band of small wavenumbers, spreads mainly towards large wavenumbers. This development is explained by a flux of energy $\Pi(k)$ on average positive.

The problem of anomalous dissipation is also found in Fourier space (see the discussion in Chapter 2). After integration over k from equation (3.28), we get ($\rho_0 = 1$):

$$\frac{1}{2}\frac{\partial \langle \mathbf{u}^2 \rangle}{\partial t} = -2\nu \int_0^{+\infty} k^2 E(k) dk \equiv -\varepsilon(\nu) \xrightarrow[\nu \to 0]{} -\varepsilon . \tag{3.33}$$

Let us now introduce an external force F, which is homogeneous, stationary, and localized at small wavenumbers around k_0. We can write:

$$\frac{\partial E(k)}{\partial t} + \frac{\partial \Pi(k)}{\partial k} = F\delta(k_0) - 2\nu k^2 E(k), \tag{3.34}$$

which gives us after integration on k and within the limit $\nu \to 0$:

$$\frac{1}{2}\frac{\partial \langle \mathbf{u}^2 \rangle}{\partial t} = F - \varepsilon . \tag{3.35}$$

The study of solutions in turbulence is generally carried out in the stationary case. We see that this necessarily implies that the mean rate of dissipation is adjusted to the mean rate of energy injection F.

Let us now return to equation (3.34) and place ourselves in the inertial range, that is, on a scale k such that: $k_0 \ll k$. We see that within the limit of a small viscosity, the dissipation term can be neglected, and consequently:

$$\frac{\partial E(k)}{\partial t} + \frac{\partial \Pi(k)}{\partial k} = 0. \qquad (3.36)$$

A stationary turbulence therefore corresponds to a constant energy flux Π. To find this value, we have to integrate equation (3.34):

$$\int_0^k \frac{\partial E(k')}{\partial t} dk' + \int_0^k \frac{\partial \Pi(k')}{\partial k'} dk' = \int_0^k F\delta(k_0) dk' - 2\nu \int_0^k k'^2 E(k') dk', \qquad (3.37)$$

with $k \gg k_0$. For large-scale stationary turbulence, the relationship is simplified:

$$\Pi = F - 2\nu \int_0^k k'^2 E(k') dk'. \qquad (3.38)$$

Within the limit of a small viscosity, we finally obtain:

$$\boxed{\Pi = F = \varepsilon}. \qquad (3.39)$$

The energy flux in spectral space is therefore precisely the mean rate of energy dissipation, which we discussed at length in Chapter 2. The flux associated with a direct cascade is therefore positive. Note that in the case of an inverse cascade with a force acting at a small scale, we can easily show from expression (3.37) that this flux is negative. From the flux-transfer relations (3.32), we see that for a fixed wavenumber k, the transfer function T is positive for wavenumbers greater than k and negative for wavenumbers smaller than k.

3.3.2 Kolmogorov's Spectrum

The Kolmogorov phenomenology developed in Chapter 2 allowed us to obtain the scaling relation ($\rho_0 = 1$):

$$\varepsilon \ell \sim u_\ell^3, \qquad (3.40)$$

valid for statistically homogeneous and isotropic turbulence. From this relationship and a simple dimensional analysis, we can deduce:

$$u_\ell^2 \sim (\varepsilon \ell)^{2/3} \sim \varepsilon^{2/3} k^{-2/3} \sim k E_k, \qquad (3.41)$$

hence, the spectral prediction (one introduces the equal sign and therefore a constant of proportionality):

$$\boxed{E_k = C_K \varepsilon^{2/3} k^{-5/3}}, \qquad (3.42)$$

with C_K the Kolmogorov constant; the measurements give $C_K \simeq 0.5$ (Sreenivasan, 1995; Welter et al., 2009). The energy flux involved here is the one that will be

3.3 Statistical Theory

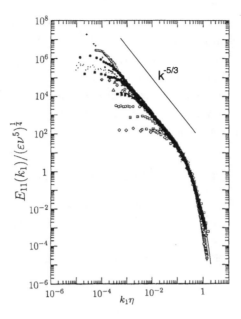

Figure 3.3 Energy spectrum found in experimental measurements: a universal law in $k^{-5/3}$ is identified over approximately four orders of magnitude. Reprinted with permission (Saddoughi and Veeravalli, 1994).

measured in the spectral space. Note that with a pure dimensional analysis involving only E_k, ε, and k, we arrive at the same scaling prediction. Although proposed for the first time by Obukhov (1941a), this spectrum is known as Kolmogorov's spectrum. It is quite remarkable to note that despite its phenomenological origin, this prediction is very clearly observed at sea (Grant et al., 1962) and in many laboratory experiments (Saddoughi and Veeravalli, 1994), with a power law in $k^{-5/3}$ over several orders of magnitude, as shown in Figure 3.3.

Using our analysis in physical space and Kolmogorov's phenomenology, we can interpret Figure 3.3 as follows. The spectrum in $k^{-5/3}$ results from a balance between the injection of energy at small wavenumbers and its dissipation at large wavenumbers. Between the two, we have an inertial range where turbulence behaves universally. This inertial range is crossed by a flux of energy that connects large to small scales: this is the region where energy cascades. The cascade process means that the energy flows continuously from one scale to another. The Kolmogorov spectrum is therefore not a set of lines, but a continuous distribution of energy. Note that the dissipative part of the spectrum has been the subject of many studies to find its shape. A simple model is the one proposed by Pao (1965) in which the spectrum is simply the $-5/3$ law weighted by an exponential function whose coefficient depends on the dissipation scale, called the Kolmogorov scale.

One of the objectives for theoreticians is to find a rigorous justification for the Kolmogorov spectrum (in the inertial range), which requires a self-consistent spectral equation. Despite numerous attempts by physicists and mathematicians, to date no solution to this problem has been found in eddy turbulence. The problem is fundamentally related to nonlinearities and the lack of closure of the hierarchy of equations. We will see in Chapter 4 that it is possible to find an analytical solution to this problem in the wave turbulence regime.

3.3.3 Infinite Hierarchy of Equations

Let us take a closer look at the closure problem. This one being fundamentally related to the nonlinearities, to simplify we will forget the viscous linear term which can be reintroduced trivially at the end of the calculation.

From the Navier–Stokes spectral equations (3.17), we obtain the equation for the second-order moment:

$$\partial_t \langle \hat{u}_i(\mathbf{k}) \hat{u}_j(\mathbf{k}') \rangle = -ik_m P_{in} \int \langle \hat{u}_j(\mathbf{k}') \hat{u}_m(\mathbf{p}) \hat{u}_n(\mathbf{q}) \rangle \delta(\mathbf{k} - \mathbf{p} - \mathbf{q}) d\mathbf{p} d\mathbf{q} \quad (3.43)$$

$$- ik'_m P'_{jn} \int \langle \hat{u}_i(\mathbf{k}) \hat{u}_m(\mathbf{p}) \hat{u}_n(\mathbf{q}) \rangle \delta(\mathbf{k}' - \mathbf{p} - \mathbf{q}) d\mathbf{p} d\mathbf{q}.$$

This equation involves a third-order moment whose evolution equation is:

$$\partial_t \langle \hat{u}_i(\mathbf{k}) \hat{u}_j(\mathbf{k}') \hat{u}_l(\mathbf{k}'') \rangle = -ik_m P_{in} \int \langle \hat{u}_j(\mathbf{k}') \hat{u}_l(\mathbf{k}'') \hat{u}_m \quad (3.44)$$

$$\times (\mathbf{p}) \hat{u}_n(\mathbf{q}) \rangle \delta(\mathbf{k} - \mathbf{p} - \mathbf{q}) d\mathbf{p} d\mathbf{q}$$

$$- ik'_m P'_{jn} \int \langle \hat{u}_i(\mathbf{k}) \hat{u}_l(\mathbf{k}'') \hat{u}_m(\mathbf{p}) \hat{u}_n(\mathbf{q}) \rangle \delta(\mathbf{k}' - \mathbf{p} - \mathbf{q}) d\mathbf{p} d\mathbf{q}$$

$$- ik''_m P''_{jn} \int \langle \hat{u}_i(\mathbf{k}) \hat{u}_j(\mathbf{k}') \hat{u}_m(\mathbf{p}) \hat{u}_n(\mathbf{q}) \rangle \delta(\mathbf{k}'' - \mathbf{p} - \mathbf{q}) d\mathbf{p} d\mathbf{q}.$$

We now need to write the moment equation for the third-order, fourth-order, and so on: we end up with an infinite hierarchy of equations. It is generally at this level that spectral closures are introduced. We will present in Sections 3.3.4–3.3.6 the main closure methods, that is, the QN, EDQN, EDQNM, then DIA approximations. There are others, like the Lagrangian DIA approximation (LHDIA; Kraichnan, 1965), or the one based on stochastic models (Kraichnan, 1961).

3.3.4 QN Closure

The first closure model was proposed by Millionschikov (1941): it is based on the quasi-normal approximation (QN), which consists in simply neglecting the contribution of the fourth-order cumulant, and therefore act as if the fourth-order moment was that of a Gaussian distribution. We recall that the fourth-order cumulant appears when rewriting the fourth-order moment as second-order moment products (see Chapter 1). With simpler notations, we have the statistical relationship:

$$\langle \hat{u}_1 \hat{u}_2 \hat{u}_3 \hat{u}_4 \rangle = \langle \hat{u}_1 \hat{u}_2 \rangle \langle \hat{u}_3 \hat{u}_4 \rangle + \langle \hat{u}_1 \hat{u}_3 \rangle \langle \hat{u}_2 \hat{u}_4 \rangle + \langle \hat{u}_1 \hat{u}_4 \rangle \langle \hat{u}_2 \hat{u}_3 \rangle \qquad (3.45)$$
$$+ \{\hat{u}_1 \hat{u}_2 \hat{u}_3 \hat{u}_4\},$$

where the term in the second line is the fourth-order cumulant. In the case of a Gaussian statistical distribution, this term is by definition zero; it is the same for all odd moments (which are also cumulants). The hypothesis made by Millionschikov (1941) is therefore not the Gaussianity hypothesis, which would trivialize the problem (with vanishing third-order moments), but a close hypothesis, hence the name of quasi-normal closure (or approximation). Schematically, we then obtain:

$$\partial_t \langle \hat{u}_i(\mathbf{k}) \hat{u}_j(\mathbf{k}') \hat{u}_l(\mathbf{k}'') \rangle = \sum \langle \hat{u}\hat{u} \rangle \langle \hat{u}\hat{u} \rangle. \qquad (3.46)$$

With this hypothesis, we see that it is possible after a time integration to obtain a self-consistent equation for the second-order moment, and therefore in particular for the energy spectrum.

Kraichnan (1957) showed, however, that this closure was inconsistent because it violated certain statistical inequalities (the feasibility conditions); then Ogura (1963) numerically demonstrated that this closure could lead to a negative energy spectrum for some wavenumbers. The energy spectrum being a definite positive quantity, this behavior constitutes a major defect of the model, which was consequently abandoned. The origin of the problem was understood later by Orszag (1970): in the absence of fourth-order cumulant, the impact of products of second-order moment on third-order moment is overestimated. In reality, as the experimental measurements show, they saturate. The fourth-order cumulant therefore has the role of damping for third-order moments.

3.3.5 EDQN and EDQNM Closure

In order to correct the defect of the QN approximation, a more sophisticated closure was proposed by Orszag (1970): it is the EDQNM (eddy-damped quasi-normal Markovian) approximation, itself based on the EDQN approximation. In this approach, the fourth-order cumulant is modeled by a linear damping term. Thus, the schematic equation (3.46) is replaced by:

$$(\partial_t + \mu_{kk'k''}) \langle \hat{u}_i(\mathbf{k}) \hat{u}_j(\mathbf{k}') \hat{u}_l(\mathbf{k}'') \rangle = \sum \langle \hat{u}\hat{u} \rangle \langle \hat{u}\hat{u} \rangle. \qquad (3.47)$$

The coefficient $\mu_{kk'k''}$ has the dimension of the inverse of a time and is called the damping rate of the third-order moment. The definition of this rate is phenomenological: we have

$$\mu_{kk'k''} = \mu_k + \mu_{k'} + \mu_{k''} \qquad (3.48)$$

and
$$\mu_k \sim \sqrt{k^3 E_k}. \tag{3.49}$$

In the isotropic case and after reintroduction of the viscosity, we schematically obtain the EDQN equation:

$$\left(\partial_t + 2\nu k^2\right) E(k,t) = \tag{3.50}$$
$$\int_0^t \int_\Delta \sum \langle \hat{u}\hat{u}\rangle \langle \hat{u}\hat{u}\rangle (\tau) e^{-[\mu_{kpq}+\nu(k^2+p^2+q^2)](t-\tau)} dp dq d\tau,$$

where Δ is the domain of integration corresponding to the triadic relation $\mathbf{k} = \mathbf{p} + \mathbf{q}$. However, the equations obtained with this closure conserve the weakness of the QN model: they do not guarantee the statistical feasibility condition in all situations. The spectrum can therefore become negative.

The last refinement brought to this spectral closure is called Markovianization (M): here, we take into account the fact that the characteristic linear damping time is shorter than the characteristic nonlinear time of variation of second-order moment products. This timescale difference is used to justify a separate average over short time of the contribution of μ_k at the level of the expression of third-order moments. Concretely, equation (3.50) becomes:

$$\left(\partial_t + 2\nu k^2\right) E(k,t) = \int_\Delta \theta_{kpq} \sum \langle \hat{u}\hat{u}\rangle \langle \hat{u}\hat{u}\rangle (t) dp dq, \tag{3.51}$$

with

$$\theta_{kpq} = \int_0^t e^{-[\mu_{kpq}+\nu(k^2+p^2+q^2)](t-\tau)} d\tau. \tag{3.52}$$

The final expression of the EDQNM spectral equation, in the case of nonhelical turbulence, is written (Lesieur, 1997):

$$\boxed{\left(\partial_t + 2\nu k^2\right) E_k = \int_\Delta b_{kpq} E_q (k^2 E_p - p^2 E_k) \theta_{kpq} dp dq}, \tag{3.53}$$

with the geometric coefficient (we consider the angles inside the triangle kpq):

$$b_{kpq} = \frac{1}{q}[\cos\widehat{(p,q)}\cos\widehat{(k,q)} + \cos^3\widehat{(k,p)}]. \tag{3.54}$$

A last simplifying assumption is made to evaluate θ_{kpq}: it is assumed that the damping rate does not vary over time integration, hence the expression:

$$\theta_{kpq} = \frac{1 - e^{-[\mu_{kpq}+\nu(k^2+p^2+q^2)]t}}{\mu_{kpq} + \nu(k^2+p^2+q^2)}. \tag{3.55}$$

As we will see in Chapter 4, the asymptotic closure of wave turbulence is based on a timescale separation between the wave period and the nonlinear time of variation of moments. There is therefore a certain proximity between the two approaches. Note that the EDQNM closure, which is subsequent to that of wave turbulence whose development dates from the mid-1960s, emanates from the thoughts of Kraichnan in the late 1950s on another closure method, called DIA.

3.3.6 DIA Closure

The direct interaction approximation (DIA) was proposed by Kraichnan (1958, 1959) in order to remedy certain shortcomings of the QN closure (Millionschikov, 1941) as the nonconservation of energy (Kraichnan, 1957).[4] This approach, which has no adjustable parameter, is based on a field theory method. The fundamental idea is that a disturbed flow over an interval of wavenumbers will see its disturbance spread over a large number of modes. Within the limit $L \to +\infty$, with L the side of the cube in which the fluid is confined, this interval becomes of infinite size, which suggests that the mode coupling becomes infinitely weak. The response to the perturbation can then be treated systematically. Under certain assumptions, two integro-differential equations are obtained for the two-point (in space and time) correlation functions and the response function (see the review by Zhou, 2021).

In practice, we are interested in the evolution of two velocity fields. The first $\hat{u}_i$ is generated by an external force $\hat{f}_i$ which is introduced to keep the turbulence stationary. The second is a small perturbation (of the speed $\hat{u}_i$) $\delta\hat{u}_i$ generated by a small perturbation (of the force $\hat{f}_i$) $\delta\hat{f}_i$. The equation associated with the perturbation is written:

$$(\partial_t + \nu k^2)\delta\hat{u}_i(\mathbf{k}, t) - \mathcal{M}_{ijm}(\hat{u}_j(\mathbf{p}, t), \delta u_m(\mathbf{q}, t)) = \delta\hat{f}_i(t). \quad (3.56)$$

If we assume that the two velocity fields are independent, then $\mathcal{M}$ is a linear operator whose form can be deduced from the Navier–Stokes spectral equations (3.17). The linearity of equation (3.56) is justified as long as the perturbations remain relatively small. The solution to this equation is the response to an infinitesimal perturbation:

$$\delta\hat{u}_i(\mathbf{k}, t) = \int_{-\infty}^{t} \hat{G}_{ij}(\mathbf{k}, t, t')\delta\hat{f}_j(\mathbf{k}, t')dt', \quad (3.57)$$

with, for $t \geq t'$:

$$(\partial_t + \nu k^2)\hat{G}_{ij}(\mathbf{k}, t, t') - \mathcal{M}(\hat{u}_i(\mathbf{p}, t), \hat{G}_{ij}(\mathbf{q}, t, t')) = P_{ij}(\mathbf{k})\delta(t - t'). \quad (3.58)$$

$\hat{G}_{ij}$ is the infinitesimal response tensor. We see that these spectral equations introduce two different times, t and t', therefore there is a temporal memory, which is not the case with the EDQNM approach.

---

[4] For a moment Kraichnan believed that his theory was exact, which is not the case as demonstrated by the non invariance of its equations by random Galilean transformation. The origin of the problem comes from the truncated series used (Kraichnan, 1975).

Although more sophisticated than the EDQNM approach, the DIA method remains an approximation. Briefly, we can say that this technique consists in developing the velocity field and the infinitesimal response tensor in power series of ϵ (hence of the Reynolds number). The zero-order term in velocity is assumed to be a random variable with a Gaussian statistics. DIA closure consists of stopping the expansion to the lowest possible nontrivial order, then taking $\epsilon = 1$, which involves subtle issues about divergence of series (Leslie, 1973; Kraichnan, 1975). We speak of direct interaction in the sense that all the equations are truncated to a given order and that we assume that there is no indirect interaction between the order of the truncation and a higher order. We finally obtain two coupled integro-differential equations (whose appearance resembles the EDQNM equations) which preserve the detailed energy conservation. The prediction for the energy spectrum, in $k^{-3/2}$, is, however, not in dimensional agreement with the theory of Kolmogorov, nor with the main spectral measurements.

Improvements were then made with a Lagrangian approach – called Lagrangian history direct interaction approximation (LHDIA) – to solve some problems like the noninvariance by random Galilean transformation (Kraichnan, 1966): this new theory can be seen as the most sophisticated model of closure. However, its degree of complexity is such that its use remains rare (Nakayama, 1999, 2001). Nowadays, the most used spectral closure method remains the EDQNM, because it is a spectral theory of reasonable complexity with which it is relatively easy to perform numerical simulation (Lesieur, 1997).

3.4 Two-Dimensional Eddy Turbulence

The motivation for the study of two-dimensional hydrodynamic turbulence stems from work showing that a two-dimensional approach could account relatively well for atmospheric dynamics (Rossby and collaborators, 1939). It is now known that the rotation (or stratification) of the Earth's atmosphere tends to confine its nonlinear dynamics to horizontal planes (see Chapter 6). Numerous works are dedicated to this subject; for fundamental aspects, we refer the reader to the reviews of Boffetta and Ecke (2012) and Alexakis and Biferale (2018).

Very early on, two-dimensional hydrodynamic turbulence was suspected to behave differently from the three-dimensional case. For example, Lee (1951) demonstrated that the existence of a direct cascade of energy would violate the conservation of enstrophy which is an invariant of the two-dimensional equations. At the end of his book, Batchelor (1953) notes that the existence of this second invariant should contribute to the emergence, by aggregation, of larger and larger eddies and concludes by asserting the very large difference between two- and three-dimensional turbulence. Using energy and enstrophy, Fjørtoft (1953), for his part, was able to demonstrate, in particular with dimensional arguments, that energy should tend to cascade preferentially towards large scales.

3.4 Two-Dimensional Eddy Turbulence

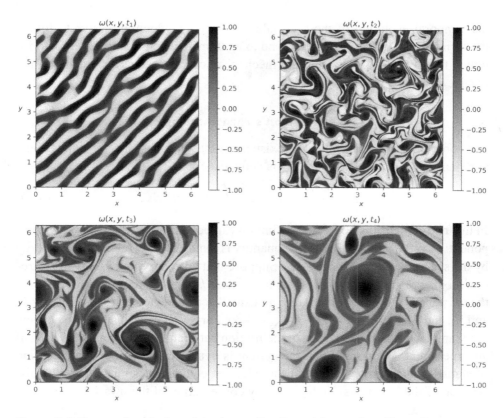

Figure 3.4 Temporal evolution of the (normalized) vorticity produced by a direct numerical simulation of the two-dimensional Navier–Stokes equations. Vortices form rapidly with a variable direction of rotation (direct in black and retrograde in white), then agglomerate to leave finally only a few structures. This merging mechanism can be seen as a signature of the inverse cascade. The numeric box used of spatial resolution 1024×1024 is periodic in both directions.

It is in this context, clearly in favor of an inverse energy cascade, that Kraichnan became interested in two-dimensional turbulence (Kraichnan and Montgomery, 1980). With the help of an analytical development of Navier–Stokes equations in Fourier space, the use of symmetries and under certain hypotheses such as the scale invariance of double and triple moments, Kraichnan (1967) brought major quantitative arguments in favor of a direct cascade of enstrophy and an inverse cascade of energy, for which the proposed spectrum is in $k^{-5/3}$. The existence in the same system of two different cascades – called dual cascade – was quite new in strong turbulence. This prediction has since been accurately verified both at the experimental (Couder, 1984; Kellay and Goldburg, 2002) and numerical levels (Leith, 1968; Pouquet et al., 1975; Chertkov et al., 2007). An illustration of what happens in physical space is given in Figure 3.4: this is a result of a direct numerical simulation of the two-dimensional Navier–Stokes equations. We see the temporal evolution of the (normalized) vorticity, which allows us to follow the formation of eddies. The latter tend to agglomerate and finally leave room for only a few structures. This merging mechanism can be seen as a signature of the inverse cascade.

3.4.1 Fjørtoft's Phenomenology

It is possible to demonstrate the existence of a dual cascade by dimensional arguments. We recall that enstrophy is defined as:

$$\Omega \equiv \frac{1}{2}\int_{\mathbb{R}^2} \mathbf{w}^2 d\mathbf{x} \equiv \frac{1}{2}\int_{\mathbb{R}^2} (\nabla \times \mathbf{u})^2 d\mathbf{x}. \tag{3.59}$$

In this two-dimensional case, the definition can be reduced to the transverse component: $\Omega \equiv \frac{1}{2}\int w_z^2(x,y)dxdy$. This quantity is dimensionally related to the energy by the spectral relation: $k^2 E_k \sim \Omega_k$. Suppose that an energy flux ε_i and an enstrophy flux η_i are injected into the system at a scale k_i. For our proof, we will assume the existence of dissipation at large and small scales, denoted respectively by k_0 and k_∞ and such that $0 < k_0 < k_i < k_\infty < +\infty$. Large-scale dissipation can be seen, for example, as the consequence of the friction between the eddies and the walls of an experiment. We assume that the system is dynamically balanced (assumption of statistical stationarity) and that the injection compensates on average the dissipation of the invariants. In this case, by conservation of the invariants we have:

$$\varepsilon_i = \varepsilon_0 + \varepsilon_\infty, \tag{3.60a}$$

$$\eta_i = \eta_0 + \eta_\infty, \tag{3.60b}$$

with ε_0, η_0, ε_∞, and η_∞ the values of the energy and entropy fluxes at the scales mentioned above. Furthermore, we have the dimensional relation:

$$k^2 \varepsilon \sim \eta, \tag{3.61}$$

which is valid at the three scales introduced above. The combination of the different relationships gives us:

$$\frac{\varepsilon_0}{\varepsilon_\infty} \sim \frac{k_\infty^2/k_i^2 - 1}{1 - k_0^2/k_i^2}, \tag{3.62a}$$

$$\frac{\eta_0}{\eta_\infty} \sim \frac{k_i^2/k_\infty^2 - 1}{1 - k_i^2/k_0^2}. \tag{3.62b}$$

In the limit of an extended inertial range, $0 < k_0 \ll k_i \ll k_\infty < +\infty$, we obtain:

$$\frac{\varepsilon_0}{\varepsilon_\infty} \sim \frac{k_\infty^2}{k_i^2} \to +\infty, \tag{3.63a}$$

$$\frac{\eta_0}{\eta_\infty} \sim \frac{k_0^2}{k_i^2} \to 0, \tag{3.63b}$$

which means that the energy and entropy fluxes have an opposite direction. In other words, we have a direct cascade of entropy and an inverse cascade of energy.

The spectral law can be obtained by the Kolmogorov phenomenology presented in Section 3.3.2, which does not take into account the direction of the cascade. In the case of energy, the prediction remains the same and we obtain a large-scale spectrum in $E_k \sim \varepsilon^{2/3} k^{-5/3}$ for the inverse cascade. We then have to adapt the phenomenology to the enstrophy. We have in the inertial range:

$$\eta \sim \frac{\Omega}{\tau} \sim \frac{\Omega}{\ell/u} \sim \Omega^{3/2} \sim (k\Omega_k)^{3/2}, \qquad (3.64)$$

hence the small-scale entropy spectrum for the direct cascade:

$$\boxed{\Omega_k \sim \eta^{2/3} k^{-1}}. \qquad (3.65)$$

Using the dimensional relationship between the two spectra, we get the energy spectrum at small scales:

$$\boxed{E_k \sim \eta^{2/3} k^{-3}}. \qquad (3.66)$$

The dependence in η is there to remind us that this spectral law is the consequence of the entropy cascade and not of energy. In Figure 3.5 we show the energy spectrum of two-dimensional turbulence, which is therefore characterized by two different spectral slopes: while the power law at large scales is the

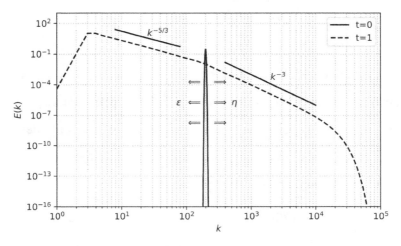

Figure 3.5 Schematic evolution (in two-dimensional turbulence) of the one-dimensional energy spectrum between the initial time $t = 0$ and $t = 1$. An external force applied at an intermediate scale injects energy flux ε and entropy flux η. A dual cascade emerges and the energy spectrum is characterized by two different spectral slopes.

consequence of the inverse cascade of energy, that at small scales must be interpreted as a signature of a direct entropy cascade. Note, to conclude, that the formation of the energy spectrum in $-5/3$ is slower for an inverse cascade than for a direct cascade. The origin of this difference is the infinite capacity of the system to accumulate energy at small wavenumbers k, while it is finite at large wavenumbers.

3.4.2 Detailed Conservation

The analytical theory proposed by Kraichnan (1967) makes it possible to go further in the description. For this we need to use the detailed energy and enstrophy conservation. Note that the demonstration that we are going to do differs slightly from that of Kraichnan because we will use the stream function ψ such that:

$$\mathbf{u} \equiv \mathbf{e_z} \times \nabla \psi, \qquad (3.67)$$

with $\mathbf{e_z}$ a unit vector in the transverse direction to the two-dimensional plan. In Fourier space, this relation is written:

$$\hat{\mathbf{u}}(\mathbf{k}) = i\hat{\psi}_\mathbf{k} \mathbf{e_z} \times \mathbf{k}. \qquad (3.68)$$

With the use of the stream function the condition of zero divergence on the velocity is automatically satisfied and the equations become simpler. We have also:

$$\mathbf{k} \cdot \hat{\mathbf{u}}(\mathbf{p}) = i\hat{\psi}_\mathbf{p} \mathbf{e_z} \cdot (\mathbf{p} \times \mathbf{k}), \qquad (3.69a)$$

$$\mathbf{k} \cdot \hat{\mathbf{u}}(\mathbf{q}) = i\hat{\psi}_\mathbf{q} \mathbf{e_z} \cdot (\mathbf{q} \times \mathbf{k}), \qquad (3.69b)$$

$$\hat{\mathbf{u}}(\mathbf{k}) \cdot \hat{\mathbf{u}}(\mathbf{p}) = -\hat{\psi}_\mathbf{k} \hat{\psi}_\mathbf{p} \mathbf{k} \cdot \mathbf{p}, \qquad (3.69c)$$

$$\hat{\mathbf{u}}(\mathbf{k}) \cdot \hat{\mathbf{u}}(\mathbf{q}) = -\hat{\psi}_\mathbf{k} \hat{\psi}_\mathbf{q} \mathbf{k} \cdot \mathbf{q}. \qquad (3.69d)$$

Therefore, expression (3.22) can be written as:

$$S(\mathbf{k}, \mathbf{p}, \mathbf{q}) = [(\mathbf{k} \cdot \mathbf{p})(\mathbf{e_z} \cdot (\mathbf{p} \times \mathbf{q})) + (\mathbf{k} \cdot \mathbf{q})(\mathbf{e_z} \cdot (\mathbf{q} \times \mathbf{p}))] \Re[\hat{\psi}_\mathbf{k} \hat{\psi}_\mathbf{p} \hat{\psi}_\mathbf{q}], \qquad (3.70)$$

with $\Re$ the real part. By using the triadic relation, the relation simplifies:

$$S(\mathbf{k}, \mathbf{p}, \mathbf{q}) = [(q^2 - p^2)(\mathbf{e_z} \cdot (\mathbf{p} \times \mathbf{q}))] \Re[\hat{\psi}_\mathbf{k} \hat{\psi}_\mathbf{p} \hat{\psi}_\mathbf{q}], \qquad (3.71)$$

with the energy conservation equation per mode:

$$\partial_t |\hat{\mathbf{u}}(\mathbf{k})|^2 + 2\nu k^2 |\hat{\mathbf{u}}(\mathbf{k})|^2 = \int_{\mathbb{R}^4} S(\mathbf{k}, \mathbf{p}, \mathbf{q}) \delta(\mathbf{k} + \mathbf{p} + \mathbf{q}) d\mathbf{p} d\mathbf{q}. \qquad (3.72)$$

Furthermore, we can demonstrate (see Exercise I.3) that the conservation of entropy per mode is written:

$$\partial_t |\hat{w}_z(\mathbf{k})|^2 + 2\nu k^2 |\hat{w}_z(\mathbf{k})|^2 = \int_{\mathbb{R}^4} k^2 S(\mathbf{k}, \mathbf{p}, \mathbf{q}) \delta(\mathbf{k} + \mathbf{p} + \mathbf{q}) d\mathbf{p} d\mathbf{q}. \qquad (3.73)$$

3.4 Two-Dimensional Eddy Turbulence

For that we have to use the relation:

$$\hat{w}_z(\mathbf{k}) = -k^2 \hat{\psi}_\mathbf{k} \mathbf{e_z}. \tag{3.74}$$

From these expressions, it is possible to demonstrate the detailed conservation of energy and enstrophy (see Exercise I.3) per triad of interaction; these conservation laws are written respectively:

$$S(\mathbf{k},\mathbf{p},\mathbf{q}) + S(\mathbf{p},\mathbf{q},\mathbf{k}) + S(\mathbf{q},\mathbf{k},\mathbf{p}) = 0, \tag{3.75a}$$

$$k^2 S(\mathbf{k},\mathbf{p},\mathbf{q}) + p^2 S(\mathbf{p},\mathbf{q},\mathbf{k}) + q^2 S(\mathbf{q},\mathbf{k},\mathbf{p}) = 0. \tag{3.75b}$$

These two relations are generalized in the case of statistically isotropic turbulence (see Exercise I.3). In this case we have:

$$\boxed{T(k,p,q) + T(p,q,k) + T(q,k,p) = 0} \tag{3.76}$$

and

$$\boxed{k^2 T(k,p,q) + p^2 T(p,q,k) + q^2 T(q,k,p) = 0}, \tag{3.77}$$

with the evolution equations of the one-dimensional energy $E(k)$ and entropy $\Omega(k)$ spectra:

$$\partial_t E(k) + 2\nu k^2 E(k) = T(k) = \int_\Delta T(k,p,q) dp dq, \tag{3.78a}$$

$$\partial_t \Omega(k) + 2\nu k^2 \Omega(k) = k^2 T(k) = \int_\Delta k^2 T(k,p,q) dp dq, \tag{3.78b}$$

where Δ means an integral verifying the triadic relation $\mathbf{k} + \mathbf{p} + \mathbf{q} = 0$ and:

$$\boxed{T(k,p,q) = 2\pi k(q^2 - p^2)\frac{pq}{\sin\theta_k} \mathbf{e_z} \cdot (\mathbf{e_p} \times \mathbf{e_q})\Re\langle \hat{\psi}_\mathbf{k} \hat{\psi}_\mathbf{p} \hat{\psi}_\mathbf{q}\rangle}, \tag{3.79}$$

with $\mathbf{e_p}$ and $\mathbf{e_q}$ two unit vectors oriented according to $\mathbf{p}$ and $\mathbf{q}$, respectively, and θ_k the opposite angle to k in the triangle $\mathbf{k} + \mathbf{p} + \mathbf{q} = 0$.

3.4.3 Zakharov Transformation and Power-Law Solutions

We will follow a method different from Kraichnan (1967) and make use of the Zakharov transformation. Usually, this transformation is only used in wave turbulence to obtain the exact solutions of the kinetic equations (see Chapter 5). We will assume the scale invariance:

$$T(ak, ap, aq) = a^x T(k, p, q) \tag{3.80}$$

and introduce the dimensionless variables: $\xi_p = p/k$ and $\xi_q = q/k$. We are first of all interested in energy; we have in the inertial range:

$$\partial_t E(k) = \frac{1}{3} \int_\Delta k^{2+x} [T(1, \xi_p, \xi_q) + T(1, \xi_p, \xi_q) + T(1, \xi_p, \xi_q)] d\xi_p d\xi_q. \tag{3.81}$$

We then apply the following transformations respectively to the last two integrands (see Figure 3.6):

$$\xi_p \to \frac{1}{\xi_p} \quad \text{and} \quad \xi_q \to \frac{\xi_q}{\xi_p}, \tag{3.82a}$$

$$\xi_p \to \frac{\xi_p}{\xi_q} \quad \text{and} \quad \xi_q \to \frac{1}{\xi_q}, \tag{3.82b}$$

which gives:

$$\partial_t E(k) = \frac{k^{2+x}}{3} \int_\Delta [T(1,\xi_p,\xi_q) \tag{3.83}$$
$$+ \xi_p^{-3-x} T(\xi_p,1,\xi_q) + \xi_q^{-3-x} T(\xi_q,\xi_p,1)] d\xi_p d\xi_q$$
$$= \frac{k^{2+x}}{3} \int_\Delta [T(1,\xi_p,\xi_q) + \xi_q^{-3-x} T(\xi_q,1,\xi_p) + \xi_p^{-3-x} T(\xi_p,\xi_q,1)] d\xi_p d\xi_q.$$

The detailed conservation laws of energy and enstrophy give us the relationships:

$$\frac{T(\xi_q,1,\xi_p)}{T(1,\xi_p,\xi_q)} = \frac{1-\xi_p^2}{\xi_p^2 - \xi_q^2}, \tag{3.84a}$$

$$\frac{T(\xi_p,\xi_q,1)}{T(1,\xi_p,\xi_q)} = \frac{\xi_q^2-1}{\xi_p^2 - \xi_q^2}. \tag{3.84b}$$

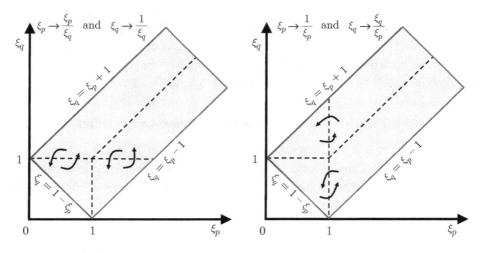

Figure 3.6 Zakharov conformal transformation for three-wave interactions. The infinitely extended gray band corresponds to the solutions of the triangular relation $\mathbf{k} + \mathbf{p} + \mathbf{q} = \mathbf{0}$ (the boundaries correspond to flattened triangles). The transformation consists of the exchange of four regions separated by dashes. The transformations (3.82a) and (3.82b) are shown on the right and on the left, respectively.

3.4 Two-Dimensional Eddy Turbulence

With the introduction of these relationships, we get:

$$\partial_t E(k) = \frac{k^{2+x}}{3} \int_\Delta T(1, \xi_p, \xi_q) \tag{3.85}$$

$$\times \left[1 + \xi_q^{-3-x} \frac{1-\xi_p^2}{\xi_p^2 - \xi_q^2} + \xi_p^{-3-x} \frac{\xi_q^2 - 1}{\xi_p^2 - \xi_q^2} \right] d\xi_p d\xi_q = \frac{k^{2+x}}{3} \int_\Delta T(1, \xi_p, \xi_q)$$

$$\times \left[\frac{(\xi_p \xi_q)^{3+x}(\xi_p^2 - \xi_q^2) + \xi_p^{3+x} - \xi_p^{5+x} + \xi_q^{5+x} - \xi_q^{3+x}}{(\xi_p \xi_q)^{3+x}(\xi_p^2 - \xi_q^2)} \right] d\xi_p d\xi_q.$$

Therefore, the exact stationary constant flux solution corresponds to $x = -3$.

What is the associated energy spectrum $E(k)$? To answer this question, it is necessary to introduce a similarity hypothesis. Dimensionally, we have:

$$[E(k)]^{3/2} k^{3/2} = [T(k)] = k^{x+2}. \tag{3.86}$$

If we assume that $E(k) \sim k^y$ we get the scaling relation: $3y = 2x + 1$. Therefore, the stationary solution corresponds to $y = -5/3$: it is the Kolmogorov spectrum. This solution is not exact because we made an assumption of similarity which is not true in turbulence, as proven by the presence of intermittency (see Chapter 2). However, the intermittency correction at this statistical level is relatively small.

Let us now turn to enstrophy; we have in the inertial range:

$$\partial_t \Omega(k) = \frac{1}{3} \int_\Delta k^{4+x} [T(1, \xi_p, \xi_q) + T(1, \xi_p, \xi_q) + T(1, \xi_p, \xi_q)] d\xi_p d\xi_q. \tag{3.87}$$

We apply the Zakharov transformation:

$$\partial_t \Omega(k) = \frac{k^{4+x}}{3} \int_\Delta [T(1, \xi_p, \xi_q) \tag{3.88}$$
$$+ \xi_q^{-3-x} T(\xi_q, 1, \xi_p) + \xi_p^{-3-x} T(\xi_p, \xi_q, 1)] d\xi_p d\xi_q.$$

Using detailed conservation laws gives:

$$\partial_t \Omega(k) = \frac{k^{4+x}}{3} \int_\Delta T(1, \xi_p, \xi_q) \tag{3.89}$$

$$\times \left[\frac{(\xi_p \xi_q)^{3+x}(\xi_p^2 - \xi_q^2) + \xi_p^{3+x} - \xi_p^{5+x} + \xi_q^{5+x} - \xi_q^{3+x}}{(\xi_p \xi_q)^{3+x}(\xi_p^2 - \xi_q^2)} \right] d\xi_p d\xi_q.$$

The exact stationary solution at constant flux therefore corresponds to $x = -5$, that is, an entrophy spectrum in $\Omega(k) \sim k^{-1}$. We recover the phenomenological prediction. Here again this result is not exact, because we have used a similarity hypothesis – not valid in the presence of intermittency.

3.4.4 Energy and Enstrophy Fluxes

Then, we can study the sign of the energy flux. We have:

$$\Pi(k) = \int_k^{+\infty} \frac{k'^{2+x}}{3} dk' I(x) = -\frac{k^{3+x}}{3(3+x)} I(x), \tag{3.90}$$

with:

$$I(x) \equiv \int_\Delta T(1, \xi_p, \xi_q) \tag{3.91}$$
$$\times \frac{(\xi_p \xi_q)^{3+x}(\xi_p^2 - \xi_q^2) + \xi_p^{3+x} - \xi_p^{5+x} + \xi_q^{5+x} - \xi_q^{3+x}}{(\xi_p \xi_q)^{3+x}(\xi_p^2 - \xi_q^2)} d\xi_p d\xi_q.$$

We implicitly assumed that the contributions of the triadic interaction for $k \to +\infty$ tend towards 0 (and therefore that $\Pi(+\infty) = 0$): this is a locality hypothesis whose validity must be verified a posteriori. The constant flux is obtained in the stationary case, that is, when $x = -3$. To take this limit, it is necessary to use L'Hôspital's rule since $I(-3) = 0$. With $z = x + 3$, we have:

$$\lim_{z \to 0} \Pi(k) = \varepsilon = -\frac{1}{3} \frac{dI(z)}{dz}\bigg|_{z=0}, \tag{3.92}$$

with:

$$\frac{dI(z)}{dz}\bigg|_{z=0} = \tag{3.93}$$
$$\int_\Delta T(1, \xi_p, \xi_q) \frac{-\xi_q^{-z} \ln(\xi_q) + \xi_p^2 \xi_q^{-z} \ln(\xi_q) - \xi_q^2 \xi_p^{-z} \ln(\xi_p) + \xi_p^{-z} \ln(\xi_p)}{\xi_p^2 - \xi_q^2} d\xi_p d\xi_q\bigg|_{z=0}$$
$$= \int_\Delta T(1, \xi_p, \xi_q) \frac{(\xi_p^2 - 1)\ln(\xi_q) - (\xi_q^2 - 1)\ln(\xi_p)}{\xi_p^2 - \xi_q^2} d\xi_p d\xi_q.$$

We get the exact expression for the energy flux:

$$\boxed{\varepsilon = \int_\Delta T(1, \xi_p, \xi_q) A(\xi_p, \xi_q) d\xi_p d\xi_q}, \tag{3.94}$$

with:

$$A(\xi_p, \xi_q) = \frac{1}{3} \frac{(\xi_p^2 - 1)\ln(\xi_q) - (\xi_q^2 - 1)\ln(\xi_p)}{\xi_q^2 - \xi_p^2}. \tag{3.95}$$

To go further, we can introduce the normalized expression (3.79):

$$T(1, \xi_p, \xi_q) = (\xi_q^2 - \xi_p^2)\xi_p \xi_q \frac{\mathbf{e_z} \cdot (\mathbf{e_p} \times \mathbf{e_q})}{\sin \theta_k} \Re[2\pi \langle \hat{\psi}_1 \hat{\psi}_{\xi_p} \hat{\psi}_{\xi_q} \rangle]. \tag{3.96}$$

A Taylor expansion in the neighborhood of the point $\xi_p = \xi_q = 1$ gives $A(\xi_p, \xi_q) \simeq -\frac{1}{3}\xi_p \xi_q$, therefore $A(1, 1) = 0$. Furthermore, we see that $A(\xi_p, \xi_q)$ is zero for $\xi_p = 1$ or $\xi_q = 1$. Thus, the gray band in Figure 3.6 on which we are can be divided into four regions with an axial symmetry along the diagonal $\xi_p = \xi_q$ on which $A = 0$. The sign of A in each region is in fact given by an estimate in the neighborhood of the point (1, 1) (see Figure 3.7).

The enstrophy flux writes:

$$\eta(k) = \int_k^{+\infty} \frac{k'^{4+x}}{3} dk' J(x) = -\frac{k^{5+x}}{3(5+x)} J(x), \tag{3.97}$$

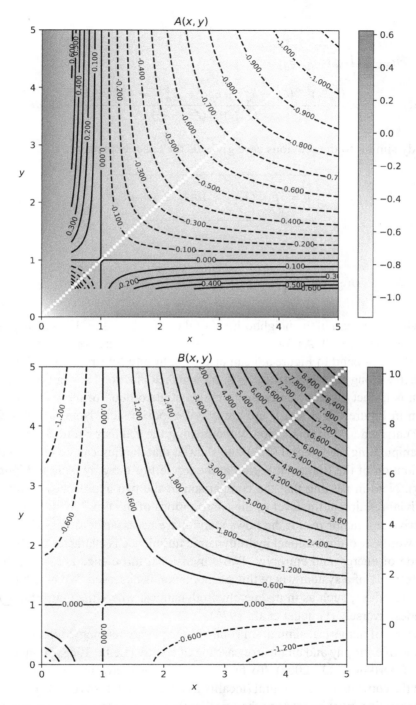

Figure 3.7 Functions $A(x,y)$ (top) and $B(x,y)$ (bottom). Only the values on the shaded band in Figure 3.6 are relevant. We can distinguish four regions on which A and B have opposite signs.

with:

$$J(x) \equiv \int_\Delta T(1, \xi_p, \xi_q) \qquad (3.98)$$
$$\times \frac{(\xi_p \xi_q)^{3+x}(\xi_p^2 - \xi_q^2) + \xi_p^{3+x} - \xi_p^{5+x} + \xi_q^{5+x} - \xi_q^{3+x}}{(\xi_p \xi_q)^{3+x}(\xi_p^2 - \xi_q^2)} d\xi_p d\xi_q.$$

A study similar to the previous case gives us the exact expression:

$$\boxed{\eta = \int_\Delta T(1, \xi_p, \xi_q) B(\xi_p, \xi_q) d\xi_p d\xi_q}, \qquad (3.99)$$

with:

$$B(\xi_p, \xi_q) = \frac{1}{3} \frac{\xi_q^2(\xi_p^2 - 1)\ln(\xi_q) - \xi_p^2(\xi_q^2 - 1)\ln(\xi_p)}{\xi_q^2 - \xi_p^2}. \qquad (3.100)$$

A Taylor expansion in the neighborhood of the point $\xi_p = \xi_q = 1$ gives $B(\xi_p, \xi_q) \simeq \frac{1}{3}\xi_p \xi_q$, so $B(1, 1) = 0$. As for A, we see that $B(\xi_p, \xi_q)$ is zero for $\xi_p = 1$ or $\xi_q = 1$. Thus the gray band in Figure 3.6 can also be split into four regions with an axial symmetry along the diagonal $\xi_p = \xi_q$ on which B is zero. The sign of B in each region is in fact given by an estimate in the neighborhood of the point $(1, 1)$. As shown in Figure 3.7, A and B are always of opposite sign. Note that Kraichnan (1967) arrived at this conclusion without using the Zakharov transformation but by manipulating the integral (3.90): he showed that the flux can be reduced to the contribution of the triangular region at the top left of the gray band in Figure 3.6 (right). Note in passing that the comparison of the two approaches highlights a nontrivial relation at the level of the contributions of the flux.[5] With these results we arrive at a major result: the flows ε and η are necessarily of opposite sign. In other words, two-dimensional hydrodynamic turbulence is characterized by a dual cascade of energy and entropy. Two-dimensional turbulence is historically the first example of a system where this phenomenon is observed. We now know that there are others, such as in magnetohydrodynamics, where the magnetic helicity cascades inversely (Pouquet et al., 1976).

The use of numerical simulation confirms the phenomenology: we have a direct cascade of entropy and an inverse cascade of energy (Leith, 1968; Pouquet et al., 1975; Chertkov et al., 2007). To finish, note that we have made an assumption about the convergence of integral (locality assumption). It turns out that a logarithmic correction must be made to the small-scale spectrum to verify this hypothesis (Kraichnan, 1971). This type of correction is not uncommon: it is also found in the wave turbulence regime (see, e.g., Düring et al., 2017).

[5] The approach developed here from the Zakharov transformation is original and has never been published.

3.5 Dual Cascade

We have just seen that two-dimensional hydrodynamic turbulence is characterized by an inverse cascade of energy, whereas it is direct in three dimensions. From a theoretical point of view, one can wonder if the transition between a direct and an inverse cascade of energy happens suddenly when the thickness of the fluid decreases, or if it is done gradually. Numerical studies on the subject show that the second scenario is the right one: there is a thickness interval on which the transition happens gradually (Benavides and Alexakis, 2017).

In the three-dimensional case, the decomposition of the velocity field on a complex helicity basis (Craya, 1954) makes it possible to separate triadic interactions by class and highlight those that contribute to positive and negative energy flux (Sahoo et al., 2017; Alexakis and Biferale, 2018). We can show that this behavior is linked to the amount of kinetic helicity injected into the system (Plunian et al., 2020). Note that the introduction of an external force, such as the Coriolis force, can fundamentally change the three-dimensional physics with the emergence of an inverse cascade (Smith and Waleffe, 1999) (see Chapter 6).

The duality – direct cascade / inverse cascade – remains a subject very much studied in hydrodynamic turbulence but also in many others systems. A review on this theme is proposed by Pouquet et al. (2017), 50 years after the discovery made by Kraichnan (1967).

3.6 Nonlinear Diffusion Model

In this section, we will present a phenomenological model of turbulence based on the idea of a turbulent diffusion in the spectral space. The approach is different from the closure methods as EDQNM in that we seek to express the cascade mechanism directly without going through a hierarchy of equations. This type of model was first proposed by Leith (1967). It has since been generalized to other systems in strong/eddy turbulence (Leith, 1968; Zhou and Matthaeus, 1990; Connaughton and Nazarenko, 2004; Matthaeus et al., 2009; Thalabard et al., 2015) and in weak/wave turbulence (Hasselmann et al., 1985; Dyachenko et al., 1992; Zakharov and Pushkarev, 1999; Boffetta et al., 2009; Galtier and Buchlin, 2010; Galtier et al., 2019). Note that in the latter case, the nonlinear diffusion equation can be obtained by a rigorous calculation from the weak wave turbulence equations in the approximation of local interactions. In Chapter 6, such a limit is discussed in detail in the context of rotating hydrodynamic turbulence.

In this approach, the cascade is modeled by a diffusion process in the spectral space. We will consider the diffusion equation of energy valid in the inertial range:

$$\partial_t E(\mathbf{k}) = -\nabla \cdot \mathbf{\Pi}(\mathbf{k}), \qquad (3.101)$$

with $\Pi(\mathbf{k})$ the energy flux. We assume a statistically isotropic turbulence, therefore we can write:

$$\partial_t E(\mathbf{k}) = -\frac{1}{k^2}\frac{\partial(k^2\Pi_k(\mathbf{k}))}{\partial k}, \qquad (3.102)$$

with Π_k the radial component of the flux in spherical coordinates. This component is modeled as follows:

$$\Pi_k = -D\frac{\partial E(\mathbf{k})}{\partial k}, \qquad (3.103)$$

where D is a diffusion coefficient. Dimensionally, we find:

$$D = \frac{k^2}{\tau_{NL}}, \qquad (3.104)$$

with τ_{NL} the characteristic transfer time of energy. Its phenomenological expression is given by the relation (see Chapter 2):

$$\tau_{NL} \sim \frac{\ell}{u_\ell}. \qquad (3.105)$$

We introduce the one-dimensional energy spectrum, $E(k) = 4\pi k^2 E(\mathbf{k})$. The diffusion coefficient can then be written:

$$D \sim k^3 \sqrt{kE(k)}. \qquad (3.106)$$

After a few manipulations, we finally obtain the following nonlinear diffusion equation, in which we have introduced a term of dissipation to include the dissipative scales:

$$\boxed{\partial_t E(k) = \frac{\partial}{\partial k}\left(k^{11/2}\sqrt{E(k)}\frac{\partial(E(k)/k^2)}{\partial k}\right) - 2\nu k^2 E(k)}. \qquad (3.107)$$

This equation models three-dimensional, homogeneous, and isotropic hydrodynamic turbulence. We could have introduced a constant of proportionality in front of the right term, however, we can always renormalize time to make this constant disappear. In other words, the characteristic timescale of this diffusion equation is normalized by the characteristic time of the cascade.

We can verify that this equation has as its solution the Kolmogorov spectrum. For that we introduce the spectrum:

$$E(k) = Ak^x, \qquad (3.108)$$

with $A > 0$ because the energy spectrum is a definite positive quantity. The expression for the flux then takes the form:

$$\Pi_k(k) = -\sqrt{A}k^{(11+x)/2}\frac{\partial(Ak^{x-2})}{\partial k} = A^{3/2}(2-x)k^{(5+3x)/2}. \qquad (3.109)$$

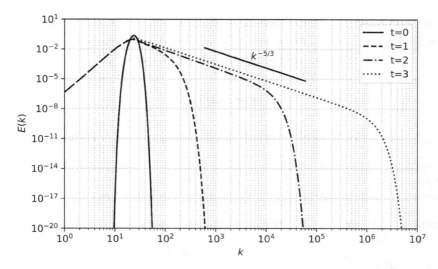

Figure 3.8 Schematic evolution of the energy spectrum with a nonlinear diffusion model.

The non-zero constant flux solution therefore corresponds to $x = -5/3$: it is indeed the Kolmogorov spectrum. We see that for this value of x, we have:

$$\Pi_k = \varepsilon = \frac{11}{3} A^{3/2}, \qquad (3.110)$$

which is positive, as expected for a direct cascade. It is not difficult to numerically simulate the nonlinear diffusion equation and obtain the Kolmogorov spectrum over several decades, as shown in Figure 3.8.

This access to an extended inertial range makes it possible to numerically address the problem of the nonstationary regime, during which the spectrum formed may not be that of the stationary regime. In the hydrodynamic case, a slightly steeper spectrum around -1.85 was measured. This spectrum is currently understood as a self-similar solution of the second kind, that is, a solution that we cannot predict exactly (Connaughton and Nazarenko, 2004; Thalabard et al., 2015). This solution is characterized by an explosive propagation of the spectrum towards large wavenumbers: we can show that the self-similar solution reaches in principle $k = +\infty$ in a finite time. We will come back to this question in Chapter 6 in the context of rotating hydrodynamics and in Chapter 8 in plasma physics.

To conclude, let us note that there is another type of turbulence model that has been often used in the past: the shell model. As for the diffusion equation, in this model the nonlinear interactions are assumed to be local. The velocity is modeled in the Fourier space by a complex scalar depending on the wavenumber and whose temporal evolution is governed by an ordinary differential equation built in an ad hoc manner. Despite its simplicity, this hydrodynamic model has nontrivial intermittency properties. In the case of magnetohydrodynamics, chaotic reversals of magnetic polarity have been observed. For more information, we refer the reader to the review articles of Biferale (2003) and Plunian et al. (2013) for hydrodynamics and magnetohydrodynamics, respectively.

References

Alexakis, A., and Biferale, L. 2018. Cascades and transitions in turbulent flows. *Phys. Rep.*, **767**, 1–101.
Batchelor, G. K. 1953. *The Theory of Homogeneous Turbulence*. Cambridge University Press.
Benavides, S. J., and Alexakis, A. 2017. Critical transitions in thin layer turbulence. *J. Fluid Mech.*, **822**, 364–385.
Biferale, L. 2003. Shell models of energy cascade in turbulence. *Ann. Rev. Fluid Mech.*, **35**(35), 441–468.
Boffetta, G., and Ecke, R. E. 2012. Two-dimensional turbulence. *Ann. Rev. Fluid Mech.*, **44**(1), 427–451.
Boffetta, G., Celani, A., Dezzani, D., Laurie, J., and Nazarenko, S. 2009. Modeling Kelvin wave cascades in superfluid helium. *J. Low Temp. Phys.*, **156**, 193–214.
Chertkov, M., Connaughton, C., Kolokolov, I., and Lebedev, V. 2007. Dynamics of energy condensation in two-dimensional turbulence. *Phys. Rev. Lett.*, **99**(8), 084501.
Connaughton, C., and Nazarenko, S. 2004. Warm cascades and anomalous scaling in a diffusion model of turbulence. *Phys. Rev. Lett.*, **92**, 044501.
Couder, Y. 1984. Two-dimensional grid turbulence in a thin liquid layer. *J. Phys. Lett.*, **45**, 353–360.
Craya, A. 1954. Contribution à l'analyse de la turbulence associée à des vitesses moyennes. *PST Ministère de l'Air*, **345**.
Düring, G., Josserand, C., and Rica, S. 2017. Wave turbulence theory of elastic plates. *Physica D*, **347**, 42–73.
Dyachenko, S., Newell, A. C., Pushkarev, A., and Zakharov, V. E. 1992. Optical turbulence: Weak turbulence, condensates and collapsing filaments in the nonlinear Schrödinger equation. *Physica D*, **57**, 96–160.
Fjørtoft, R. 1953. On the changes in the spectral distribution of kinetic energy for two-dimensional, non-divergent flow. *Tellus*, **5**, 225–230.
Galtier, S., and Buchlin, E. 2010. Nonlinear diffusion equations for anisotropic magneto-hydrodynamic turbulence with cross-helicity. *Astrophys. J.*, **722**(2), 1977–1983.
Galtier, S., Nazarenko, S. V., Buchlin, E., and Thalabard, S. 2019. Nonlinear diffusion models for gravitational wave turbulence. *Physica D*, **390**, 84–88.
Grant, H. L., Stewart, R. W., and Moilliet, A. 1962. Turbulence spectra from a tidal channel. *J. Fluid Mech.*, **12**(2), 241–268.
Hasselmann, S., Hasselmann, K., Allender, J. H., and Barnett, T. P. 1985. Computations and parameterizations of the nonlinear energy transfer in a gravity-wave spectrum. Part II. *J. Phys. Oceano.*, **15**, 1378–1391.
Kellay, H., and Goldburg, W. I. 2002. Two-dimensional turbulence: A review of some recent experiments. *Rep. Prog. Phys.*, **65**(5), 845–894.
Kraichnan, R. H. 1957. Relation of fourth-order to second-order moments in stationary isotropic turbulence. *Phys. Rev.*, **107**(6), 1485–1490.
Kraichnan, R. H. 1958. Irreversible statistical mechanics of incompressible hydromagnetic turbulence. *Phys. Rev.*, **109**(5), 1407–1422.
Kraichnan, R. H. 1959. The structure of isotropic turbulence at very high Reynolds numbers. *J. Fluid Mech.*, **5**, 497–543.

Kraichnan, R. H. 1961. Dynamics of nonlinear stochastic systems. *J. Math. Phys.*, **2**(1), 124–148.

Kraichnan, R. H. 1965. Lagrangian-history closure approximation for turbulence. *Phys. Fluids*, **8**(4), 575–598.

Kraichnan, R. H. 1966. Isotropic turbulence and inertial-range structure. *Phys. Fluids*, **9**(9), 1728–1752.

Kraichnan, R. H. 1967. Inertial ranges in two-dimensional turbulence. *Phys. Fluids*, **10**(7), 1417–1423.

Kraichnan, R. H. 1971. Inertial-range transfer in two- and three-dimensional turbulence. *J. Fluid Mech.*, **47**, 525–535.

Kraichnan, R. H. 1975. Remarks on turbulence theory. *Adv. Math.*, **16**, 305–331.

Kraichnan, R. H., and Montgomery, D. 1980. Two-dimensional turbulence. *Rep. Prog. Phys.*, **43**(5), 547–619.

Lee, T. D. 1951. Difference between turbulence in a two-dimensional fluid and in a three-dimensional fluid. *J. Appl. Phys.*, **22**(4), 524 (1 page).

Leith, C. E. 1967. Diffusion approximation to inertial energy transfer in isotropic turbulence. *Phys. Fluids*, **10**(7), 1409–1416.

Leith, C. E. 1968. Diffusion approximation for two-dimensional turbulence. *Phys. Fluids*, **11**(3), 671–672.

Lesieur, M. 1997. *Turbulence in Fluids*. Kluwer Academic Publishers.

Leslie, D. C. 1973. Review of developments in turbulence theory. *Rep. Prog. Phys.*, **36**(11), 1365–1424.

Matthaeus, W. H., Oughton, S., and Zhou, Y. 2009. Anisotropic magnetohydrodynamic spectral transfer in the diffusion approximation. *Phys. Rev. E*, **79**(3), 035401.

Millionschikov, M. D. 1941. Theory of homogeneous isotropic turbulence. *Dokl. Akad. Nauk SSSR*, **22**, 241–242.

Nakayama, K. 1999. Statistical theory of anisotropic MHD turbulence: An approach to strong shear Alfvén turbulence by DIA. *Astrophys. J.*, **523**(1), 315–327.

Nakayama, K. 2001. Statistical theory of anisotropic MHD turbulence. II. Lagrangian theory of strong shear Alfvén turbulence. *Astrophys. J.*, **556**(2), 1027–1037.

Obukhov, A. M. 1941a. On the distribution of energy in the spectrum of turbulent flow. *Dokl. Akad. Nauk SSSR*, **32**, 22–24.

Obukhov, A. M. 1941b. Spectral energy distribution in a turbulent flow. *Izv. Akad. Nauk SSSR Ser. Geogr. Geofiz.*, **5**, 453–466.

Ogura, Y. 1963. A consequence of the zero-fourth-cumulant approximation in the decay of isotropic turbulence. *J. Fluid Mech.*, **16**, 33–40.

Orszag, S. A. 1970. Analytical theories of turbulence. *J. Fluid Mech.*, **41**, 363–386.

Pao, Y.-H. 1965. Structure of turbulent velocity and scalar fields at large wavenumbers. *Phys. Fluids*, **8**(6), 1063–1075.

Plunian, F., Stepanov, R., and Frick, P. 2013. Shell models of magnetohydrodynamic turbulence. *Phys. Rep.*, **523**(1), 1–60.

Plunian, F., Teimurazov, A., Stepanov, R., and Verma, M. K. 2020. Inverse cascade of energy in helical turbulence. *J. Fluid Mech.*, **895**, A13.

Pouquet, A., Lesieur, M., Andre, J. C., and Basdevant, C. 1975. Evolution of high Reynolds number two-dimensional turbulence. *J. Fluid Mech.*, **72**, 305–319.

Pouquet, A., Frisch, U., and Leorat, J. 1976. Strong MHD helical turbulence and the nonlinear dynamo effect. *J. Fluid Mech.*, **77**, 321–354.

Pouquet, A., Marino, R., Mininni, P. D., and Rosenberg, D. 2017. Dual constant-flux energy cascades to both large scales and small scales. *Phys. Fluids*, **29**(11), 111108.

Rossby, C. G., and collaborators. 1939. Relation between variations in the intensity of the zonal circulation of the atmosphere and the displacements of the semi-permanent centers of action. *J. Marine Res.*, **2**, 38–55.

Saddoughi, S. G., and Veeravalli, S. V. 1994. Local isotropy in turbulent boundary layers at high Reynolds number. *J. Fluid Mech.*, **268**, 333–372.

Sahoo, G., Alexakis, A., and Biferale, L. 2017. Discontinuous transition from direct to inverse cascade in three-dimensional turbulence. *Phys. Rev. Lett.*, **118**(16), 164501.

Smith, L. M., and Waleffe, F. 1999. Transfer of energy to two-dimensional large scales in forced, rotating three-dimensional turbulence. *Phys. Fluids*, **11**(6), 1608–1622.

Sreenivasan, K. R. 1995. On the universality of the Kolmogorov constant. *Phys. Fluids*, **7**(11), 2778–2784.

Thalabard, S., Nazarenko, S., Galtier, S., and Medvedev, S. 2015. Anomalous spectral laws in differential models of turbulence. *J. Phys. A: Math. & Theo.*, **48**, 285501.

Welter, G. S., Wittwer, A. R., Degrazia, G. A. et al. 2009. Measurements of the Kolmogorov constant from laboratory and geophysical wind data. *Phys. A: Stat. Mech. Appl.*, **388**(18), 3745–3751.

Zakharov, V. E., and Pushkarev, A. N. 1999. Diffusion model of interacting gravity waves on the surface of deep fluid. *Nonlin. Proc. Geophys.*, **6**, 1–10.

Zhou, Y. 2021. Turbulence theories and statistical closure approaches. *Phys. Rep.*, **935**, 1–117.

Zhou, Y., and Matthaeus, W. H. 1990. Models of inertial range spectra of interplanetary magnetohydrodynamic turbulence. *J. Geophys. Res.*, **95**(A9), 14881–14892.

Exercises I

1.1 1D HD Turbulence: Burgers' Equation

Burgers' equation (Burgers, 1948) is often regarded as a one-dimensional model of compressible hydrodynamic turbulence. This equation is written:

$$\frac{\partial u}{\partial t} + u\frac{\partial u}{\partial x} = \nu \frac{\partial^2 u}{\partial x^2},$$

with u a scalar velocity and ν a viscosity. Burgers' equation is a simple model often used to test new ideas in turbulence. Here we will study the properties of dissipation and then intermittency using the tools introduced for the Navier–Stokes equations.

(1) Demonstrate that:

$$u(x,t) = \frac{1}{t}\left[x - L\tanh\left(\frac{xL}{2\nu t}\right)\right]$$

is an exact solution.

(2) Find the limit of this solution when $\nu \to 0$.

(3) Calculate the mean rate of energy dissipation ε.

(4) Calculate the mean rate of viscous dissipation ε_ν. Conclude.

(5) Find the expression of the smoothed Burgers equation by introducing an anomalous dissipation $\mathcal{D}_I^\ell$.

(6) Calculate the expression of the inertial dissipation $\mathcal{D}_I$ with the exact solution. Conclude.

(7) By using the exact solution in the limit $\nu \to 0$, find the exponents ζ_p by distinguishing the case where $p < 1$ from the case where $p \geq 1$.

1.2 Structure Function and Spectrum

We are interested in the relationship between the one-dimensional energy spectrum $E_{1d}(k)$ and the second-order structure function $S_2(r)$ in the case of three-dimensional homogeneous isotropic hydrodynamic turbulence.

(1) Let $R(\mathbf{r})$ be the two-point correlation function of the velocity, that is, $R(\mathbf{r}) = \langle \mathbf{u} \cdot \mathbf{u}' \rangle$. Write down the general relation between this function and the three-dimensional energy spectrum $E_{3d}(\mathbf{k})$.

(2) Focusing on the isotropic case, demonstrate the relationship:

$$S_2(r) = 4\int_0^{+\infty}\left(1 - \frac{\sin kr}{kr}\right)E_{1d}(k)dk, \tag{3.111}$$

where $S_2(r) = \langle (\delta \mathbf{u})^2 \rangle$.

(3) It is assumed that the one-dimensional energy spectrum is given by the relation $E_{1d}(k) = C_K \varepsilon^{2/3} k^{-5/3}$, where C_K is the Kolmogorov constant. Find the relation:

$$S_2(r) = C_2 \varepsilon^{2/3} r^{2/3}, \qquad (3.112)$$

where C_2 is a constant that will be given. Compare your results with the experimental measurements $C_K \simeq 0.5$ and $C_2 \simeq 2.5$ (Sreenivasan, 1995; Welter et al., 2009).

I.3 2D HD Turbulence: Detailed Conservation

In the two-dimensional case, the Navier–Stokes equations become simpler and it is possible to demonstrate analytically the existence of a dual cascade of energy and enstrophy (see Chapter 3). For this, we must use the detailed conservation laws for these invariants. The aim of this exercise is to obtain these two laws. We will introduce the stream function ψ such that $\mathbf{u} \equiv \mathbf{e}_z \nabla \psi$. With the use of this function the zero velocity divergence condition is automatically satisfied and the calculations are simplified.

(1) Write the spectral expression of enstrophy conservation.

(2) Demonstrate the detailed conservation of enstrophy using the relationships in the triangle formed by $\mathbf{k}$, $\mathbf{p}$, and $\mathbf{q}$.

(3) Same question for energy.

(4) Generalize the result in the case of statistically isotropic turbulence.

References

Burgers, J. M. 1948. A mathematical model illustrating the theory of turbulence. *Adv. Appl. Mech.*, **1**, 171–199.

Sreenivasan, K. R. 1995. On the universality of the Kolmogorov constant. *Phys. Fluids*, **7**(11), 2778–2784.

Welter, G. S., Wittwer, A. R., Degrazia, G. A. et al. 2009. Measurements of the Kolmogorov constant from laboratory and geophysical wind data. *Phys. A: Stat. Mech. Appl.*, **388**(18), 3745–3751.

Part II
Wave Turbulence

4

Introduction

Waves and turbulence are the two pillars of this second part. As we shall see, the wave turbulence regime offers the possibility to develop an analytical theory. Beyond its mathematical beauty, the spectral theory obtained makes it possible to understand in depth a weakly nonlinear system and to develop an intuition on the physics of strong wave turbulence. In essence, the wave turbulence regime only concerns systems in which waves can be excited, which excludes the Navier–Stokes equations of incompressible hydrodynamics.[1] We have seen in Part I that these equations are fundamental in turbulence: indeed, it is from them and from laboratory experiments based on water or air that the first theoretical advances (concepts, exact laws) have been made. This is one of the reasons why the literature on turbulence focuses mainly – if not entirely – on incompressible hydrodynamics, thus excluding a whole range of applications where waves make a major contribution to the dynamics.[2] The present book is an exception, as it is the first to devote significant attention to (strong) eddy turbulence and (weak) wave turbulence. As we shall see in this part, turbulent systems that have waves within them are very numerous. In contrast, incompressible hydrodynamics (pure eddy turbulence) appear quite singular.

We can distinguish two regimes in wave turbulence: one where the waves are of weak amplitude and one where they are not. The regime that concerns us here is mainly that of weak turbulence (first case). When there will be no ambiguity, we will follow the usages on the subject and speak of wave turbulence in the sense of weak turbulence. The existence of a small parameter – the amplitude of the waves – allows a systematic approach to the problem. The general idea is that the weak amplitude of the waves makes it possible to distinguish between the temporal evolution of the amplitude and that of the phase, the former evolving

---

[1] We mean here the unmodified Navier–Stokes equations. If, for example, the Coriolis force is added, inertial waves appear. In the limit of a small Rossby number, we obtain the inertial wave turbulence regime, which is the subject of Chapter 6.
[2] Another reason is that books on turbulence are often written by fluid mechanists and not by physicists.

Figure 4.1 Wave turbulence: Spatial deformation of the free surface of a liquid produced by direct numerical simulation (see Chapter 5).

more slowly than the latter. This separation of scales in time makes it possible to obtain a uniform asymptotic closure of the hierarchy of equations.

A brief history will allow us to understand how the theory of wave turbulence was constructed. We will apply, in a simple example and then in a nonlinear equation model, the method of multiple scales, which allows us to justify the uniformity of the asymptotic development. The systematic method of wave turbulence will then be presented exhaustively in what is probably the simplest example, that of capillary waves (Chapter 5). Chapter 5 is a technical chapter, but necessary for those who wish to make the method their own. Chapters 6 and 7 will be devoted to hydrodynamics in rapid rotation, then to magnetohydrodynamics (MHD) in a strong magnetic field. A synthetic presentation of the main properties will be given. Chapter 8 will look at wave turbulence in a compressible plasma at sub-MHD scales. The final Chapter 9 deals with a subject at the limits of our knowledge: gravitational wave turbulence, which could play a fundamental role in the mechanism at the origin of the cosmological inflation that appeared in the first second of the universe, just after the Big Bang.

4.1 Brief History

4.1.1 Prehistory

It is essentially in the field of oceanography and the study of surface waves on the sea that we find the beginnings of the theory of wave turbulence (Phillips, 1981). This work began in the late 1950s, at a time when surface waves were treated essentially in a linear manner, the spectra as a superposition of linear waves, and nonlinear effects restricted, for example, to periodic wave distortion phenomena described a long time previously by Stokes (Stokes, 1847). Gravity waves are by definition on the surface of the ocean; inside the ocean they are called "internal gravity waves." A second type of surface wave exists: capillary waves, which are discussed in detail in the Chapter 5. The Navier–Stokes equations, modified by the force of gravity, are the theoretical framework from which the first results emerged. To simplify the problem, the movements are generally assumed to be irrotational: this corresponds to an air–water interface disturbed by a unidirectional blowing wind – a typical condition encountered in the open sea. The problem is then reduced to Bernoulli's equation applied to the free surface of the fluid, to which is added a Lagrangian equation to describe the deformation of the fluid surface and another to take into account an additional hypothesis: the deep-water hypothesis (see Chapter 5).

4.1.2 Resonant Wave Interactions

Work in the late 1950s and early 1960s led to a major theoretical breakthrough with the discovery of the existence of resonant interactions between nonlinear waves of weak amplitude.[3] The waves in question are, of course, gravity waves. The idea has its origin in the work of Phillips (1960), Longuet-Higgins (1962), and Hasselmann (1962). who tried to understand how the nonlinear interactions between gravity waves could redistribute the energy initially present in these modes.[4] The underlying idea is that in the initial phase of the development of turbulence it is the resonant interactions that provide the dominant mechanism for transferring energy from one wave to another. The ultimate phase of development had already been studied previously by Phillips (1958) who, using dimensional analysis, was able to make a spectral prediction of strong (wave) turbulence which is still in use today (see Chapter 7 on critical balance).

After long calculations based on a classical perturbative development, it was possible to demonstrate that the energy redistribution was generally negligible for interactions involving less than four waves, but that it became important for four waves if the following resonance relationship was satisfied:

$$\begin{cases} \mathbf{k}_1 + \mathbf{k}_2 = \mathbf{k}_3 + \mathbf{k}, \\ \omega_1 + \omega_2 = \omega_3 + \omega, \end{cases} \quad (4.1)$$

with the dispersion relation $\omega^2 = g|\mathbf{k}|$ (g being the acceleration of gravity). The experimental demonstration of the existence of resonant interactions was then published by Longuet-Higgins and Smith (1966) and McGoldrick et al. (1966).

---

[3] Note that earlier studies on three- and four-wave resonance had already been carried out, for example, in the field of thermal conduction in crystals (Peierls, 1929). See also Nordheim (1928), who wrote the first kinetic equation: it was for four-wave interactions in the context of the electron theory of conductivity.

[4] K. Hasselmann was awarded the Nobel Prize in Physics in 2021 "for the physical modelling of Earth's climate, quantifying variability and reliably predicting global warming."

Although the existence of resonant wave interactions appeared to be real, their interpretation was then subject to debate because the energy exchange between resonant waves also led to a linear growth (in time) of the amplitude of the waves, ultimately breaking the hypothesis underlying the existence of resonant interactions, that is, the presence of waves of weak amplitude; in this case, the perturbation is said to be nonuniform.[5] It was therefore not clear whether this energy transfer could be really effective. An answer to this question was first proposed statistically by Hasselmann (1962), who considered a random set of gravity waves. The hypothesis of weak nonlinearities was then used to justify another hypothesis, that of a Gaussian statistics which greatly simplifies the calculations by eliminating in particular moments of odd order (see also Drummond and Pines, 1962).

4.1.3 Multiple Scale Method

A second major theoretical breakthrough came with Benney's work, which showed that the problem of wave turbulence is similar to that of weakly coupled oscillators in mechanics (Akylas, 2020). With a new mathematical technique for the analysis of dispersive wave packets involving two timescale, Benney (1962) obtained a relatively simple form for the (discrete and nonstatistical) equations that govern the temporal evolution of resonant modes. In this approach, the secular terms disappear: they are somehow absorbed by the slow variation of the wave amplitude. This work shows, in a simple way, how the exchange of energy between four gravity waves is carried out: this exchange takes place while conserving energy. Here we find a property of strong turbulence, that of detailed conservation (see Chapter 3).

This approach paved the way for the use of a new mathematical technique called the multiple scale method (Sturrock, 1957; Nayfeh, 2004), which made it possible to demonstrate that the (continuous) statistical equations of wave turbulence have a natural closure due to the separation of scales in time. As shown by Benney and Saffman (1966) for quadratic nonlinearities and Benney (1967) for cubic nonlinearities, the equations of wave turbulence are asymptotically valid over long periods of time and do not require the statistical assumption of Gaussianity made by Hasselmann (1962). The method of multiple timescale offers a systematic and consistent theoretical framework in which the procedure expansion allows one in principle to determine the slow rate of change of the amplitude of the waves at any order in ϵ (Benney and Newell, 1967, 1969). The so-called kinetic equation of gravity wave turbulence then takes the following (schematic) integro-differential form in d-dimension:

---

[5] We are talking about secular terms, an illustration of which is given in Section 4.2 from the Duffing equation. This growth can also be nonlinear in t^α ($\alpha > 0$) for higher-order corrective terms.

$$\frac{\partial N(\mathbf{k})}{\partial t} = \epsilon^4 \int_{\mathbb{R}^{3d}} S(\mathbf{k}, \mathbf{k}_1, \mathbf{k}_2, \mathbf{k}_3)[N_1 N_2 (N + N_3) - N_3 N(N_1 + N_2)]$$
$$\delta(\omega_1 + \omega_2 - \omega_3 - \omega)\delta(\mathbf{k}_1 + \mathbf{k}_2 - \mathbf{k}_3 - \mathbf{k}) d\mathbf{k}_1 d\mathbf{k}_2 d\mathbf{k}_3, \quad (4.2)$$

with $N \equiv N(\mathbf{k}) = E(\mathbf{k})/\omega$ the wave action spectrum and $E(\mathbf{k})$ the energy spectrum.[6] The Dirac functions translate the resonance condition (4.1) discussed in Section 4.1.2. The presence of the ϵ^4 factor means that the timescale (normalized to the wave period $\sim 1/\omega$) on which the spectra are modified by the nonlinear dynamics is of the order of $\mathcal{O}(1/\epsilon^4)$. Therefore, it is a relatively slow process.

4.1.4 Kolmogorov–Zakharov Spectrum

Parallel to the work carried out in the Western world, major theoretical advances were also made in the East. From the beginning of the 1960s, the Soviet school became interested in wave turbulence, mainly through plasma physics (Sagdeev and Galeev, 1966; Vedenov, 1967), from which certain notations and vocabulary were borrowed (one speaks, for example, of the kinetic equation or collision integral). In passing, it is curious to note that this work was carried out simultaneously by the two parts of the world without much communication between them. In particular, by the method known as random phase approximation, kinetic equations of wave turbulence were proposed in a primitive form by Kadomtsev and Petviashvili (1963) for a problem of plasma physics, then in a modern form by Zakharov and his collaborators (Zakharov, 1965, 1967; Zakharov and Filonenko, 1966, 1967). This work is generally based on a Hamiltonian approach to the problem, whereas it is the Eulerian approach that was mainly used in the West.

The random phase approximation leads, in practice, to the same kinetic equations as by the (more rigorous) method of multiple scales. From these integro-differential equations, a major breakthrough was achieved with the discovery of a conformal transformation to extract from the nonlinear kinetic equations the exact power-law solutions. This transformation – now called the Zakharov transformation – was first proposed for capillary wave turbulence involving triadic interactions (Zakharov, 1967), then for Langmuir wave turbulence, where the interactions are quartic (Zakharov, 1967; Kaner and Yakovenko, 1970).[7] For more details on the Zakharov transformation, we refer the reader to the Chapter 5, on capillary wave turbulence, in which this transformation is used. However,

---

[6] By analogy with plasma physics, wave action is often associated with particles. This is a quantity that is sometimes conserved in four-wave processes. This is the case, for example, in gravitational wave turbulence (see Chapter 9).

[7] Zakharov was a student of Sagdeev. He defended his PhD thesis on surface waves in 1966 with several fundamental results in wave turbulence to his credit, such as the discovery of exact solutions to the kinetic equations. This discovery is reported in the article by Zakharov (1965) in which the author first verified that the collision integral (obtained from a relatively simple ad hoc model of three-wave interactions) tended towards $\pm\infty$ for two power-law exponents. He then demonstrated that the solution (the energy spectrum) associated with the index exactly in the middle of this interval cancels nontrivially the collision integral: Zakharov had just discovered an exact stationary solution. The discovery of the so-called Zakharov transformation came shortly after (Zakharov and Filonenko, 1966).

we can already note that there are two types of solution: the zero flux solution (which was the regime studied by Nordheim, 1928) and the nonzero constant flux solution. The first case corresponds to the thermodynamic solution (constant entropy) and the second to the Kolmogorov–Zakharov spectrum, which is the most interesting solution because it is nontrivial. Note that we have already made use of the Zakharov transformation in the study of two-dimensional hydrodynamic turbulence (see Chapter 3).

4.1.5 Applications of Wave Turbulence

There are many examples of the application of wave turbulence. Below is a nonexhaustive list of applications with some references.

- **Surface waves:** These are the first applications of wave turbulence. They include capillary waves and gravity waves. The former involve three-wave interactions, the theory of which was published in English by Zakharov (1967) in the deep water limit. This is the subject of Chapter 5, in which many references are given. Note that the shallow water limit is also the subject of studies (see, e.g., Clark di Leoni et al., 2014). Gravity wave turbulence is a problem that needs to be addressed at the level of four waves (Hasselmann, 1962). This regime has been well reproduced in the laboratory or by direct numerical simulations (see, e.g., Deike et al., 2011; Zhang and Pan, 2022). Its detection at sea is more difficult, but not impossible (see, e.g., Hwang et al., 2000). The two subjects being linked, several experiments deal with the interaction between gravity and capillary waves (see Chapter 5). All in all, it is a subject that is still very much under study (see the review by Falcon and Mordant, 2022).
- **Internal gravity waves:** These waves are a variant of the previous ones in the sense that we are interested here in gravity waves under the surface of the ocean. These waves contribute dynamically to the turbulent transport of heat, which is important to understand in order to properly evaluate the impact of oceans on the climate (MacKinnon, 2017). Internal gravity wave turbulence is an anisotropic three-wave problem for which an Eulerian theory has been developed by Caillol and Zeitlin (2000), as well as several experiments (see, e.g., Savaro et al., 2020).
- **Inertial waves:** This is the closest example to the standard case of eddy turbulence, in the sense that the equations are those of Navier–Stokes which are modified simply by adding the Coriolis force. Inertial wave turbulence is a three-wave anisotropic problem, whose theory has been published by Galtier (2003). Chapter 6 is devoted to this subject: the main properties of this regime are described, and a review of the numerous numerical and experimental works is given (see, e.g., Monsalve et al., 2020).

- **Rossby waves:** These waves appear in a situation of differential rotation. They are used in the modeling of planetary atmospheres, the interiors of gaseous planets, and stars; these are often referred to as planetary waves (Longuet-Higgins and Gill, 1967). In wave turbulence, the dynamics is driven by three-wave processes with a dominance of nonlocal interactions (Balk et al., 1990a,b).
- **Plasma waves:** As mentioned in Section 4.1.4, we find the beginnings of wave turbulence in the field of plasma physics (Sagdeev and Galeev, 1966; Vedenov, 1967). In this very vast domain, waves are legion. In Chapter 7 we present the special case of Alfvén wave turbulence, which is based on the simplest model of plasma physics. The three-wave theory was published by Galtier et al. (2000): it is a case where the anisotropy is so strong that the cascade is completely inhibited in the direction of the strong applied magnetic field. Further examples of plasma physics are mentioned and accompanied by references in Chapter 8, which is devoted to compressible plasmas.
- **Geodynamo waves:** Geodynamo waves are defined as waves present in the earth's outer liquid core. These waves take part in the dynamo effect, that is, the physical mechanism that maintains the earth's magnetic field (Finlay, 2008). These are magnetostrophic waves and inertial waves. The three-wave theory models a homogeneous medium with a small Rossby number (Galtier, 2014): in this framework, an anisotropic inverse cascade of hybrid helicity is predicted, which could be at the origin of the regeneration of the magnetic field on a large scale (Menu et al., 2019).
- **Acoustic waves:** Acoustic wave turbulence is driven by three-wave interactions, but these waves are not dispersive. This is a critical situation for the application of wave turbulence, because the uniformity of the development is not guaranteed (it depends on the dimension of the problem). The first works on the subject date back to the early 1970s (Zakharov and Sagdeev, 1970). The way in which the asymptotic is subtly modified is discussed by Newell and Aucoin (1971), and later by L'vov et al. (1997). It can be shown for this regime that energy is at best redistributed according to rays (in three-dimensional Fourier space).
- **Elastic waves:** This is a turbulence produced by a thin elastic plate or by the introduction of long polymer molecules into a liquid (Steinberg, 2021). In the former case, we are in a very different situation from the traditional turbulence produced by a fluid. The vibrations of a plate can in theory produce weak or strong wave turbulence, depending on the forces acting on it. The theory of wave turbulence has been published by Düring et al. (2006): it is a four-wave problem characterized by a direct energy cascade. Although the wave action is not conserved in this problem, direct numerical simulations show an inverse cascade which seems to be established in an explosive way (Düring et al., 2015). Laboratory experiments using steel plates have been carried out in order

to reproduce, with varied success, the theoretical predictions (Boudaoud et al., 2008; Mordant, 2008; Cobelli et al., 2009; Mordant, 2010).
- **Optical waves:** Wave turbulence is also found in the field of nonlinear optics. The multidimensional nonlinear Schrödinger equation can be used to describe the evolution of quasi-monochromatic plane wave envelopes (Sulem and Sulem, 1999). The theory is explained by Dyachenko et al. (1992): this wave turbulence is governed by four-wave interactions, and an intermittency mechanism characterized by a collapse phenomenon in physical space can occur. In the case of a reduction of the problem to one dimension, it can be shown that the dominant resonant interactions are six-wave interactions, involving a much slower nonlinear dynamics (Laurie et al., 2012). All this work is part of the study of nonlinear effects on the propagation of incoherent optical beams (Mitchell et al., 1996; Picozzi et al., 2014).
- **Quantum turbulence:** This subject is very close to the previous one, in the sense that the model used is a variant of the nonlinear Schrödinger equation: by changing the sign of the nonlinear term, the interaction becomes repulsive and the physics of turbulence is modified. The associated equation – also called the Gross–Pitaevskii equation – describes a Bose gas at very low temperatures. Note that the emergence of quantum turbulence in an oscillating Bose–Enstein condensate has been experimentally demonstrated (Henn et al., 2009). The wave turbulence regime is described by Dyachenko et al. (1992): it is shown that four-wave interactions conserve energy and wave action. The inverse cascade associated with the latter leads to the formation of a stable condensate, that is, the accumulation of wave action in the $k = 0$ mode. Direct numerical simulations in two (Nazarenko and Onorato, 2006; Nazarenko, 2007) and three (Proment et al., 2009, 2012) dimensions illustrate this regime.
- **Kelvin waves:** Kelvin waves are studied in the context of superfluids whose temperature is close the absolute zero. These waves can propagate along filaments of vorticity and modify the dynamics of turbulence (Vinen, 2000; Kivotides et al., 2001). Kelvin's theory of wave turbulence has been proposed by Kozik and Svistunov (2004): it involves six-wave processes that conserve both energy and wave action. As these cascade processes are extremely slow, it is particularly interesting to consider a local model of nonlinear diffusion (Nazarenko, 2006).
- **Gravitational waves:** This example illustrates the wide range of possible applications of wave turbulence. It relates to cosmology, and more precisely the birth of the universe: primordial gravitational waves could be at the origin of the inflation phase of the universe, a question that is still open to this day and that touches the limits of our knowledge (Galtier et al., 2020). The theory has been published by Galtier and Nazarenko (2017): the dynamics is governed by four-wave processes sufficiently symmetric to conserve wave action. The latter is characterized by an explosive inverse cascade. It is interesting to note that

this regime is close to that of elastic waves in the strong tension limit, as shown by the theoretical, numerical, and experimental work of Hassaini et al. (2019). Chapter 9 is devoted to this cosmological subject.

To go further, the reader is invited to consult the books of Zakharov et al. (1992) and Nazarenko (2011), in which other examples are presented.

4.2 Multiple Scale Method

4.2.1 Duffing's Equation

Classical Perturbation Development

The aim of this section is not to provide a rigorous and comprehensive demonstration of the theoretical validity of wave turbulence, but rather to show the reader the main steps in a schematic manner. To illustrate our point, we will first consider the Duffing equation:

$$\frac{d^2 f}{dt^2} + f = -\epsilon f^3, \qquad (4.3)$$

with ϵ a small parameter ($0 < \epsilon \ll 1$) which measures the intensity of the nonlinearity and f a function of time only. The main solution to this equation is a harmonic oscillator. This solution will, however, be slightly modified over time by the presence of the small nonlinear perturbation (right-hand-side term). Let us first use the standard perturbation theory and introduce the following power series development:

$$f = \sum_{i=0}^{+\infty} \epsilon^i f_i. \qquad (4.4)$$

We then obtain an infinite system of equations which, for the first three orders, can be written:

$$\mathcal{O}(\epsilon^0): \quad \frac{d^2 f_0}{dt^2} + f_0 = 0, \qquad (4.5a)$$

$$\mathcal{O}(\epsilon^1): \quad \frac{d^2 f_1}{dt^2} + f_1 = -f_0^3, \qquad (4.5b)$$

$$\mathcal{O}(\epsilon^2): \quad \frac{d^2 f_2}{dt^2} + f_2 = -3 f_0^2 f_1. \qquad (4.5c)$$

We can see that the solution to a given order will affect the solution to the higher order. The first solution is trivially $f_0 = A\cos(t + \phi)$. By introducing this one in the second equation, we can deduce the exact solution for f_1. If we stop at this order, the solution reads:

$$f(t) = A\cos(t+\phi) + \epsilon A^3 \left[-\frac{3}{8} t \sin(t+\phi) + \frac{1}{32} \cos(3t+3\phi) \right] + \mathcal{O}(\epsilon^2). \quad (4.6)$$

As a first approximation, this system behaves well as a harmonic oscillator. The condition to be checked is that the nonlinear perturbation remains of weak amplitude and the time considered relatively short. On the other hand, over long periods of the order of $\mathcal{O}(1/\epsilon)$, this solution is modified in a nonnegligible way. The term that is at the origin of it, in ϵt, is said to be secular. For even longer periods of time the development diverges; this divergence is all the more so as certain secular terms of a higher order reinforce it. Then, the uniformity of the development is broken.

We can pursue the analysis to evaluate the period of oscillations. To do this, the Duffing equation, which has been previously multiplied by the derivative of f, is integrated:

$$\frac{1}{2}\left(\frac{df}{dt}\right)^2 + \frac{1}{2}f^2 + \frac{1}{4}\epsilon f^4 = E, \quad (4.7)$$

where E is a constant. One obtains:

$$dt = \frac{df}{\sqrt{2E - f^2 - \frac{1}{2}\epsilon f^4}}. \quad (4.8)$$

This equation can be integrated over a period T. By posing $f = a\sin\theta$, we obtain the following expression, which is exact in its integral form:

$$T = 4\int_0^{\pi/2} \frac{d\theta}{\sqrt{1 + \epsilon a^2(1 - \frac{1}{2}\cos^2\theta)}} = 2\pi\left(1 - \frac{3}{8}\epsilon a^2 + \mathcal{O}(\epsilon^2 a^4)\right). \quad (4.9)$$

A Taylor expansion is used to evaluate the integral. It is found that the 2π period of the harmonic oscillator is corrected by the nonlinear perturbation. The smaller the nonlinearity, the smaller the correction.

What information is useful for our problem? This very simple system illustrates schematically the problems that we can encounter in a classical perturbation development. Such a development should allow us, in principle, to follow the evolution of the dynamics of wave turbulence at various orders in ϵ. The main problem is to ensure that the development remains neat (or uniform). In other words, this means making sure that no secular term of the order n comes to interfere in the dynamics at an order m, such as $m < n$, in order not to find oneself in the same situation as Hasselmann (1962), that is, in the presence of a secular term. This type of term constitutes an obstacle to the closure of the hierarchy of equations. The more sophisticated method of multiple scales will allow us to solve this problem.

Multiple Scale Method

Let us take the Duffing equation again and introduce the following independent time variables:

$$T_n \equiv \epsilon^n t \quad \text{with} \quad n = 0, 1, 2, \ldots . \tag{4.10}$$

The power series development of the variable f is then written:

$$f = \sum_{i=0}^{+\infty} \epsilon^i f_i(t, T_1, T_2, \ldots), \tag{4.11}$$

and the time derivative becomes a sum of partial derivatives:

$$\frac{d}{dt} = \sum_{n=0}^{+\infty} \epsilon^n \frac{\partial}{\partial T_n}. \tag{4.12}$$

The introduction of relations (4.10)–(4.12) into the Duffing equation gives:

$$\left(\frac{\partial}{\partial t} + \epsilon \frac{\partial}{\partial T_1} + \epsilon^2 \frac{\partial}{\partial T_2} + \ldots \right)^2 (f_0 + \epsilon f_1 + \epsilon^2 f_2 + \ldots) \tag{4.13}$$
$$+ (f_0 + \epsilon f_1 + \epsilon^2 f_2 + \ldots) = -\epsilon (f_0 + \epsilon f_1 + \epsilon^2 f_2 + \ldots)^3 .$$

We obtain an infinite system of equations which, for the first two orders, can be written:

$$\mathcal{O}(\epsilon^0): \quad \frac{\partial^2 f_0}{\partial t^2} + f_0 = 0, \tag{4.14a}$$

$$\mathcal{O}(\epsilon^1): \quad \frac{\partial^2 f_1}{\partial t^2} + f_1 + 2 \frac{\partial^2 f_0}{\partial t \partial T_1} = -f_0^3. \tag{4.14b}$$

We can see the presence of a new term in the time evolution equation of f_1, whose importance emerges over long times. We will look for a solution of the form:

$$f_0 = A(T_1, T_2, \ldots) \cos(t + \phi(T_1, T_2, \ldots)), \tag{4.15}$$

with an amplitude A and a phase ϕ which can vary slowly over time. The introduction of the previous expression in (4.14b) gives the equation:

$$\frac{\partial^2 f_1}{\partial t^2} + f_1 = 2 \frac{\partial A}{\partial T_1} \sin(t + \phi) + \left(2A \frac{\partial \phi}{\partial T_1} - \frac{3}{4} A^3 \right) \cos(t + \phi)$$
$$- \frac{A^3}{4} \cos(3(t + \phi)). \tag{4.16}$$

To ensure that the f_1 solution does not include a secular term, the following conditions must be imposed:

$$\frac{\partial A}{\partial T_1} = 0 \quad \text{and} \quad \frac{\partial \phi}{\partial T_1} = \frac{3}{8} A^2, \tag{4.17}$$

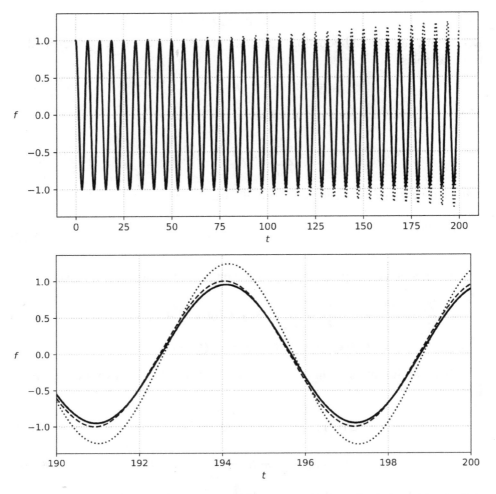

Figure 4.2 Variation of $f(t)$ (solid line) obtained by a numerical simulation of the Duffing equation with $\epsilon = 0.01$. The solutions (4.6) and (4.18) are plotted in dotted and dashed lines, respectively.

which gives $\phi = (3/8)A^2\epsilon t + \theta$. The solution is then written:

$$f(t) = A\cos\left(t + \frac{3}{8}A^2\epsilon t + \theta\right) + \frac{\epsilon A^3}{32}\cos\left(3t + \frac{9}{8}A^2\epsilon t + 3\theta\right) + \mathcal{O}(\epsilon^2), \quad (4.18)$$

with $A(T_2, T_3, \ldots)$ and $\theta(T_2, T_3, \ldots)$. In the order of the truncation, we can therefore consider that A and θ are constant. Note that this development is compatible with the previous one (4.6) for $t \leq \mathcal{O}(\epsilon)$. This example illustrates the fact that we can achieve a uniform development in a systematic way using the multiple scale method.

We show in Figure 4.2 the result of a numerical simulation of the Duffing equation with $\epsilon = 0.01$. The function $f(t)$ is plotted in solid line. At the top, we see the evolution for $t \in [0.200]$ and at the bottom, we show an enlargement of the end of the simulation. The mathematical solutions (4.6) and (4.18) are plotted in dotted and dashed lines, respectively. The divergence with t of the solution (4.6) whose origin is the secular term can be seen. As expected, this divergence occurs over a time t of the order of $\mathcal{O}(1/\epsilon)$ (the reference time is the wave period). On the other hand, the solution (4.18) remains close to the real solution until the final time.

4.3 Weakly Nonlinear Model

4.3.1 Fundamental Equation

To go further, let us now consider the following inviscid model:

$$\frac{\partial u(\mathbf{x},t)}{\partial t} = \mathcal{L}(u) + \epsilon \mathcal{N}(u,u), \qquad (4.19)$$

where u is a null mean random function, $\mathcal{L}$ is a linear operator which ensures that waves are linear solutions to the problem, and $\mathcal{N}$ is a nonlinear quadratic operator thus implying triadic interactions. The coefficient ϵ is a small parameter ($0 < \epsilon \ll 1$) which measures the amplitude of the nonlinearities. Direct and inverse Fourier transforms in d-dimension (see Chapter 3) are introduced:

$$u(\mathbf{x},t) \equiv \int_{\mathbb{R}^d} A(\mathbf{k},t) e^{i\mathbf{k}\cdot\mathbf{x}} d\mathbf{k}, \qquad (4.20a)$$

$$A(\mathbf{k},t) \equiv \frac{1}{(2\pi)^d} \int_{\mathbb{R}^d} u(\mathbf{x},t) e^{-i\mathbf{k}\cdot\mathbf{x}} d\mathbf{x}. \qquad (4.20b)$$

The Fourier transform of equation (4.19) takes the following schematic form:

$$\left(\frac{\partial}{\partial t} + i\omega\right) A(\mathbf{k},t) = \epsilon \int_{\mathbb{R}^{2d}} \mathcal{H}_{\mathbf{kpq}} A(\mathbf{p},t) A(\mathbf{q},t) \delta(\mathbf{k}-\mathbf{p}-\mathbf{q}) d\mathbf{p} d\mathbf{q}. \qquad (4.21)$$

On the left we find ω, which is fixed by the dispersion relation, whereas on the right $\mathcal{H}_{\mathbf{kpq}}$ is an operator which depends on the shape of the nonlinearities and is symmetrical in $\mathbf{p}$ and $\mathbf{q}$. We recall that the presence of the Dirac function finds its origin in the convolution product, and the quadratic nature of the nonlinearities leads to triadic interactions. The equation is simplified by making the following change of variables (writing is simplified in passing):

$$A(\mathbf{k},t) = \sum_s a^s(\mathbf{k},t) e^{-is\omega_k t} = \sum_s a_k^s e^{-is\omega_k t}, \qquad (4.22)$$

with $s = \pm$. One obtains the fundamental equation:

$$\frac{\partial a_k^s}{\partial t} = \epsilon \sum_{s_p s_q} \int_{\mathbb{R}^{2d}} \mathcal{H}_{\mathbf{kpq}} a_p^{s_p} a_q^{s_q} e^{i\Omega_{k,pq} t} \delta_{k,pq} d\mathbf{p} d\mathbf{q}, \qquad (4.23)$$

with $\delta_{k,pq} \equiv \delta(\mathbf{k}-\mathbf{p}-\mathbf{q})$ and $\Omega_{k,pq} \equiv s\omega_k - s_p\omega_p - s_q\omega_q$. This equation highlights the temporal evolution of the amplitude of the wave: this evolution is relatively slow since it induces a nonlinear term proportional to ϵ. We will find this property by the multiple scale method. The presence of the complex exponential is fundamental for the asymptotic closure: since we are interested in the dynamics over a long time compared to the period of the waves, the contribution of this exponential is essentially zero. Only certain terms will survive: those for which $\Omega_{k,pq} = 0$. With the condition imposed by the Dirac, we obtain the resonance condition:

$$\begin{cases} \mathbf{k} = \mathbf{p} + \mathbf{q}, \\ s\omega_k = s_p\omega_p + s_q\omega_q. \end{cases} \quad (4.24)$$

4.3.2 Dispersion Relation and Resonance

The triadic resonance condition (4.24) does not always have a solution. To realize this, we can represent this condition geometrically in the two-dimensional and isotropic case: Figure 4.3 shows a (axisymmetric) dispersion relation of the type $\omega_k \sim k^x$, with $x > 1$ at the top and $0 < x < 1$ at the bottom. The solutions of the resonance condition (4.24) correspond to the intersection between the surface $\omega(\mathbf{p})$ (which is identified with that of $\omega(\mathbf{k})$) and the surface $\omega(\mathbf{q})$. We find that this intersection exists only for a convex dispersion relation ($x > 1$). The concave case corresponding to $0 < x < 1$ is the one encountered with gravity waves.

The special case $x = 1$ is that of nondispersive waves. We can easily see that the only possible solution is that where all three wavevectors are aligned. Although a solution to the triadic resonance condition exists, we are not sure that there is a an asymptotic closure of the hierarchy of equations. Physically, we can easily understand this problem: two nondispersive waves propagate at the same speed, therefore, if they initially overlap their interactions will be strong, otherwise they will never interact. Using the multiscale method presented in the Section 4.3.3, it can be shown in 1D that a contribution from nonlinear terms is possible on a timescale of the order of T_1, while the closure is made on a longer timescale in T_2 (Benney and Saffman, 1966). This problem is found in the case of acoustic wave turbulence, discussed briefly in Section 4.1.5. It should be noted that in the case of four-wave interactions, the situation is less constraining because new possibilities exist. Thus, gravity waves can be analyzed at this order. In the case of a four-wave problem of a nondispersive nature, it is also possible to develop a theory of wave turbulence, because the associated four wavevectors are not necessarily collinear. An example will be discussed in Chapter 9: these are gravitational waves for which the dispersion relation is $\omega_k = ck$, with c the speed of light.

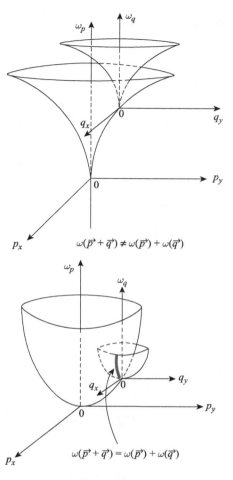

Figure 4.3 Resonance condition for a dispersion relation of the type $\omega_k \sim k^x$ with $x > 1$ (top) and $0 < x < 1$ (bottom).

4.3.3 Uniform Asymptotic Development

Following our analysis of the Duffing equation, we wish to make a power series development of equation (4.23) following the method of multiple timescales. A series of timescales are introduced, $T_0, T_1, T_2, \ldots$, which will be treated as independent variables, with:

$$T_0 \equiv t, \quad T_1 \equiv \epsilon t, \quad T_2 \equiv \epsilon^2 t, \ldots . \tag{4.25}$$

We obtain:

$$\left(\frac{\partial}{\partial t} + \epsilon \frac{\partial}{\partial T_1} + \epsilon^2 \frac{\partial}{\partial T_2} + \ldots\right) a_k^s = \epsilon \sum_{s_p s_q} \int_{\mathbb{R}^{2d}} \mathcal{H}_{\mathbf{kpq}} a_p^{s_p} a_q^{s_q} e^{i\Omega_{k,pq}t} \delta_{k,pq} d\mathbf{p} d\mathbf{q} . \tag{4.26}$$

The variable a_k^s must also be expanded to the power of ϵ to make the various scales appear in time:

$$a_k^s = \sum_{n=0}^{+\infty} \epsilon^n a_{k,n}^s(t, T_1, T_2, \ldots) = a_{k,0}^s + \epsilon a_{k,1}^s + \epsilon^2 a_{k,2}^s + \ldots . \quad (4.27)$$

This expression must then be introduced into the fundamental equation (4.26). We obtain for the first three terms:

$$\frac{\partial a_{k,0}^s}{\partial t} = 0, \quad (4.28a)$$

$$\frac{\partial a_{k,1}^s}{\partial t} = -\frac{\partial a_{k,0}^s}{\partial T_1} + \sum_{s_p s_q} \int_{\mathbb{R}^{2d}} \mathcal{H}_{\mathbf{kpq}} a_{p,0}^{s_p} a_{q,0}^{s_q} e^{i\Omega_{k,pq}t} \delta_{k,pq} d\mathbf{p} d\mathbf{q}, \quad (4.28b)$$

$$\frac{\partial a_{k,2}^s}{\partial t} = -\frac{\partial a_{k,1}^s}{\partial T_1} - \frac{\partial a_{k,0}^s}{\partial T_2} + 2\sum_{s_p s_q} \int_{\mathbb{R}^{2d}} \mathcal{H}_{\mathbf{kpq}} a_{p,0}^{s_p} a_{q,1}^{s_q} e^{i\Omega_{k,pq}t} \delta_{k,pq} d\mathbf{p} d\mathbf{q}. \quad (4.28c)$$

To lighten the writing, the time dependency of the variables has been omitted. Let us look at the solutions; we get after integration on t:

$$a_{k,0}^s = a_{k,0}^s(k, T_1, T_2, \ldots), \quad (4.29a)$$

$$a_{k,1}^s = -t\frac{\partial a_{k,0}^s}{\partial T_1} + b_{k,1}^s, \quad (4.29b)$$

$$a_{k,2}^s = -t\frac{\partial a_{k,1}^s}{\partial T_1} - t\frac{\partial a_{k,0}^s}{\partial T_2} - \frac{t^2}{2}\frac{\partial^2 a_{k,0}^s}{\partial T_1^2} + b_{k,2}^s, \quad (4.29c)$$

with

$$b_{k,1}^s \equiv \sum_{s_p s_q} \int_{\mathbb{R}^{2d}} \mathcal{H}_{\mathbf{kpq}} a_{p,0}^{s_p} a_{q,0}^{s_q} \Delta(\Omega_{k,pq}) \delta_{k,pq} d\mathbf{p} d\mathbf{q}, \quad (4.30)$$

and $b_{k,2}^s$, a term whose integral form will not be shown because it is not very useful for our purpose. One can rewrite expression (4.30) within the long time limit by using the Riemann–Lebesgue lemma for the distributions (which corresponds to a phase mixing):

$$\Delta(X) = \int_0^t e^{iXt} dt$$

$$= \frac{e^{iXt} - 1}{iX} \xrightarrow{t \to +\infty} \pi\delta(X) + i\mathcal{P}\left(\frac{1}{X}\right). \quad (4.31)$$

This long time limit means that we are considering a much longer timescale than the period of linear waves ($\sim 1/\omega$); however, this timescale remains shorter than any interaction time. We notice that the first solution (4.29a) is compatible with the hypothesis that amplitude and phase do not evolve on the same timescale: the phase evolves on a timescale t shorter than the amplitude timescale (which is on

T_1 or longer). In relation (4.22), amplitude and phase are therefore two variables separated in time.

We now need to define the conditions that will enable us to ensure the uniformity of the development. Here, the development we are interested in concerns statistical objects, that is, cumulants and moments. In practice, we wish to have a uniform development at long times for the cumulants. We obtain from expression (4.29b):

$$\langle a^s_{k,1} a^{s'}_{k',0} \rangle + \langle a^s_{k,0} a^{s'}_{k',1} \rangle = -t \frac{\partial \langle a^s_{k,0} a^{s'}_{k',0} \rangle}{\partial T_1} + \langle a^{s'}_{k',0} b^s_{k,1} + a^s_{k,0} b^{s'}_{k',1} \rangle, \quad (4.32)$$

with $\langle \rangle$ the ensemble average. We will assume statistically homogeneous turbulence. The terms on the left are moments (or cumulants) at two points. They are quantities which are physically bounded (we can see these terms as energy spectra). The second term on the right is also bounded in the long time limit. Therefore, to ensure the uniformity of the development we must eliminate the potentially secular term and impose the condition:

$$\frac{\partial \langle a^s_{k,0} a^{s'}_{k',0} \rangle}{\partial T_1} = 0. \quad (4.33)$$

In other words, this means that the cumulants (in two points) of zero order evolve on a timescale longer than T_1.

The analysis must continue at the higher order with the solution (4.29c). We can obtain:

$$\langle a^s_{k,2} a^{s'}_{k',0} \rangle + \langle a^s_{k,0} a^{s'}_{k',2} \rangle = -t \left\langle a^s_{k,0} \frac{\partial a^{s'}_{k',1}}{\partial T_1} + a^{s'}_{k',0} \frac{\partial a^s_{k,1}}{\partial T_1} \right\rangle - t \frac{\partial \langle a^s_{k,0} a^{s'}_{k',0} \rangle}{\partial T_2}$$

$$- \frac{t^2}{2} \left\langle a^s_{k,0} \frac{\partial^2 a^{s'}_{k',0}}{\partial T_1^2} + a^{s'}_{k',0} \frac{\partial^2 a^s_{k,0}}{\partial T_1^2} \right\rangle + \langle a^{s'}_{k',0} b^s_{k,2} + a^s_{k,0} b^{s'}_{k',2} \rangle. \quad (4.34)$$

A condition on the uniformity of the development emerges after a few additional – sometimes subtle – manipulations, and reuse of expression (4.29b) (see Benney and Saffman, 1966).[8] We can then show that the previous expression is simplified (for example, the first term in the second line is null). The terms on the left (cumulants) are considered to be physically bounded at long time, which finally leads us to a nontrivial uniformity condition:

$$\frac{\partial \langle a^s_{k,0} a^{s'}_{k',0} \rangle}{\partial T_2} = C_t \langle b^s_{k,1} b^{s'}_{k',1} \rangle + C_t \langle a^{s'}_{k',0} b^s_{k,2} + a^s_{k,0} b^{s'}_{k',2} \rangle, \quad (4.35)$$

[8] Several relationships that mix, for example, products (within the long time limit) of Δ introduced above must be used. For example, there is the Poincaré–Bertrand formula, or the relation $\Delta(X)\Delta(-X) \sim 2\pi t \delta(X) + 2\mathcal{P}\left(\frac{1}{X}\right)\frac{\partial}{\partial X}$, in which we have a secular contribution. For more formulas, see Benney and Newell (1967).

where $\mathcal{C}_t f$ means the coefficient of f proportional to t. This relation implies, in particular, that the energy spectrum (two-point cumulant at zero order) evolves on the timescale T_2: it is on this timescale that the modes can thus exchange energy.

The next part of the study consists of evaluating the right-hand-side term of equation (4.35). To understand the principle, we will only focus on the contribution of the first term on the right of this expression. This term shows four-point moments of zero order. These moments are decomposed into sums of products of two-point cumulants (of zero order) and four-point cumulants (of zero order). Therefore, we have:

$$\mathcal{C}_t \langle b^s_{k,1} b^{s'}_{k',1} \rangle = \mathcal{C}_t \sum_{s_i} \int_{\mathbb{R}^{4d}} \mathcal{H}_{\mathbf{kpq}} \mathcal{H}_{\mathbf{k'rs}} \langle a^{s_p}_{p,0} a^{s_q}_{q,0} a^{s_r}_{r,0} a^{s_s}_{s,0} \rangle \Delta(\Omega_{k,pq}) \Delta(\Omega_{k',rs})$$
$$\delta_{k,pq} \delta_{k',rs} d\mathbf{p} d\mathbf{q} d\mathbf{r} d\mathbf{s}, \qquad (4.36)$$

with:

$$\langle a^{s_p}_{p,0} a^{s_q}_{q,0} a^{s_r}_{r,0} a^{s_s}_{s,0} \rangle = q_0^{s_p s_q s_r s_s}(\mathbf{p}, \mathbf{q}, \mathbf{r}, \mathbf{s})\delta(\mathbf{p}+\mathbf{q}+\mathbf{r}+\mathbf{s}) \qquad (4.37)$$
$$+ q_0^{s_p s_q}(\mathbf{p}, \mathbf{q}) q_0^{s_r s_s}(\mathbf{r}, \mathbf{s})\delta(\mathbf{p}+\mathbf{q})\delta(\mathbf{r}+\mathbf{s})$$
$$+ q_0^{s_p s_r}(\mathbf{p}, \mathbf{r}) q_0^{s_q s_s}(\mathbf{q}, \mathbf{s})\delta(\mathbf{p}+\mathbf{r})\delta(\mathbf{q}+\mathbf{s})$$
$$+ q_0^{s_p s_s}(\mathbf{p}, \mathbf{s}) q_0^{s_q s_r}(\mathbf{q}, \mathbf{r})\delta(\mathbf{p}+\mathbf{s})\delta(\mathbf{q}+\mathbf{r}).$$

In this writing $q_0^{ss'}$ is a two-point cumulant of order 0 such that $q_0^{ss'}(\mathbf{k}, \mathbf{k}')\delta(\mathbf{k}+\mathbf{k}') = \langle a^s_{k,0} a^{s'}_{k',0} \rangle$. We can then show by manipulating Dirac products that the contribution of the cumulant $q_0^{s_p s_q s_r s_s}$ to the dynamics is null, that is, it is not possible to generate from this term a linear contribution in t. Indeed, both Δ functions are independent, so the relationship (4.31) must be used as a product. On the other hand, a nonzero contribution can come from products of two-point cumulants: more precisely, these are products of the type $\Delta(X)\Delta(-X)$, which give a linear contribution in t. An analysis of the second right-hand-side term of equation (4.35) comes to the same conclusion. Therefore, the contribution proportional to t that we are looking for does not involve the fourth-order cumulant. It should be noted that for a Gaussian distribution the fourth-order cumulant is zero. The situation we have here is thus similar in appearance to the one of a Gaussian distribution, however, we have not made the assumption of Gaussianity. This result does not preclude a contribution to the dynamics of the non-Gaussian part of the statistic: in principle, it can intervene in the dynamics by slowly changing the spectrum, but at a higher order in ϵ. The demonstration presented here is only sketched. For more information, one can refer to the work of Benney and collaborators (Benney and Saffman, 1966; Benney, 1967; Benney and Newell, 1967, 1969), where the case of four-wave interactions is also studied. In particular, it is shown for three-wave interactions that the development is uniformly valid for times $\omega t \ll \mathcal{O}(1/\epsilon^4)$, with the wave period as the reference time.

4.3 Weakly Nonlinear Model

In summary, we can say that wave turbulence is characterized by a dynamics on two timescale. Over short times, of the order of the wave period, we have a phase mixing which leads, because of the dispersive nature of the waves, to the decoupling of the correlations initially present and to a statistics that is close to Gaussianity, as expected from the central limit theorem.[9] (It is assumed, however, that initially there are no coherent structures leading to too strong correlation as can be the case in strong turbulence.) On a longer timescale, the nonlinear coupling – weak over short times – becomes nonnegligible because of the resonance mechanism. This coupling leads to a regeneration of the cumulants via the product of lower-order cumulants. It is these terms that are at the origin of the energy transfer mechanism. Their contribution is such that an asymptotic closure uniform in time is achievable. As we will see in the Chapter 5, on capillary wave turbulence – for which the interactions are three-wave – the analytical development can be done in practice directly from the fundamental equation (4.23) by assuming a scale separation in time, between phase and amplitude. However, it should be borne in mind that this is a demonstration by the multiscale method of the uniformity of the asymptotic development, which allows the theory of wave turbulence to be rigorously justified.

In this book, we implicitly assume that the systems studied are of infinite size and that they can therefore be treated as continuous. Note that numerical simulation with its grid of points escapes this description. Effects (freezing of the cascade) related to the discretization of the Fourier space may emerge because the resonance conditions are necessarily more difficult to satisfy (see, e.g., Connaughton et al. (2001) for capillary waves). In theory, the weaker the nonlinearities, the stronger these effects. For example, in inertial wave turbulence Bourouiba (2008) has shown that discretization effects become preponderant when the Rossby number, R_o, is smaller than 10^{-3}. Above this value, but still for a small R_o, these effects are negligible because of the quasi-resonances which, together with the resonances, contribute to the transfer of energy.

Finally, we give the schematic form of the kinetic equation of (dispersive) wave turbulence for three-wave interactions:

$$\frac{\partial N(\mathbf{k})}{\partial t} = \quad (4.38)$$

$$\epsilon^2 \int_{\mathbb{R}^{2d}} S(\mathbf{k}, \mathbf{p}, \mathbf{q})(N_p N_q - N N_p - N N_q) \delta(\omega - \omega_p - \omega_q) \delta(\mathbf{k} - \mathbf{p} - \mathbf{q}) d\mathbf{p} d\mathbf{q},$$

with $N \equiv N(\mathbf{k}) = E(\mathbf{k})/\omega$ the wave action spectrum and $E(\mathbf{k})$ the energy spectrum. This schematic equation describes the dynamics on a timescale $\omega t \sim \mathcal{O}(1/\epsilon^2)$. This timescale is therefore shorter than in the case of four-wave interactions. New contributions of higher orders can emerge over longer timescales (from

---

[9] Note that in a multidimensional problem, the dispersive nature of the waves is required for three-wave interactions but not for four-wave interactions, as shown in Chapter 9 on gravitational waves.

$\mathcal{O}(1/\epsilon^4)$), whereas over shorter times the dynamics of wave turbulence does not have time to develop. Note that the wave action is not conserved for three-wave interactions: it can be conserved only if the interactions are even (four waves, six waves, etc.).

In nature we often encounter three-wave problems. Chapters 5, 6, 7, and 8 correspond to this type of situation, while Chapter 9 presents a four-wave problem. As mentioned in Section 4.1.5, sometimes higher-level problems (up to six waves) can occur. Despite the diversity of situations, the kinetic equation takes a universal form in the sense that it is written (at the main order):

$$\frac{\partial N(\mathbf{k})}{\partial t} = \epsilon^{2n-4} T(\mathbf{k}), \qquad (4.39)$$

for n-wave interactions, with T the transfer function (or collision integral). The higher the degree of interaction, the longer the transfer time τ_{tr}. This time is characterized by the relation:

$$\omega \tau_{tr} \sim \mathcal{O}(1/\epsilon^{2n-4}). \qquad (4.40)$$

Since the characteristic time of the waves is $\tau_\omega \sim 1/\omega$ and the small parameter is none other than the ratio between this time and the nonlinear time τ_{NL}, we arrive at the following phenomenological expression for the transfer time:

$$\tau_{tr} \sim \frac{\tau_\omega}{\epsilon^{2n-4}} \sim \frac{\tau_{NL}^{2n-4}}{\tau_\omega^{2n-5}}. \qquad (4.41)$$

For three-wave interactions, we will use the expression: $\tau_{tr} \sim \omega \tau_{NL}^2$. In Chapter 7, we will show how this characteristic time can be found with a phenomenology based on collisions of wave packets.

References

Akylas, T. R. 2020. David J. Benney: Nonlinear wave and instability processes in fluid flows. *Ann. Rev. Fluid Mech.*, **52**(1), 010518–040240.
Balk, A. M., Zakharov, V. E., and Nazarenko, S. V. 1990a. Nonlocal turbulence of drift waves. *J. Exp. Theor. Phys.*, **98**, 446–467.
Balk, A. M., Nazarenko, S. V., and Zakharov, V. E. 1990b. On the nonlocal turbulence of drift type waves. *Phys. Lett. A*, **146**(4), 217–221.
Benney, D. J. 1962. Non-linear gravity wave interactions. *J. Fluid Mech.*, **14**, 577–584.
Benney, D. J. 1967. Asymptotic behavior of nonlinear dispersive waves. *J. Math. Phys.*, **46**(2), 115–132.
Benney, D. J., and Newell, A. C. 1967. Sequential time closures for interacting random waves. *J. Math. Phys.*, **46**(4), 363–392.
Benney, D. J., and Newell, A. C. 1969. Random wave closures. *Stud. App. Maths.*, **48**(1), 29–53.
Benney, D. J., and Saffman, P. G. 1966. Nonlinear interactions of random waves in a dispersive medium. *Proc. R. Soc. Lond. A*, **289**(1418), 301–320.

Boudaoud, A., Cadot, O., Odille, B., and Touzé, C. 2008. Observation of wave turbulence in vibrating plates. *Phys. Rev. Lett.*, **100**(23), 234504.

Bourouiba, L. 2008. Discreteness and resolution effects in rapidly rotating turbulence. *Phys. Rev. E*, **78**, 056309.

Caillol, P., and Zeitlin, V. 2000. Kinetic equations and stationary energy spectra of weakly nonlinear internal gravity waves. *Dyn. Atmos. Oceans*, **32**(2), 81–112.

Clark di Leoni, P., Cobelli, P. J., and Mininni, P. D. 2014. Wave turbulence in shallow water models. *Phys. Rev. E*, **89**(6), 063025.

Cobelli, P., Petitjeans, P., Maurel, A., Pagneux, V., and Mordant, N. 2009. Space-time resolved wave turbulence in a vibrating plate. *Phys. Rev. Lett.*, **103**(20), 204301.

Connaughton, C., Nazarenko, S., and Pushkarev, A. 2001. Discreteness and quasiresonances in weak turbulence of capillary waves. *Phys. Rev. E*, **63**, 046306.

Deike, L., Laroche, C., and Falcon, E. 2011. Experimental study of the inverse cascade in gravity wave turbulence. *EPL (Europhys. Lett.)*, **96**(3), 34004.

Drummond, W. E., and Pines, D. 1962. Non-linear stability of plasma oscillations. *Nuclear Fusion Supp.*, **3**, 1049–1057.

Düring, G., Josserand, C., and Rica, S. 2006. Weak turbulence for a vibrating plate: Can one hear a Kolmogorov spectrum? *Phys. Rev. Lett.*, **97**(2), 025503.

Düring, G., Josserand, C., and Rica, S. 2015. Self-similar formation of an inverse cascade in vibrating elastic plates. *Phys. Rev. E*, **91**(5), 052916.

Dyachenko, S., Newell, A. C., Pushkarev, A., and Zakharov, V. E. 1992. Optical turbulence: Weak turbulence, condensates and collapsing filaments in the nonlinear Schrödinger equation. *Physica D*, **57**, 96–160.

Falcon, E., and Mordant, N. 2022. Experiments in surface gravity-capillary wave turbulence. *Ann. Rev. Fluid Mech.*, **54**(1), 1–25.

Finlay, C. C. 2008. Waves in the presence of magnetic fields, rotation and convection. In P. Cardin and L. F. Cugliandolo (eds.) *Dynamos: Lecture Notes of the Les Houches Summer School 2007* (pages 403–450). Elsevier.

Galtier, S. 2003. Weak inertial-wave turbulence theory. *Phys. Rev. E*, **68**(1), 015301.

Galtier, S. 2014. Weak turbulence theory for rotating magnetohydrodynamics and planetary flows. *J. Fluid Mech.*, **757**, 114–154.

Galtier, S., and Nazarenko, S. V. 2017. Turbulence of weak gravitational waves in the early universe. *Phys. Rev. Lett.*, **119**(22), 221101.

Galtier, S., Nazarenko, S. V., Newell, A. C., and Pouquet, A. 2000. A weak turbulence theory for incompressible magnetohydrodynamics. *J. Plasma Phys.*, **63**, 447–488.

Galtier, S., Laurie, J., and Nazarenko, S. V. 2020. A plausible model of inflation driven by strong gravitational wave turbulence. *Universe*, **6**(7), 98 (16 pages).

Hassaini, R., Mordant, N., Miquel, B., Krstulovic, G., and Düring, G. 2019. Elastic weak turbulence: From the vibrating plate to the drum. *Phys. Rev. E*, **99**(3), 033002.

Hasselmann, K. 1962. On the non-linear energy transfer in a gravity-wave spectrum. Part 1. General theory. *J. Fluid Mech.*, **12**, 481–500.

Henn, E. A. L., Seman, J. A., Roati, G., Magalhaes, K. M. F., and Bagnato, V. S. 2009. Emergence of turbulence in an oscillating Bose–Einstein condensate. *Phys. Rev. Lett.*, **103**(4), 045301.

Hwang, P. A., Wang, D. W., Walsh, E. J., Krabill, W. B., and Swift, R. N. 2000. Airborne measurements of the wavenumber spectra of ocean surface waves. Part I: Spectral slope and dimensionless spectral coefficient. *J. Phys. Ocean.*, **30**(11), 2753.

Kadomtsev, B. B., and Petviashvili, V. I. 1963. Weakly turbulent plasma in a magnetic field. *J. Exp. Theor. Phys.*, **16**, 1578–1585.

Kaner, É. A., and Yakovenko, V. M. 1970. Weak turbulence spectrum and second sound in a plasma. *J. Exp. Theor. Phys.*, **31**, 316–330.

Kivotides, D., Vassilicos, J. C., Samuels, D. C., and Barenghi, C. F. 2001. Kelvin waves cascade in superfluid turbulence. *Phys. Rev. Lett.*, **86**(14), 3080–3083.

Kozik, E., and Svistunov, B. 2004. Kelvin-wave cascade and decay of superfluid turbulence. *Phys. Rev. Lett.*, **92**(3), 035301.

Laurie, J., Bortolozzo, U., Nazarenko, S., and Residori, S. 2012. One-dimensional optical wave turbulence: Experiment and theory. *Phys. Rep.*, **514**(4), 121–175.

Longuet-Higgins, M. S. 1962. Resonant interactions between two trains of gravity waves. *J. Fluid Mech.*, **12**, 321–332.

Longuet-Higgins, M. S., and Gill, A. E. 1967. Resonant interactions between planetary waves. *Proc. Royal Soc. Lond. A*, **299**(1456), 120–140.

Longuet-Higgins, M. S., and Smith, N. D. 1966. An experiment on third-order resonant wave interactions. *J. Fluid Mech.*, **25**, 417–435.

L'vov, V. S., L'vov, Yu., Newell, A. C., and Zakharov, V. E. 1997. Statistical description of acoustic turbulence. *Phys. Rev. E*, **56**(1), 390–405.

MacKinnon, J. A., Zhao, Z., Whalen, C. B. et al. 2017. Climate process team on internal wave-driven ocean mixing. *Bull. Am. Meteo. Soc.*, **98**(11), 2429–2454.

McGoldrick, L. F., Phillips, O. M., Huang, N. E., and Hodgson, T. H. 1966. Measurements of third-order resonant wave interactions. *J. Fluid Mech.*, **25**, 437–456.

Menu, M. D., Galtier, S., and Petitdemange, L. 2019. Inverse cascade of hybrid helicity in $B_0 - \Omega$ MHD turbulence. *Phys. Rev. Fluids*, **4**(7), 073701.

Mitchell, M., Chen, Z., Shih, M.-F., and Segev, M. 1996. Self-trapping of partially spatially incoherent light. *Phys. Rev. Lett.*, **77**(3), 490–493.

Monsalve, E., Brunet, M., Gallet, B., and Cortet, P.-P. 2020. Quantitative experimental observation of weak inertial-wave turbulence. *Phys. Rev. Lett.*, **125**, 254502.

Mordant, N. 2008. Are there waves in elastic wave turbulence? *Phys. Rev. Lett.*, **100**(23), 234505.

Mordant, N. 2010. Fourier analysis of wave turbulence in a thin elastic plate. *Europ. Phys. J. B*, **76**(4), 537–545.

Nayfeh, A. H. 2004. *Perturbation Methods*. Weinheim, Germany: Wiley-VCH Verlag GmbH & Co. KGaA.

Nazarenko, S. 2006. Differential approximation for Kelvin wave turbulence. *JETP Lett.*, **83**(5), 198–200.

Nazarenko, S. 2011. *Wave Turbulence*. Lecture Notes in Physics, vol. 825. Springer Verlag.

Nazarenko, S., and Onorato, M. 2006. Wave turbulence and vortices in Bose–Einstein condensation. *Physica D*, **219**(1), 1–12.

Nazarenko, S., and Onorato, M. 2007. Freely decaying turbulence and Bose–Einstein condensation in Gross–Pitaevski model. *J. Low Temp. Phys.*, **146**(1–2), 31–46.

Newell, A. C., and Aucoin, P. J. 1971. Semi-dispersive wave systems. *J. Fluid Mech.*, **49**, 593–609.

Nordheim, L. W. 1928. On the kinetic method in the new statistics and its application in the electron theory of conductivity. *Proc. Royal Soc. London Series A*, **119**(783), 689–698.

Peierls, R. 1929. Zur kinetischen Theorie der Wärmeleitung in Kristallen. *Annalen der Physik*, **395**(8), 1055–1101.

Phillips, O. M. 1958. The equilibrium range in the spectrum of wind-generated waves. *J. Fluid Mech.*, **4**, 426–434.

Phillips, O. M. 1960. On the dynamics of unsteady gravity waves of finite amplitude. Part 1. The elementary interactions. *J. Fluid Mech.*, **9**, 193–217.

Phillips, O. M. 1981. Wave interactions – The evolution of an idea. *J. Fluid Mech.*, **106**, 215–227.

Picozzi, A., Garnier, J., Hansson, T. et al. 2014. Optical wave turbulence: Towards a unified nonequilibrium thermodynamic formulation of statistical nonlinear optics. *Phys. Rep.*, **542**(1), 1–132.

Proment, D., Nazarenko, S., and Onorato, M. 2009. Quantum turbulence cascades in the Gross–Pitaevskii model. *Phys. Rev. A*, **80**(5), 051603.

Proment, D., Nazarenko, S., and Onorato, M. 2012. Sustained turbulence in the three-dimensional Gross–Pitaevskii model. *Physica D*, **241**(3), 304–314.

Sagdeev, R. Z., and Galeev, A. A. 1966. *Lectures on the Non-linear Theory of Plasma*. Trieste: International Center for Theoretical Physics.

Savaro, C., Campagne, A., Linares, M. C. et al. 2020. Generation of weakly nonlinear turbulence of internal gravity waves in the Coriolis facility. *Phys. Rev. Fluids*, **5**(7), 073801.

Steinberg, V. 2021. Elastic turbulence: An experimental view on inertialess random flow. *Ann. Rev. Fluid Mech.*, **53**(1), 010719–060129.

Stokes, G. G. 1847. On the theory of oscillatory waves. *Trans. Camb. Philos. Soc.*, **8**, 441–455.

Sturrock, P. A. 1957. Non-linear effects in electron plasmas. *Proc. R. Soc. Lond. A*, **242**(1230), 277–299.

Sulem, C., and Sulem, P.-L. 1999. *The Nonlinear Schrödinger equation: Self-focusing and wave collapse*. Applied Mathematical Sciences, vol. 139. Springer-Verlag.

Vedenov, A. A. 1967. Theory of a weakly turbulent plasma. *Rev. Plasma Physics*, **3**, 229–276.

Vinen, W. F. 2000. Classical character of turbulence in a quantum liquid. *Phys. Rev. B*, **61**(2), 1410–1420.

Zakharov, V. E. 1965. Weak turbulence in media with a decay spectrum. *J. Appl. Mech. Tech. Phys.*, **6**, 22–24.

Zakharov, V. E. 1967. Weak-turbulence spectrum in a plasma without a magnetic field. *J. Exp. Theor. Phys.*, **24**, 455–459.

Zakharov, V. E., and Filonenko, N. N. 1966. The energy spectrum for stochastic oscillations of a fluid surface. *Doclady Akad. Nauk. SSSR*, **170**, 1292–1295.

Zakharov, V. E., and Filonenko, N. N. 1967. Weak turbulence of capillary waves. *J. Appl. Mech. Tech. Phys.*, **8**(5), 37–40.

Zakharov, V. E., and Sagdeev, R. Z. 1970. Spectrum of acoustic turbulence. *Sov. Phys. Dok.*, **15**, 439–440.

Zakharov, V. E., L'Vov, V. S., and Falkovich, G. 1992. *Kolmogorov Spectra of Turbulence*, vol. 1: *Wave Turbulence*. Springer Series in Nonlinear Dynamics. Springer.

Zhang, Z., and Pan, Y. 2022. Numerical investigation of turbulence of surface gravity waves. *J. Fluid Mech.*, **933**, A58.

5

Theory for Capillary Wave Turbulence

5.1 Introduction

Together with gravity waves, capillary waves are the main surface waves encountered in nature. The latter have an advantage over the former in that they are easier to treat analytically in the nonlinear regime. This is the choice we make in this chapter: we present capillary wave turbulence in detail (see Figure 5.1). It is a technical chapter which is, however, necessary for those who wish to master the theory of wave turbulence. For others, the calculation steps can be ignored.

We will consider an incompressible fluid ($\nabla \cdot \mathbf{u} = 0$) (such as water) subject to irrotational movements ($\mathbf{u} = \nabla \phi$ with ϕ the potential). This condition is well justified when the air–water interface is perturbed by a unidirectional wind (a typical condition encountered in the open sea). The nonlinear equations that describe the dynamics of capillary waves are obtained by firstly noting that the deformation of the fluid at the air–water interface verifies the exact Lagrangian relationship:

$$\frac{d\eta}{dt} = u_z = \frac{\partial \phi}{\partial z}\Big|_\eta, \tag{5.1}$$

where $\eta(x, y, t)$ is the deformation and $\phi(x, y, z, t)$ the potential (see Figure 5.2 for an illustration). Bernoulli's equation applied to the free surface of the fluid (at $z = \eta$) is written:

$$\frac{\partial \phi}{\partial t}\Big|_\eta = -\frac{1}{2}(\nabla \phi)^2\Big|_\eta + \sigma \Delta \eta, \tag{5.2}$$

where $\sigma = \gamma/\rho_{water}$ with γ the surface tension coefficient (for the air–water interface $\gamma \simeq 0.07$ N/m) and ρ_{water} the mass density of the water. Note that the mass density of air is negligible compared to that of water. The surface tension term is obtained by assuming that the deformation is relatively small: $|\nabla \eta| \ll 1$. This tension is responsible for a discontinuity between the pressure of the fluid at its free surface P_f and the pressure of the atmosphere P_a; it is modeled by the relation $P_f - P_a = \sigma/R$, with R the radius of curvature of the free surface. The

Figure 5.1 Superposition of capillary waves (ripples) on gravity waves. Photo taken near Cargèse (Corsica).

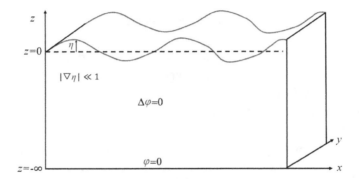

Figure 5.2 Schematic section of a capillary wave in deep water. It is assumed that the deformation η of the air–water interface is on average at altitude $z = 0$ and is such that $|\nabla \eta| \ll 1$, that is, of weak amplitude. In addition, we will assume that the fluid is incompressible ($\nabla \cdot \mathbf{u} = 0$) and irrotational ($\mathbf{u} = \nabla \phi$): in this case $\Delta \phi = 0$. The deep-water hypothesis means that the potential ϕ is nul at altitude $z = -\infty$. A more realistic illustration produced by a direct numerical simulation can be seen in Figure 5.3.

assumption of weak deformation (or weak curvature) simplifies the modeling. After developing the first equation, we obtain the following system:

$$\frac{\partial \eta}{\partial t} = -\nabla_\perp \phi|_\eta \cdot \nabla_\perp \eta + \frac{\partial \phi}{\partial z}\bigg|_\eta, \tag{5.3a}$$

$$\frac{\partial \phi}{\partial t}\bigg|_\eta = -\frac{1}{2}(\nabla \phi)^2|_\eta + \sigma \Delta \eta, \tag{5.3b}$$

where the symbol $\perp$ means that we only take derivatives in the x and y directions. This system of equations is the one used by Zakharov (1967) to develop the theory that interests us in this chapter. Basically, it involves the use of the potential ϕ in $z = \eta$, which can be tricky to handle. One way to get around this problem is to use a Taylor development in quadratic order to express ϕ in $z = \eta$ from its (Eulerian) value in $z = 0$. We obtain (Benney, 1962; Case and Chiu, 1977):

5.1 Introduction

$$\frac{\partial \eta}{\partial t} = -\nabla_\perp \phi|_0 \cdot \nabla_\perp \eta + \frac{\partial \phi}{\partial z}|_0 + \eta \frac{\partial^2 \phi}{\partial z^2}|_0, \tag{5.4a}$$

$$\frac{\partial \phi}{\partial t}|_0 + \eta \frac{\partial^2 \phi}{\partial z \partial t}|_0 = -\frac{1}{2}(\nabla \phi)^2|_0 + \sigma \Delta \eta. \tag{5.4b}$$

We have limited ourselves to quadratic nonlinearities because the problem of capillary wave turbulence can be solved at this level, that is, for three-wave interactions (McGoldrick, 1965). This situation differs from that of gravity waves, which must be treated at the cubic level (four-wave interactions). Equations (5.4a) and (5.4b) are the ones we will consider to develop the theory of capillary wave turbulence. They are completed by the incompressible and irrotational conditions of the fluid:

$$\Delta \phi = 0. \tag{5.5}$$

Under the deep-water hypothesis, that is, $\phi = 0$ in $z = -\infty$, we obtain a function of form:

$$\phi(x, y, z, t) = \psi(x, y, t) e^{kz}, \tag{5.6}$$

with k the wavevector norm, $\mathbf{k} \equiv (k_x, k_y)$.

The linearization of the system (5.4a)–(5.4b) will give us the dispersion relation. We obtain after Fourier transform:

$$-i\omega_k \hat{\eta}_k = k\hat{\phi}_k, \tag{5.7a}$$

$$-i\omega_k \hat{\phi}_k = -\sigma k^2 \hat{\eta}_k, \tag{5.7b}$$

with by definition:

$$\hat{\eta}_k \equiv \hat{\eta}(k_x, k_y) = \frac{1}{(2\pi)^2} \int_{\mathbb{R}^2} \eta(\mathbf{x}) \exp^{-i\mathbf{k}\cdot\mathbf{x}} d\mathbf{x}, \tag{5.8a}$$

$$\hat{\phi}_k \equiv \hat{\phi}(k_x, k_y) = \frac{1}{(2\pi)^2} \int_{\mathbb{R}^2} \phi(\mathbf{x}) \exp^{-i\mathbf{k}\cdot\mathbf{x}} d\mathbf{x}. \tag{5.8b}$$

Finally, the dispersion relationship is obtained:

$$\boxed{\omega_k^2 = \sigma k^3}. \tag{5.9}$$

Note that the presence of gravity at the linear level corrects this relation in $\omega_k^2 = \sigma k^3 + gk$. Therefore, our study is valid in the case where $k \gg k_*$ with $k_* \equiv \sqrt{g\rho_{water}/\gamma}$ (we have explicitly written the surface tension coefficient to obtain a numerical value). This corresponds to a critical wavelength $\lambda_* \simeq 1.7$ cm for the air–water interface. As a result, capillary waves appear at small scales. They are dispersive with a phase velocity v_ϕ that increases with the wavenumber ($v_\phi \propto \sqrt{k}$). This property can be observed by slightly disturbing the water surface: short wavelength waves are the fastest to escape from the disturbed area. Let us note in passing that in the case of gravity waves we have the inverse situation (easily verifiable by experiment): long wavelength gravity waves are the fastest to escape from the disturbed region (but they are preceded by capillary waves).

5.2 Phenomenology

Phenomenological analysis plays a fundamental role in turbulence because in the regime of strong turbulence it is the method used to arrive at a spectral prediction, for example, for energy. In the case of wave turbulence, it is possible to obtain this solution analytically; however, phenomenological analysis remains indispensable in order to, on the one hand, rapidly arrive at a first prediction and, on the other hand, be able to explain simply how the solution sought emerges. If we consider the nonlinear contribution of equation (5.4b), we arrive at the following phenomenological expression:

$$\frac{\phi}{\tau_{NL}} \sim k^2 \phi^2, \tag{5.10}$$

where τ_{NL} is the nonlinear time with $\partial \phi/\partial t \sim \phi/\tau_{NL}$ and $(\nabla \phi)^2 \sim k^2 \phi^2$. We then obtain:

$$\tau_{NL} \sim \frac{1}{k^2 \phi}. \tag{5.11}$$

A similar analysis using equation (5.4a) leads to the same expression. We can note, however, that the nonlinear term $\eta \partial^2 \phi/(\partial t \partial z)|_0$ of equation (5.4b) gives us an additional expression when this is balanced with the time derivative term (expression (5.11) should also be used), that is:

$$k\phi^2 \sim k^2 \eta^2. \tag{5.12}$$

This can be interpreted as an equipartition relation between the kinetic $k\phi^2$ and surface potential $k^2\eta^2$ energies.

To arrive at a prediction for the total energy spectrum, we need to introduce the mean rate of energy transfer ε in the inertial range:

$$\varepsilon \sim \frac{kE_k}{\tau_{tr}} \sim \frac{kE_k}{\omega \tau_{NL}^2} \sim \frac{kE_k k^4 \phi^2}{k^{3/2}} \sim k^{7/2} E_k^2, \tag{5.13}$$

with τ_{tr} the transfer (or cascade) time of capillary wave turbulence and E_k the one-dimensional spectrum of total energy (we assume that turbulence is statistically isotropic). We use here the expression τ_{tr} for triadic interactions (see Chapter 4 or 7 for a justification). One gets the relationship:

$$\boxed{E_k \sim \sqrt{\varepsilon} k^{-7/4}}. \tag{5.14}$$

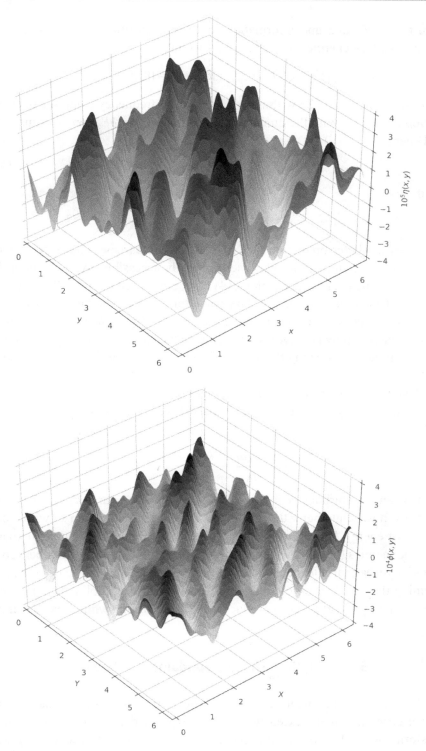

Figure 5.3 Spatial variations of the surface deformation $\eta(x,y)$ (top) and velocity potential $\phi(x,y)$ (bottom) in the wave turbulence regime. This direct numerical simulation is presented in Section 5.9.

From this prediction and information on the equipartition between kinetic and potential surface energies, we obtain the spectra:

$$E_k^\phi \equiv |\hat{\phi}_k|^2 \sim \sqrt{\varepsilon} k^{-11/4} \quad \text{and} \quad E_k^\eta \equiv |\hat{\eta}_k|^2 \sim \sqrt{\varepsilon} k^{-15/4}. \tag{5.15}$$

These spectra can also be written as a function of frequency using the dispersion relation $\omega \sim k^{3/2}$. With the dimensional relation $kE_k \sim \omega E_\omega$, one obtains for the total energy:

$$E_\omega \sim \sqrt{\varepsilon} \omega^{-3/2}, \tag{5.16}$$

and then:

$$E_\omega^\phi \equiv |\hat{\phi}_\omega|^2 \sim \sqrt{\varepsilon} \omega^{-13/6} \quad \text{and} \quad E_\omega^\eta \equiv |\hat{\eta}_\omega|^2 \sim \sqrt{\varepsilon} \omega^{-17/6}. \tag{5.17}$$

It is often the latter prediction that is used for comparison with experiment (or direct numerical simulation) because it is easily accessible. We will see in Section 5.6 that the energy spectrum (5.14) can be obtained analytically as an exact solution of the equations of capillary wave turbulence. The analytical approach also makes it possible to demonstrate that the energy cascade is direct (with a flux of energy strictly positive) and to estimate the so-called Kolmogorov constant of proportionality, allowing the sign "$\sim$" to be substituted for "$=$" in expression (5.14).

A last comment can be made on the capillary wave turbulence regime if we express the ratio χ between the wave period and the nonlinear time. With the prediction (5.15), we get:

$$\chi = \frac{1/\omega}{\tau_{NL}} \sim \frac{k^2 \phi}{\omega} \sim k^{-3/8}, \tag{5.18}$$

which means that turbulence is weaker at small scales (i.e. large k). In other words, if at a given wavenumber k the turbulence is weak, it will remain so with a direct cascade of energy. However, this property is not generic: for example, in magnetohydrodynamics (Chapter 7) the ratio χ increases with the wavenumbers so that an initially weak turbulence inevitably becomes strong at small scales (of course, assuming that small-scale dissipation effects remain negligible). This is also the case for gravity waves that can produce wave breaking and whitecaps (Nazarenko, 2011).

5.3 Analytical Theory: Fundamental Equation

For the nonlinear treatment of capillary wave turbulence, we will pass through Fourier space and make extensive use of the properties of the (spatial) Fourier transform. We will follow the Eulerian method proposed by Galtier (2021), which is the only complete demonstration published to date; the article of Zakharov (1967) only gives an overview of the demonstration, and it is the same for the later Pushkarev and Zakharov (1996, 2000), in which, however, there are some additional steps; note that these articles follow a Hamiltonian approach. The system (5.4a)–(5.4b) can be rewritten:

$$\frac{\partial \hat{\eta}_k}{\partial t} - k\hat{\phi}_k = \int_{\mathbb{R}^4} [(\mathbf{p}\cdot\mathbf{q})\hat{\phi}_p\hat{\eta}_q + p^2\hat{\phi}_p\hat{\eta}_q]\delta(\mathbf{k}-\mathbf{p}-\mathbf{q})d\mathbf{p}d\mathbf{q}, \qquad (5.19a)$$

$$\frac{\partial \hat{\phi}_k}{\partial t} + \sigma k^2 \hat{\eta}_k = \frac{1}{2}\int_{\mathbb{R}^4}[(\mathbf{p}\cdot\mathbf{q}-pq)\hat{\phi}_p\hat{\phi}_q + 2\sigma p^3 \hat{\eta}_p \hat{\eta}_q]\delta(\mathbf{k}-\mathbf{p}-\mathbf{q})d\mathbf{p}d\mathbf{q}. \qquad (5.19b)$$

The convolution product is expressed through the presence of the Dirac $\delta(\mathbf{k} - \mathbf{p} - \mathbf{q})$. One can notice at this level of analysis a difference with the equations of Zakharov (1967) whose attribution can be given to our initial choice to use a Taylor development around the equilibrium position. We now introduce the canonical variables A_k^s of this system:

$$\hat{\eta}_k \equiv \left(\frac{4}{\sigma k}\right)^{1/4} \sum_s A_k^s, \qquad (5.20a)$$

$$\hat{\phi}_k \equiv -i(4\sigma k)^{1/4} \sum_s s A_k^s, \qquad (5.20b)$$

with $s = \pm$. As the functions η and ϕ are real, we have:

$$\hat{\eta}_{-k}^* = \hat{\eta}_k, \quad \hat{\phi}_{-k}^* = \hat{\phi}_k, \qquad (5.21)$$

which gives us the remarkable relationship: $A_k^{s*} = A_{-k}^{-s}$ (with * the complex conjugate). The introduction of relations (5.20a)–(5.20b) in (5.19a)–(5.19b) gives (we also use the triadic relation $\mathbf{q} = \mathbf{k} - \mathbf{p}$ to simplify partly the nonlinear expression):

$$\frac{\partial A_k^s}{\partial t} + is\omega_k A_k^s = \frac{1}{2}\left(\frac{\sigma k}{4}\right)^{1/4}\int_{\mathbb{R}^4}(\mathbf{k}\cdot\mathbf{p})\hat{\phi}_p\hat{\eta}_q\delta(\mathbf{k}-\mathbf{p}-\mathbf{q})d\mathbf{p}d\mathbf{q} \qquad (5.22)$$

$$+ \frac{is}{4}\left(\frac{1}{4\sigma k}\right)^{1/4}\int_{\mathbb{R}^4}[(\mathbf{p}\cdot\mathbf{q}-pq)\hat{\phi}_p\hat{\phi}_q$$

$$+ 2\sigma p^3 \hat{\eta}_p\hat{\eta}_q]\delta(\mathbf{k}-\mathbf{p}-\mathbf{q})d\mathbf{p}d\mathbf{q},$$

with $\omega_k = \sqrt{\sigma k^3}$. The relevance of the choice of definitions of the canonical variables is immediately visible at the linear level: the term on the left makes the dispersion relationship explicitly apparent. The introduction of these variables at the nonlinear level gives:

$$\frac{\partial A_k^s}{\partial t} + is\omega_k A_k^s = \frac{-i\sigma^{1/4}}{\sqrt{2}}\int_{\mathbb{R}^4}\sum_{s_p s_q} s_p(\mathbf{k}\cdot\mathbf{p})\left(\frac{pk}{q}\right)^{1/4} \qquad (5.23)$$

$$\times A_p^{s_p} A_q^{s_q}\delta(\mathbf{k}-\mathbf{p}-\mathbf{q})d\mathbf{p}d\mathbf{q} - \frac{is\sigma^{1/4}}{2\sqrt{2}}\int_{\mathbb{R}^4}\sum_{s_p s_q}\bigg[s_p s_q(\mathbf{p}\cdot\mathbf{q}-pq)$$

$$\times \left(\frac{pq}{k}\right)^{1/4} - \frac{2p^3}{(kpq)^{1/4}}\bigg]A_p^{s_p}A_q^{s_q}\delta(\mathbf{k}-\mathbf{p}-\mathbf{q})d\mathbf{p}d\mathbf{q}.$$

This expression cannot be used as such for a statistical development: it needs to be simplified and above all to make it as symmetrical as possible in order to facilitate subsequent work. This is a fundamental work on which one can spend more time than the statistical development – which is merely the application of sophisticated but systematic techniques. The objective here is, moreover, to find the expression proposed by Zakharov (1967). This guide will help us to simplify the expression. The first remark concerns the first and last nonlinear terms, which can become symmetric by exchanging the wavevectors **p** and **q**, and the associated polarizations s_p and s_q; this gives us

$$\frac{\partial A_k^s}{\partial t} + is\omega_k A_k^s = -\frac{i\sigma^{1/4}}{2\sqrt{2}} \int_{\mathbb{R}^4} \sum_{s_p s_q} s_p s_q A_p^{s_p} A_q^{s_q} \delta(\mathbf{k} - \mathbf{p} - \mathbf{q}) \quad (5.24)$$

$$\times \left[s(\mathbf{p}\cdot\mathbf{q} - pq)\left(\frac{pq}{k}\right)^{1/4} + s_q(\mathbf{k}\cdot\mathbf{p})\left(\frac{pk}{q}\right)^{1/4} \right.$$

$$\left. + s_p(\mathbf{k}\cdot\mathbf{q})\left(\frac{qk}{p}\right)^{1/4} - \frac{ss_p s_q (p^3+q^3)}{(kpq)^{1/4}} \right] d\mathbf{p}d\mathbf{q}.$$

Then we introduce and subtract several terms and get:

$$\frac{\partial A_k^s}{\partial t} + is\omega_k A_k^s = -\frac{i\sigma^{1/4}}{2\sqrt{2}} \int_{\mathbb{R}^4} \sum_{s_p s_q} s_p s_q A_p^{s_p} A_q^{s_q} \delta(\mathbf{k} - \mathbf{p} - \mathbf{q}) \quad (5.25)$$

$$\times \left[s(\mathbf{p}\cdot\mathbf{q} + pq)\left(\frac{pq}{k}\right)^{1/4} + s_q(\mathbf{k}\cdot\mathbf{p} - kp)\left(\frac{pk}{q}\right)^{1/4} \right.$$

$$+ s_p(\mathbf{k}\cdot\mathbf{q} - kq)\left(\frac{qk}{p}\right)^{1/4} - 2spq\left(\frac{pq}{k}\right)^{1/4} + s_q kp\left(\frac{pk}{q}\right)^{1/4}$$

$$\left. + s_p kq\left(\frac{qk}{p}\right)^{1/4} - \frac{ss_p s_q (p^3+q^3)}{(kpq)^{1/4}} \right] d\mathbf{p}d\mathbf{q}.$$

We will see that the terms of the last line do not ultimately contribute to the nonlinear dynamics over long periods of time. For this, we introduce the angular frequency $\omega_k = \sqrt{\sigma k^3}$; we can then show that:

$$s_p s_q \left[-2spq\left(\frac{pq}{k}\right)^{1/4} + s_q kp\left(\frac{pk}{q}\right)^{1/4} + s_p kq\left(\frac{qk}{p}\right)^{1/4} - \frac{ss_p s_q (p^3+q^3)}{(kpq)^{1/4}} \right]$$

$$= \frac{s(s_p \omega_p + s_q \omega_q)(s\omega_k - s_p \omega_p - s_q \omega_q)}{\sigma (kpq)^{1/4}}. \quad (5.26)$$

Finally, we get:

$$\frac{\partial A_k^s}{\partial t} + is\omega_k A_k^s = -\frac{i\sigma^{1/4}}{2\sqrt{2}} \int_{\mathbb{R}^4} \sum_{s_p s_q} s_p s_q A_p^{s_p} A_q^{s_q} \delta(\mathbf{k} - \mathbf{p} - \mathbf{q}) \quad (5.27)$$

$$\times \left[s(\mathbf{p}\cdot\mathbf{q} + pq)\left(\frac{pq}{k}\right)^{1/4} + s_q(\mathbf{k}\cdot\mathbf{p} - kp)\left(\frac{pk}{q}\right)^{1/4} \right.$$

$$+ s_p(\mathbf{k}\cdot\mathbf{q} - kq)\left(\frac{qk}{p}\right)^{1/4}\Bigg] d\mathbf{p}d\mathbf{q} - \frac{i\sigma^{1/4}}{2\sqrt{2}}\int_{\mathbb{R}^4}$$

$$\times \sum_{s_p s_q} \frac{(s_p\omega_p + s_q\omega_q)(s\omega_k - s_p\omega_p - s_q\omega_q)}{\sigma(kpq)^{1/4}} A_p^{s_p} A_q^{s_q}$$

$$\times \delta(\mathbf{k} - \mathbf{p} - \mathbf{q}) d\mathbf{p}d\mathbf{q}.$$

Since the amplitude of the waves is assumed to be weak, the linear term will first dominate the dynamics, and the phase of the waves will evolve while maintaining the amplitude. Over longer periods of time, nonlinear terms will no longer be negligible and will modify the amplitude of the waves. Under these conditions, it is useful for canonical variables to separate the amplitude from the phase (which we have already done to find the dispersion relationship). We recall that the theory of wave turbulence consists of an asymptotic development whose uniformity was discussed in Chapter 4. A small parameter $0 < \epsilon \ll 1$ is introduced, and we write:

$$A_k^s \equiv \epsilon a_k^s e^{-is\omega_k t}, \tag{5.28}$$

hence the expression:

$$\frac{\partial a_k^s}{\partial t} = -\frac{i\epsilon\sigma^{1/4}}{2\sqrt{2}}\int_{\mathbb{R}^4}\sum_{s_p s_q} s_p s_q a_p^{s_p} a_q^{s_q} \delta(\mathbf{k}-\mathbf{p}-\mathbf{q}) e^{i(s\omega_k - s_p\omega_p - s_q\omega_q)t} \tag{5.29}$$

$$\times\Bigg[s(\mathbf{p}\cdot\mathbf{q} + pq)\left(\frac{pq}{k}\right)^{1/4} + s_q(\mathbf{k}\cdot\mathbf{p} - kp)\left(\frac{pk}{q}\right)^{1/4}$$

$$+ s_p(\mathbf{k}\cdot\mathbf{q} - kq)\left(\frac{qk}{p}\right)^{1/4}\Bigg] d\mathbf{p}d\mathbf{q} - \frac{i\epsilon s\sigma^{1/4}}{2\sqrt{2}}\int_{\mathbb{R}^4}\sum_{s_p s_q}$$

$$\times \frac{(s_p\omega_p + s_q\omega_q)(s\omega_k - s_p\omega_p - s_q\omega_q)}{\sigma(kpq)^{1/4}} a_p^{s_p} a_q^{s_q} e^{i(s\omega_k - s_p\omega_p - s_q\omega_q)t}$$

$$\times \delta(\mathbf{k}-\mathbf{p}-\mathbf{q})d\mathbf{p}d\mathbf{q}.$$

We are going to look at the nonlinear dynamics that emerges over a long time. By long time we mean a time τ much longer than the period of the waves, that is, $\tau \gg 1/\omega_k$. It is clear that the relevant contributions are those that cancel the coefficient in the exponential. Therefore, the secular contributions will not be provided by the second integral of equation (5.29), which precisely cancels for this condition. We will therefore neglect this term later on. Finally, we obtain the following nonlinear equation for the evolution of the amplitude of capillary waves:

$$\boxed{\frac{\partial a_k^s}{\partial t} = i\epsilon \int_{\mathbb{R}^4}\sum_{s_p s_q} L_{-kpq}^{-ss_ps_q} a_p^{s_p} a_q^{s_q} e^{i\Omega_{k,pq}t} \delta_{k,pq} d\mathbf{p}d\mathbf{q}}, \tag{5.30}$$

with by definition $\Omega_{k,pq} \equiv s\omega_k - s_p\omega_p - s_q\omega_q$, $\delta_{k,pq} \equiv \delta(\mathbf{k} - \mathbf{p} - \mathbf{q})$ and

$$L_{kpq}^{ss_ps_q} \equiv \frac{s_ps_q\sigma^{1/4}}{2\sqrt{2}} \left[s(\mathbf{p}\cdot\mathbf{q}+pq)\left(\frac{pq}{k}\right)^{1/4} + s_p(\mathbf{k}\cdot\mathbf{q}+kq)\left(\frac{qk}{p}\right)^{1/4} \right.$$
$$\left. + s_q(\mathbf{k}\cdot\mathbf{p}+kp)\left(\frac{pk}{q}\right)^{1/4} \right]. \tag{5.31}$$

Equation (5.30) governs the slow temporal evolution of capillary waves of weak amplitude. It is a quadratic nonlinear equation: these nonlinearities correspond to the interactions between waves propagating in the directions $\mathbf{p}$ and $\mathbf{q}$, and in the positive ($s_p, s_q > 0$) or negative direction ($s_p, s_q < 0$). Equation (5.30) is fundamental for our problem since it is from it that we will make a statistical development over asymptotically long times. This development is based on the symmetries of the fundamental equation: a lack of symmetry is a source of additional algebraic complexity (in particular for its statistical processing). Moreover, simplifications generally appear more easily on symmetrical equations. In our case, the interaction coefficient verifies the following symmetries (of which the number is sufficient, as we will see in practice):

$$L_{kpq}^{ss_ps_q} = L_{kqp}^{ss_qs_p}, \tag{5.32a}$$

$$L_{0pq}^{ss_ps_q} = 0, \tag{5.32b}$$

$$L_{-k-p-q}^{ss_ps_q} = L_{kpq}^{ss_ps_q}, \tag{5.32c}$$

$$L_{kpq}^{-s-s_p-s_q} = -L_{kpq}^{ss_ps_q}, \tag{5.32d}$$

$$ss_q L_{qpk}^{s_qs_ps} = L_{kpq}^{ss_ps_q}, \tag{5.32e}$$

$$ss_p L_{pkq}^{s_pss_q} = L_{kpq}^{ss_ps_q}. \tag{5.32f}$$

5.4 Analytical Theory: Statistical Approach

We now move on to a statistical description. We use the ensemble average $\langle\rangle$ and define the following spectral correlators (cumulants) for homogeneous turbulence (we will also assume $\langle a_k^s \rangle = 0$):

$$\langle a_k^s a_{k'}^{s'} \rangle = q_{kk'}^{ss'}(\mathbf{k},\mathbf{k}')\delta(\mathbf{k}+\mathbf{k}'), \tag{5.33a}$$

$$\langle a_k^s a_{k'}^{s'} a_{k''}^{s''} \rangle = q_{kk'k''}^{ss's''}(\mathbf{k},\mathbf{k}',\mathbf{k}'')\delta(\mathbf{k}+\mathbf{k}'+\mathbf{k}''), \tag{5.33b}$$

$$\langle a_k^s a_{k'}^{s'} a_{k''}^{s''} a_{k'''}^{s'''} \rangle = q_{kk'k''k'''}^{ss's''s'''}(\mathbf{k},\mathbf{k}',\mathbf{k}'',\mathbf{k}''')\delta(\mathbf{k}+\mathbf{k}'+\mathbf{k}''+\mathbf{k}''') \tag{5.33c}$$
$$+ q_{kk'}^{ss'}(\mathbf{k},\mathbf{k}')q_{k''k'''}^{s''s'''}(\mathbf{k}'',\mathbf{k}''')\delta(\mathbf{k}+\mathbf{k}')\delta(\mathbf{k}''+\mathbf{k}''')$$
$$+ q_{kk''}^{ss''}(\mathbf{k},\mathbf{k}'')q_{k'k'''}^{s's'''}(\mathbf{k}',\mathbf{k}''')\delta(\mathbf{k}+\mathbf{k}'')\delta(\mathbf{k}'+\mathbf{k}''')$$
$$+ q_{kk'''}^{ss'''}(\mathbf{k},\mathbf{k}''')q_{k'k''}^{s's''}(\mathbf{k}',\mathbf{k}'')\delta(\mathbf{k}+\mathbf{k}''')\delta(\mathbf{k}'+\mathbf{k}'').$$

5.4 Analytical Theory: Statistical Approach

From the fundamental equation (5.30), we get:

$$\frac{\partial \langle a_k^s a_{k'}^{s'} \rangle}{\partial t} = \left\langle \frac{\partial a_k^s}{\partial t} a_{k'}^{s'} \right\rangle + \left\langle a_k^s \frac{\partial a_{k'}^{s'}}{\partial t} \right\rangle \qquad (5.34)$$

$$= i\epsilon \int_{\mathbb{R}^4} \sum_{s_p s_q} L_{-kpq}^{-s s_p s_q} \langle a_{k'}^{s'} a_p^{s_p} a_q^{s_q} \rangle e^{i\Omega_{k,pq}t} \delta_{k,pq} d\mathbf{p} d\mathbf{q}$$

$$+ i\epsilon \int_{\mathbb{R}^4} \sum_{s_p s_q} L_{-k'pq}^{-s' s_p s_q} \langle a_k^s a_p^{s_p} a_q^{s_q} \rangle e^{i\Omega_{k',pq}t} \delta_{k',pq} d\mathbf{p} d\mathbf{q}.$$

At the next order we have:

$$\frac{\partial \langle a_k^s a_{k'}^{s'} a_{k''}^{s''} \rangle}{\partial t} = \left\langle \frac{\partial a_k^s}{\partial t} a_{k'}^{s'} a_{k''}^{s''} \right\rangle + \left\langle a_k^s \frac{\partial a_{k'}^{s'}}{\partial t} a_{k''}^{s''} \right\rangle + \left\langle a_k^s a_{k'}^{s'} \frac{\partial a_{k''}^{s''}}{\partial t} \right\rangle \qquad (5.35)$$

$$= i\epsilon \int_{\mathbb{R}^4} \sum_{s_p s_q} L_{-kpq}^{-s s_p s_q} \langle a_{k'}^{s'} a_{k''}^{s''} a_p^{s_p} a_q^{s_q} \rangle e^{i\Omega_{k,pq}t} \delta_{k,pq} d\mathbf{p} d\mathbf{q}$$

$$+ i\epsilon \int_{\mathbb{R}^4} \sum_{s_p s_q} L_{-k'pq}^{-s' s_p s_q} \langle a_k^s a_{k''}^{s''} a_p^{s_p} a_q^{s_q} \rangle e^{i\Omega_{k',pq}t} \delta_{k',pq} d\mathbf{p} d\mathbf{q}$$

$$+ i\epsilon \int_{\mathbb{R}^4} \sum_{s_p s_q} L_{-k''pq}^{-s'' s_p s_q} \langle a_k^s a_{k'}^{s'} a_p^{s_p} a_q^{s_q} \rangle e^{i\Omega_{k'',pq}t} \delta_{k'',pq} d\mathbf{p} d\mathbf{q}.$$

Here we face the classic problem of closure: a hierarchy of statistical equations of increasingly higher order emerges. In contrast to the strong turbulence regime, in the weak wave turbulence regime we can use the scale separation in time to achieve a natural closure of the system. Expressions (5.33a)–(5.33c) are introduced into equation (5.35):

$$\frac{\partial q_{kk'k''}^{ss's''}(\mathbf{k}, \mathbf{k}', \mathbf{k}'')}{\partial t} \delta(\mathbf{k} + \mathbf{k}' + \mathbf{k}'') = i\epsilon \int_{\mathbb{R}^4} \sum_{s_p s_q} L_{-kpq}^{-s s_p s_q} \qquad (5.36)$$

$$\times [q_{k'k''pq}^{s's''s_p s_q}(\mathbf{k}', \mathbf{k}'', \mathbf{p}, \mathbf{q})\delta(\mathbf{k}' + \mathbf{k}'' + \mathbf{p} + \mathbf{q})$$
$$+ q_{k'k''}^{s's''}(\mathbf{k}', \mathbf{k}'') q_{pq}^{s_p s_q}(\mathbf{p}, \mathbf{q})\delta(\mathbf{k}' + \mathbf{k}'')\delta(\mathbf{p} + \mathbf{q})$$
$$+ q_{k'p}^{s' s_p}(\mathbf{k}', \mathbf{p}) q_{k''q}^{s'' s_q}(\mathbf{k}'', \mathbf{q})\delta(\mathbf{k}' + \mathbf{p})\delta(\mathbf{k}'' + \mathbf{q})$$
$$+ q_{k'q}^{s' s_q}(\mathbf{k}', \mathbf{q}) q_{k''p}^{s'' s_p}(\mathbf{k}'', \mathbf{p})\delta(\mathbf{k}' + \mathbf{q})$$
$$\times \delta(\mathbf{k}'' + \mathbf{p})] e^{i\Omega_{k,pq}t} \delta_{k,pq} d\mathbf{p} d\mathbf{q}$$
$$+ i\epsilon \int_{\mathbb{R}^4} \left\{ (\mathbf{k}, s) \leftrightarrow (\mathbf{k}', s') \right\} d\mathbf{p} d\mathbf{q}$$
$$+ i\epsilon \int_{\mathbb{R}^4} \left\{ (\mathbf{k}, s) \leftrightarrow (\mathbf{k}'', s'') \right\} d\mathbf{p} d\mathbf{q},$$

where the last two lines correspond to the exchange at the notation level between $\mathbf{k}, s$ in the expanded expression and $\mathbf{k}', s'$ (penultimate line), then $\mathbf{k}'', s''$ (last line).

We are now going to integrate expression (5.36) both on **p** and **q**, and on time, by considering a long integrated time compared to the reference time (i.e. the period of the capillary wave). The presence of several Dirac functions leads to the conclusion that the second term on the right (in the main expression) gives no contribution, since it corresponds to $k = 0$, for which the interaction coefficient is null. It is a property of statistical homogeneity. The last two terms on the right (always in the main expression) lead to a strong constraint on wavevectors **p** and **q** which must be equal to $-\mathbf{k}'$ or $-\mathbf{k}''$. For the fourth-order cumulant, the constraint is much less strong since only the sum of **p** and **q** is imposed. The consequence is that for long times this term will not contribute to the nonlinear dynamics (see the discussion on uniformity of the development in Chapter 4). Finally, for long times the second-order cumulants are only relevant when the associated polarities have different signs. In order to understand this, it is necessary to go back to the definition of the moment, $\langle A_k^s A_{k'}^{s'} \rangle = \epsilon^2 \langle a_k^s a_{k'}^{s'} \rangle \exp(-i(s\omega_k + s'\omega_{k'})t)$, from which we see that a nonzero contribution is possible for homogeneous turbulence ($\mathbf{k} = -\mathbf{k}'$) only if $s = -s'$ (then the coefficient of the exponential is cancelled). We finally get:

$$q_{kk'k''}^{ss's''}(\mathbf{k},\mathbf{k}',\mathbf{k}'')\delta(\mathbf{k}+\mathbf{k}'+\mathbf{k}'') = i\epsilon\Delta(\Omega_{kk'k''})\delta(\mathbf{k}+\mathbf{k}'+\mathbf{k}'') \quad (5.37)$$

$$\times \left\{ \left[L_{-k-k'-k''}^{-s-s'-s''} + L_{-k-k''-k'}^{-s-s''-s'} \right] q_{k''-k''}^{s''-s''} \right.$$

$$\times (\mathbf{k}'',-\mathbf{k}'') q_{k'-k'}^{s'-s'}(\mathbf{k}',-\mathbf{k}')$$

$$+ \left[L_{-k'-k-k''}^{-s'-s-s''} + L_{-k'-k''-k}^{-s'-s''-s} \right] q_{k''-k''}^{s''-s''}$$

$$\times (\mathbf{k}'',-\mathbf{k}'') q_{k-k}^{s-s}(\mathbf{k},-\mathbf{k})$$

$$+ \left[L_{-k''-k'-k}^{-s''-s'-s} + L_{-k''-k-k'}^{-s''-s-s'} \right] q_{k-k}^{s-s}$$

$$\left. \times (\mathbf{k},-\mathbf{k}) q_{k'-k'}^{s'-s'}(\mathbf{k}',-\mathbf{k}') \right\},$$

with:

$$\Delta(\Omega_{kk'k''}) = \int_0^{t\gg 1/\omega} e^{i\Omega_{kk'k''}t'} dt' = \frac{e^{i\Omega_{kk'k''}t} - 1}{i\Omega_{kk'k''}}. \quad (5.38)$$

We can now write without ambiguity: $q_{k-k}^{s-s}(\mathbf{k},-\mathbf{k}) = q_k^s(\mathbf{k})$. Using the symmetry relations of the interaction coefficient, we obtain

$$q_{kk'k''}^{ss's''}(\mathbf{k},\mathbf{k}',\mathbf{k}'')\delta(\mathbf{k}+\mathbf{k}'+\mathbf{k}'') = -2i\epsilon\Delta(\Omega_{kk'k''})\delta(\mathbf{k}+\mathbf{k}'+\mathbf{k}'') \quad (5.39)$$

$$\left[L_{kk'k''}^{ss's''} q_{k''}^{s''}(\mathbf{k}'') q_{k'}^{s'}(\mathbf{k}') + L_{k'kk''}^{s'ss''} q_{k''}^{s''}(\mathbf{k}'') q_k^s(\mathbf{k}) + L_{k''k'k}^{s''s's} q_k^s(\mathbf{k}) q_{k'}^{s'}(\mathbf{k}') \right],$$

and then:

$$q_{kk'k''}^{ss's''}(\mathbf{k},\mathbf{k}',\mathbf{k}'')\delta(\mathbf{k}+\mathbf{k}'+\mathbf{k}'') = -2i\epsilon\Delta(\Omega_{kk'k''})\delta(\mathbf{k}+\mathbf{k}'+\mathbf{k}'') \quad (5.40)$$

$$L_{kk'k''}^{ss's''} \left[q_{k''}^{s''}(\mathbf{k}'') q_{k'}^{s'}(\mathbf{k}') + ss' q_{k''}^{s''}(\mathbf{k}'') q_k^s(\mathbf{k}) + ss'' q_k^s(\mathbf{k}) q_{k'}^{s'}(\mathbf{k}') \right].$$

The effective long time limit (which introduces irreversibility) gives us (Riemann–Lebesgue lemma):

$$\Delta(x) \to \pi\delta(x) + i\mathcal{P}(1/x), \tag{5.41}$$

with $\mathcal{P}$ the principal value integral.

The so-called kinetic equation is obtained by injecting expression (5.40) into equation (5.34) and integrating on $\mathbf{k}'$ (with the relation $q_{-\mathbf{k}}^{-s}(-\mathbf{k}) = q_{\mathbf{k}}^{s}(\mathbf{k})$):

$$\frac{\partial q_{\mathbf{k}}^{s}(\mathbf{k})}{\partial t} = 2\epsilon^{2}\int_{\mathbb{R}^{4}}\sum_{s_{p}s_{q}}|L_{-kpq}^{-ss_{p}s_{q}}|^{2}(\pi\delta(\Omega_{-kpq})+i\mathcal{P}(1/\Omega_{-kpq}))e^{i\Omega_{k,pq}t}\delta_{k,pq}$$

$$\times s_{p}s_{q}\left[s_{p}s_{q}q_{q}^{s_{q}}(\mathbf{q})q_{p}^{s_{p}}(\mathbf{p}) - ss_{q}q_{q}^{s_{q}}(\mathbf{q})q_{k}^{s}(\mathbf{k}) - ss_{p}q_{k}^{s}(\mathbf{k})q_{p}^{s_{p}}(\mathbf{p})\right]d\mathbf{p}d\mathbf{q}$$

$$+ 2\epsilon^{2}\int_{\mathbb{R}^{4}}\sum_{s_{p}s_{q}}|L_{kpq}^{ss_{p}s_{q}}|^{2}(\pi\delta(\Omega_{kpq})+i\mathcal{P}(1/\Omega_{kpq}))e^{i\Omega_{kpq}t}\delta_{kpq} \tag{5.42}$$

$$\times s_{p}s_{q}\left[s_{p}s_{q}q_{q}^{s_{q}}(\mathbf{q})q_{p}^{s_{p}}(\mathbf{p}) + ss_{q}q_{q}^{s_{q}}(\mathbf{q})q_{k}^{s}(\mathbf{k}) + ss_{p}q_{k}^{s}(\mathbf{k})q_{p}^{s_{p}}(\mathbf{p})\right]d\mathbf{p}d\mathbf{q}.$$

By changing the sign of the (dummy) variables $\mathbf{p}$ and $\mathbf{q}$ of integration, and the associated polarities, the principal values are eliminated. Using the symmetries of the interaction coefficient, we finally arrive at the following expression after simplification:

$$\frac{\partial q_{\mathbf{k}}^{s}(\mathbf{k})}{\partial t} = 4\pi\epsilon^{2}\int_{\mathbb{R}^{4}}\sum_{s_{p}s_{q}}|L_{kpq}^{ss_{p}s_{q}}|^{2}\delta(s\omega_{k}+s_{p}\omega_{p}+s_{q}\omega_{q})\delta(\mathbf{k}+\mathbf{p}+\mathbf{q}) \tag{5.43}$$

$$s_{p}s_{q}\left[s_{p}s_{q}q_{q}^{s_{q}}(\mathbf{q})q_{p}^{s_{p}}(\mathbf{p}) + ss_{q}q_{q}^{s_{q}}(\mathbf{q})q_{k}^{s}(\mathbf{k}) + ss_{p}q_{k}^{s}(\mathbf{k})q_{p}^{s_{p}}(\mathbf{p})\right]d\mathbf{p}d\mathbf{q},$$

with:

$$|L_{kpq}^{ss_{p}s_{q}}|^{2} \equiv \frac{\sqrt{\sigma}}{8}\left[s(\mathbf{p}\cdot\mathbf{q}+pq)\left(\frac{pq}{k}\right)^{1/4} + s_{p}(\mathbf{k}\cdot\mathbf{q}+kq)\left(\frac{qk}{p}\right)^{1/4}\right.$$

$$\left. + s_{q}(\mathbf{k}\cdot\mathbf{p}+kp)\left(\frac{pk}{q}\right)^{1/4}\right]^{2}. \tag{5.44}$$

Expression (5.43) is the kinetic equation of capillary wave turbulence first obtained by Zakharov & Filonenko in 1967.[1] The presence of the small parameter $\epsilon \ll 1$ means that the amplitude of the quadratic nonlinearities is weak and that, consequently, the characteristic time over which we place ourselves to measure these effects is of the order of $1/\epsilon^{2}$. As we have seen in practice, the obtaining of this expression was only possible thanks to the numerous symmetries of the interaction coefficient which we have used several times.

It is from the invariants of the system that we can bring out the main properties of wave turbulence. Energy has, therefore, a privileged role since it is always

---

[1] To be totally convinced of the equivalence of the two expressions, it remains to develop the integrand according to the values of s_{p} and s_{q} by eliminating the special case $s_{p} = s_{q} = s$, which has no solution at the resonance.

conserved. Other (inviscid) invariants can appear as the kinetic helicity in incompressible hydrodynamics subject to rapid rotation (limit of low Rossby numbers), a condition for being in the wave turbulence regime (see Chapter 6). In the capillary wave turbulence regime, we will consider the only relevant invariant, energy, for which the detailed conservation property will be demonstrated.

5.5 Detailed Energy Conservation

The kinetic equation (5.43) describes the temporal evolution of capillary wave turbulence over asymptotically long times compared to the reference time, which is the period of the waves. It is an equation involving three-wave interactions that give a nonzero contribution only when the following resonance condition is verified:

$$\begin{cases} s\omega_k + s_p\omega_p + s_q\omega_q = 0, \\ \mathbf{k} + \mathbf{p} + \mathbf{q} = 0. \end{cases} \quad (5.45)$$

In the case of capillary waves, the resonance condition has solutions, but this is not always the case. For example, for gravity waves the dispersion relation $\omega_k \propto \sqrt{k}$ does not allow us to get a solution (see Chapter 4). In this case, it is necessary to consider nonlinear contributions to the next order in the development, that is, four-wave interactions; the analytical approach then becomes more cumbersome.

A remarkable property verified by the kinetic equation is the detailed conservation of energy (see also Chapter 3 on two-dimensional eddy turbulence). To demonstrate this result, the kinetic equation must be rewritten for the polarized energy spectrum:

$$e^s(\mathbf{k}) \equiv \omega_k q_k^s(\mathbf{k}). \quad (5.46)$$

One notices in particular that: $e^s(\mathbf{k}) = e^{-s}(-\mathbf{k})$. After a few manipulations, we get:

$$\frac{\partial e^s(\mathbf{k})}{\partial t} = \frac{\pi\epsilon^2}{2\sigma} \int_{\mathbb{R}^4} \sum_{s_p s_q} |\tilde{L}_{kpq}^{ss_ps_q}|^2 \delta(s\omega_k + s_p\omega_p + s_q\omega_q)\delta(\mathbf{k}+\mathbf{p}+\mathbf{q}) \quad (5.47)$$

$$s\omega_k \left[\frac{s\omega_k}{e^s(\mathbf{k})} + \frac{s_p\omega_p}{e^{s_p}(\mathbf{p})} + \frac{s_q\omega_q}{e^{s_q}(\mathbf{q})}\right] e^s(\mathbf{k})e^{s_p}(\mathbf{p})e^{s_q}(\mathbf{q})d\mathbf{p}d\mathbf{q},$$

with:

$$|\tilde{L}_{kpq}^{ss_ps_q}|^2 \equiv \left[\frac{\mathbf{p}\cdot\mathbf{q}+pq}{sk\sqrt{pq}} + \frac{\mathbf{k}\cdot\mathbf{q}+kq}{s_pp\sqrt{kq}} + \frac{\mathbf{k}\cdot\mathbf{p}+kp}{s_qq\sqrt{kp}}\right]^2. \quad (5.48)$$

By considering the integral in $\mathbf{k}$ of the total energy spectrum, $e^+(\mathbf{k}) + e^-(\mathbf{k})$, and then playing on the permutation of the wavevectors (first we decompose the expression into three identical integrals), we can show that:

$$\frac{\partial \int_{\mathbb{R}^2} \sum_s e^s(\mathbf{k}) d\mathbf{k}}{\partial t} \tag{5.49}$$

$$= \frac{\pi\epsilon^2}{6\sigma} \int_{\mathbb{R}^6} \sum_{ss_p s_q} |\tilde{L}_{kpq}^{ss_p s_q}|^2 \delta(s\omega_k + s_p\omega_p + s_q\omega_q) \delta(\mathbf{k}+\mathbf{p}+\mathbf{q})(s\omega_k + s_p\omega_p + s_q\omega_q)$$

$$\times \left[\frac{s\omega_k}{e^s(\mathbf{k})} + \frac{s_p\omega_p}{e^{s_p}(\mathbf{p})} + \frac{s_q\omega_q}{e^{s_q}(\mathbf{q})}\right] e^s(\mathbf{k}) e^{s_p}(\mathbf{p}) e^{s_q}(\mathbf{q}) d\mathbf{k} d\mathbf{p} d\mathbf{q} = 0.$$

This means that energy is conserved by triadic interaction: the redistribution of energy takes place within a triad satisfying the resonance condition (5.45). This is a general property of wave turbulence that can be used to verify (in part) the accuracy of the kinetic equation obtained after a long computation.

5.6 Exact Solutions and Zakharov's Transformation

In this section, we look for the exact solutions of the nonlinear kinetic equation (5.47). These solutions can be obtained by applying the Zakharov transformation, which we have already introduced in Chapter 3, which is a conformal transformation applied to the integral (see Figure 5.5). Note that Balk (2000) has proposed another method to obtain these solutions without using a conformal transformation.

We will make the simplifying hypothesis that capillary wave turbulence is statistically isotropic. This assumption is reasonable, since we do not have a source of anisotropy as may be the case in other problems (e.g. in plasma physics, the strong uniform magnetic field supporting the plasma waves induces a strong anisotropy – see Chapters 7 and 8). The isotropic spectrum is therefore introduced:

$$E(k) \equiv E_k = 2\pi k \sum_s e^s(|\mathbf{k}|). \tag{5.50}$$

We will rewrite the kinetic equation using the triangular relation (see Figure 5.4):

$$q^2 = k^2 + p^2 - 2kp\cos\theta, \tag{5.51}$$

from which we deduce (at fixed k and p): $qdq = kp\sin\theta d\theta$. This relation will then be used to rewrite the kinetic equation. We use Al Kashi's formulas on the triangle to finally obtain the following expanded expression:

$$\frac{\partial E_k}{\partial t} = \frac{\epsilon^2}{2\sigma} \int_\Delta \sum_{ss_p s_q} s\omega_k |\tilde{L}_{kpq}^{ss_p s_q}|^2 \delta(s\omega_k + s_p\omega_p + s_q\omega_q) \tag{5.52}$$

$$\frac{sk\omega_k E_p E_q + s_p p\omega_p E_k E_q + s_q q\omega_q E_k E_p}{\sqrt{4k^2 p^2 - (k^2 + p^2 - q^2)^2}} dpdq,$$

with Δ the integration domain (the shaded band in Figure 5.5) and:

$$|\tilde{L}_{kpq}^{ss_ps_q}|^2 = \left[\frac{k^2 - p^2 - q^2 + 2pq}{2sk\sqrt{pq}} \right. \tag{5.53}$$
$$\left. + \frac{p^2 - k^2 - q^2 + 2kq}{2s_pp\sqrt{kq}} + \frac{q^2 - k^2 - p^2 + 2kp}{2s_qq\sqrt{kp}}\right]^2.$$

We notice that the new expression no longer uses wavevectors but only wavenumbers. A further simplification is made by introducing the adimensional wavenumbers $\xi_p \equiv p/k$ and $\xi_q \equiv q/k$; we get the expression:

$$\frac{\partial E_k}{\partial t} = \frac{\epsilon^2 k^{5/2}}{2\sqrt{\sigma}} \int_\Delta \sum_{ss_ps_q} s|\tilde{L}_{1\xi_p\xi_q}^{ss_ps_q}|^2 \delta(s + s_p\xi_p^{3/2} + s_q\xi_q^{3/2}) \tag{5.54}$$
$$\frac{sE_{k\xi_p}E_{k\xi_q} + s_p\xi_p^{5/2}E_kE_{k\xi_q} + s_q\xi_q^{5/2}E_kE_{k\xi_p}}{\sqrt{4\xi_p^2 - (1 + \xi_p^2 - \xi_q^2)^2}} d\xi_p d\xi_q,$$

with:

$$|\tilde{L}_{1\xi_p\xi_q}^{ss_ps_q}|^2 = \left[\frac{1 - \xi_p^2 - \xi_q^2 + 2\xi_p\xi_q}{2s\sqrt{\xi_p\xi_q}} \right. \tag{5.55}$$
$$\left. + \frac{\xi_p^2 - 1 - \xi_q^2 + 2\xi_q}{2s_p\xi_p\sqrt{\xi_q}} + \frac{\xi_q^2 - 1 - \xi_p^2 + 2\xi_p}{2s_q\xi_q\sqrt{\xi_p}}\right]^2.$$

We will apply the Zakharov transformation on this last expression by assuming a power-law form for the energy spectrum, $E_k = Ck^x$. In practice, we divide the integral into three equal parts and apply on two of the three integrals a different transformation, keeping the third integral intact. The Zakharov transformations are:

$$\xi_p \to \frac{1}{\xi_p}, \quad \xi_q \to \frac{\xi_q}{\xi_p}, \quad \text{(TZ1)} \tag{5.56}$$

and

$$\xi_p \to \frac{\xi_p}{\xi_q}, \quad \xi_q \to \frac{1}{\xi_q}. \quad \text{(TZ2)} \tag{5.57}$$

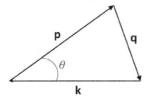

Figure 5.4 Triadic interaction.

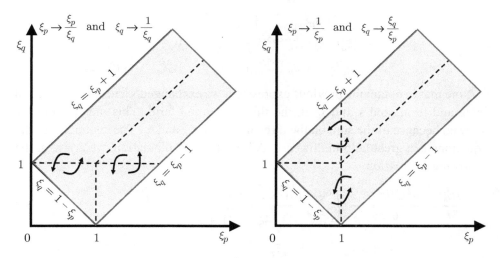

Figure 5.5 Zakharov conformal transformation for three-wave interactions. The infinitely long gray band corresponds to the solutions of the triangular relation $\mathbf{k}+\mathbf{p}+\mathbf{q}=\mathbf{0}$ (boundaries correspond to flattened triangles). The transformation consists in the exchange of the four regions separated by dashes. The transformations (5.56) and (5.57) are shown to the right and left, respectively.

They correspond to a conformal transformation of the region of integration in (ξ_p,ξ_q) space. In the case of triadic interactions, this region is an infinitely long band, as shown in Figure 5.5. The reader will be able to verify that these transformations give:

$$|\tilde{L}^{ss_ps_q}_{1\xi_p\xi_q}|^2 \xrightarrow{TZ1} |\tilde{L}^{s_pss_q}_{1\xi_p\xi_q}|^2, \tag{5.58a}$$

$$|\tilde{L}^{ss_ps_q}_{1\xi_p\xi_q}|^2 \xrightarrow{TZ2} |\tilde{L}^{s_qs_ps}_{1\xi_p\xi_q}|^2, \tag{5.58b}$$

$$\delta(s + s_p\xi_p^{3/2} + s_p\xi_q^{3/2}) \xrightarrow{TZ1} \xi_p^{3/2}\delta(s_p + s\xi_p^{3/2} + s_q\xi_q^{3/2}), \tag{5.58c}$$

$$\delta(s + s_p\xi_p^{3/2} + s_p\xi_q^{3/2}) \xrightarrow{TZ2} \xi_q^{3/2}\delta(s_q + s_p\xi_p^{3/2} + s\xi_q^{3/2}), \tag{5.58d}$$

$$1/\sqrt{4\xi_p^2 - (1+\xi_p^2-\xi_q^2)^2} \xrightarrow{TZ1} \xi_p^2/\sqrt{4\xi_p^2 - (1+\xi_p^2-\xi_q^2)^2}, \tag{5.58e}$$

$$1/\sqrt{4\xi_p^2 - (1+\xi_p^2-\xi_q^2)^2} \xrightarrow{TZ2} \xi_q^2/\sqrt{4\xi_p^2 - (1+\xi_p^2-\xi_q^2)^2}, \tag{5.58f}$$

$$d\xi_p d\xi_q \xrightarrow{TZ1} \xi_p^{-3} d\xi_p d\xi_q, \tag{5.58g}$$

$$d\xi_p d\xi_q \xrightarrow{TZ2} \xi_q^{-3} d\xi_p d\xi_q. \tag{5.58h}$$

This transforms the energy spectrum equation:

$$\frac{\partial E_k}{\partial t} = \frac{\epsilon^2 C^2 k^{2x+5/2}}{6\sqrt{\sigma}} \int_\Delta \sum_{ss_ps_q} \frac{|\tilde{L}^{ss_ps_q}_{1\xi_p\xi_q}|^2 \delta(s + s_p\xi_p^{3/2} + s_q\xi_q^{3/2})}{\sqrt{4\xi_p^2 - (1+\xi_p^2-\xi_q^2)^2}} \tag{5.59}$$

$$[\xi_p^x \xi_q^x (s + s_p \xi_p^{-x+5/2} + s_q \xi_q^{-x+5/2})s$$
$$+ \xi_p^x \xi_q^x (s + s_p \xi_p^{-x+5/2} + s_q \xi_q^{-x+5/2}) s_p \xi_p^{-2x-2}$$
$$+ \xi_p^x \xi_q^x (s + s_p \xi_p^{-x+5/2} + s_q \xi_q^{-x+5/2}) s_q \xi_q^{-2x-2}] d\xi_p d\xi_q.$$

Note that to obtain the previous expression, we exchanged s and s_p in the second integrand term, and s and s_q in the third integrand term. This manipulation is allowed because of the sum on the three indices s, s_p, and s_q. The symmetry of the equations is a great help in this case. After a final manipulation, we arrive at the following expression:

$$\frac{\partial E_k}{\partial t} = \frac{\epsilon^2 C^2 k^{2x+5/2}}{6\sqrt{\sigma}} \int_\Delta \sum_{s s_p s_q} \frac{|\tilde{L}_{1\xi_p \xi_q}^{s s_p s_q}|^2 \delta(s + s_p \xi_p^{3/2} + s_q \xi_q^{3/2})}{\sqrt{4\xi_p^2 - (1 + \xi_p^2 - \xi_q^2)^2}} \qquad (5.60)$$
$$\xi_p^x \xi_q^x [s + s_p \xi_p^{-x+5/2} + s_q \xi_q^{-x+5/2}][s + s_p \xi_p^{-2x-2} + s_q \xi_q^{-2x-2}] d\xi_p d\xi_q.$$

The stationary solutions for which the right-hand term cancels out correspond to:

$$\boxed{x = 1 \quad \text{and} \quad x = -7/4}. \qquad (5.61)$$

Indeed, for these two values one can make appear the expression $s + s_p \xi_p^{3/2} + s_q \xi_q^{3/2}$ (in the second line of (5.60)), which cancels exactly at resonance (a condition imposed by the Dirac).

The solution $x = 1$ corresponds to the thermodynamic equilibrium for which the energy flux is zero: in this case each of the three right-hand terms of expression (5.59) is null and therefore no energy transfer is possible. The solution $x = -7/4$ is more interesting, because the cancellation of the right-hand term of (5.59) is obtained by a subtle balance of its three contributions: it is the finite energy flux solution called the Kolmogorov–Zakharov spectrum. In this case, a final calculation must be performed to justify the relevance of this spectrum: this is a question of verifying the convergence of integrals in the case of strongly nonlocal interactions (this must be done before the application of the Zakharov transformation – see Figure 5.6). This corresponds to the regions close to the two right angles of the bands in Figure 5.5 as well as to the infinitely distant region of the origin. In the case of a divergence, the solution found is simply not relevant (however, sometimes a logarithmic correction is required for the convergence) and only the numerical simulation of the wave turbulence equation can be used to estimate the shape of the spectrum. In capillary wave turbulence, the locality of the Kolmogorov–Zakharov spectrum has been verified by Zakharov (1967). Contrary to their study, our solution was obtained from an equation written in wavenumber; it is therefore necessary to properly establish the area of convergence. An elementary calculation (for example, we can write the variables ξ_p and ξ_q in polar coordinates) brings us to the condition:

$$-7/2 < x < +5/4, \qquad (5.62)$$

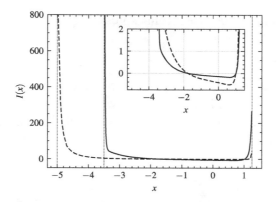

Figure 5.6 Numerical estimation of the collision integral $I(x)$ before the application of the Zakharov transform (solid line) and after (dashed line). We use expression (5.54) multiplied by a factor of 40 in the first case, while it is expression (5.60) or (5.63) in the second case. The vertical lines $x = -5$, $x = -7/2$, and $x = 5/4$ define the limits of the domains of convergence. Insert: enlargement to show the stationary Kolmogorov–Zakharov solution at $x = -7/4$. Note that the broader domain of convergence corresponds to the result published by Pushkarev and Zakharov (2000) with $]-5, -5/6[$, but in fact it should be reduced to $]-4, -5/6[$. Figure made by V. David.

where the upper bound is given by the convergence condition for a point located at infinity on the bands of Figure 5.5. This result thus demonstrates the relevance of the Kolmogorov–Zakharov spectrum. The convergence condition actually corresponds to a criterion of locality of interactions. In this sense, we find a hypothesis made by Kolmogorov to obtain the exact four-thirds (or four-fifths) law presented in Chapter 2, namely that the inertial range describes a universal physics independent of mechanisms (e.g. of forcing and dissipation) operating at the largest and smallest scales. This is the reason why the Kolmogorov–Zakharov spectrum is relevant here.

5.7 Nature of the Exact Solutions

Zakharov's transformation allowed us to obtain the exact stationary solutions in power law and to highlight two types of spectrum: the thermodynamic solution, and the dynamic solution corresponding to a cascade of energy. However, the nature of this cascade remains unknown: is it a direct or an inverse cascade? The answer to this question requires further analysis.

We will use the relation linking the energy flux Π_k to the energy spectrum:

$$\frac{\partial E_k}{\partial t} = -\frac{\partial \Pi_k}{\partial k} = \frac{\epsilon^2 C^2}{\sqrt{\sigma}} k^{2x+5/2} I(x), \qquad (5.63)$$

where $I(x)$ is the integral deduced from expression (5.60). We get:

$$\Pi_k = -\frac{\epsilon^2 C^2}{\sqrt{\sigma}} \frac{k^{2x+7/2} I(x)}{2x+7/2}. \tag{5.64}$$

The direction of the cascade will be given by the sign of the energy flux in the special case where $x = -7/4$. But in this case, we see that the denominator cancels. Fortunately, so does the numerator, since it is precisely for this value that $I(x) = 0$. L'Hôspital's rule allows us to write (we use the notations introduced in Chapter 3 on energy flux):

$$\lim_{x \to -7/4} \Pi_k = \varepsilon = -\frac{\epsilon^2 C^2}{\sqrt{\sigma}} \lim_{x \to -7/4} \frac{I(x)}{2x+7/2} = -\frac{\epsilon^2 C^2}{\sqrt{\sigma}} \lim_{y \to 3/2} \frac{I(y)}{3/2 - y},$$

$$= \frac{\epsilon^2 C^2}{\sqrt{\sigma}} \lim_{y \to 3/2} \frac{\partial I(y)}{\partial y}, \tag{5.65}$$

with $y = -2x - 2$ and:

$$\frac{\partial I(y)}{\partial y}\Big|_{3/2} = \frac{1}{6} \int_\Delta \sum_{s s_p s_q} \frac{|\tilde{L}^{s s_p s_q}_{1\xi_p \xi_q}|^2}{\sqrt{4\xi_p^2 - (1 + \xi_p^2 - \xi_q^2)^2}} \delta(s + s_p \xi_p^{3/2} + s_q \xi_q^{3/2})$$

$$\xi_p^x \xi_q^x [s + s_p \xi_p^{-x+5/2} + s_q \xi_q^{-x+5/2}] \frac{\partial [s + s_p \xi_p^{\xi_y} + s_q \xi_q^{\xi_y}]}{\partial y} d\xi_p d\xi_q \Big|_{3/2}$$

$$= \frac{1}{6} \int_\Delta \sum_{s s_p s_q} \frac{|\tilde{L}^{s s_p s_q}_{1\xi_p \xi_q}|^2}{\sqrt{4\xi_p^2 - (1 + \xi_p^2 - \xi_q^2)^2}} \delta(s + s_p \xi_p^{3/2} + s_q \xi_q^{3/2}) \tag{5.66}$$

$$\xi_p^{-7/4} \xi_q^{-7/4} [s + s_p \xi_p^{17/4} + s_q \xi_q^{17/4}][s_p \xi_p^{3/2} \ln(\xi_p) + s_q \xi_q^{3/2} \ln(\xi_q)] d\xi_p d\xi_q.$$

The sign of the previous expression can be found numerically. A positive sign was obtained (Pushkarev and Zakharov, 2000), demonstrating that the energy cascade of capillary wave turbulence is direct. It is also possible to find the numerical value of the Kolmogorov constant C_K, whose expression is deduced from C in (5.65); finally, we obtain:[2]

$$\boxed{E_k = \frac{\sigma^{1/4}}{\epsilon} C_K \sqrt{\varepsilon} k^{-7/4}}, \tag{5.67}$$

with:

$$C_K = \frac{1}{\sqrt{\partial I(y)/\partial y|_{3/2}}}. \tag{5.68}$$

The value reported by Pushkarev and Zakharov (2000) with their approach is $C_K \simeq 9.85$ but this value has been corrected by Pan and Yue (2017), who found 6.97. In the present case, the value obtained numerically for the (total) energy spectrum is $C_K \simeq 3.48$. We see that the analytical study allowed us to find the main properties of capillary wave turbulence. The strength of these results is that they are analytical.

[2] See also Chapter 7, where the Kolmogorov constant is obtained analytically and numerically.

5.8 Comparison with Experiments

Capillary waves have been studied for a long time as, demonstrated by the work of Longuet-Higgins (1963) and McGoldrick (1970): in these examples, the aim was to understand the role of resonant wave interactions and the mechanism of capillary wave generation from gravity waves. On the other hand, the experimental study of capillary wave turbulence is more recent and is still the subject of much work today. For example, Holt and Trinh (1996) have been able to produce a sea of capillary waves at the surface of a drop (of about 5.0 mm in diameter) in levitation and consisting mainly of water. The experiment shows the rapid formation (in less than one second) by a direct cascade of a spectrum in $f^{-3.58}$ (with f the frequency) for the fluctuations of the surface deformation η. The difference with the theoretical prediction (5.17) could have its origin in nonnegligible visco-elastic effects. In the article by Wright et al. (1996), a new measurement technique based on the scattering of visible light on spheres in polystyrene (with a diameter of the order of μm) is used to obtain the height variations of the water surface. From these data the authors were able to calculate the wavenumber spectrum: a narrow power law with an exponent around -4 was measured. Subsequent measurements with the same technique (but with semi-skimmed milk) gave results consistent with the prediction of $-17/6$ for the frequency spectrum (see Section 5.2) (Henry et al., 2000).

Capillary wave turbulence has also been studied using liquid hydrogen (maintained at 15–16 K). The interest of this kind of study is that the liquid has a lower kinematic viscosity than water, allowing the size of the inertial zone to be increased (Brazhnikov et al., 2001, 2002). Note, however, that in this problem the inertial range is also limited by the surface tension coefficient. The experiment shows that the theoretically predicted frequency spectra can be reproduced fairly well over more than an order of magnitude, despite high noise. Since the imagination of physicists is vast, the problem has also been addressed using the properties of fluorescence (essentially localized on the surface of the liquid) of a solution added to water. Using a blue laser projected on to the surface of the liquid, the authors (Lommer and Levinsen, 2002) measured the power law with precision, and this time the Zakharov–Filonenko solution was very well reproduced over two decades of frequencies.

The low-gravity capillary wave turbulence regime (~ 0.05 g) was achieved during parabolic (Airbus A320) flights with a duration of 22 seconds (Falcón et al., 2009). The motivation to develop such an experiment is to limit the effects of gravity waves and thus extend the purely capillary inertial range (see Figure 5.7). Two decades of frequency power law have been measured, with an exponent close

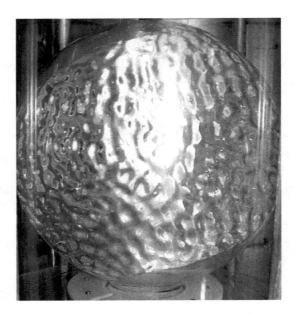

Figure 5.7 Capillary wave turbulence on the surface of a spherical container (of 15 cm diameter) filled with 20 cl of water. This result is obtained during a parabolic flight in free fall (Falcón et al., 2009). The fluid wets the inner surface of the sphere and forms a layer of spherical fluid. Purely capillary waves are then generated by vibrating the container. Credit: E. Falcon.

to the expected value.[3] Note that gravito-capillary waves have been the subject of several experimental studies in which, in particular, the transition between the regime dominated by large-scale gravity waves and small-scale capillary waves has been well highlighted, as well as the possible (nonlocal) interactions between these two types of waves (with collinear wavevectors) (see, e.g., Falcon et al., 2007; Deike et al., 2012; Berhanu and Falcon, 2013; Aubourg and Mordant, 2015, 2016; Berhanu et al., 2018; Hassaini and Mordant, 2018; Cazaubiel et al., 2019).

Direct numerical simulation is an essential complement to experimental studies because it allows a more detailed analysis of capillary wave turbulence: in principle it gives access to all fields and therefore allows precise signal processing. Little work has been done on this compared to the numerous experimental studies. These simulations were first made by Pushkarev and Zakharov (1996, 2000), then more recently by, for example, Deike et al. (2014) and Pan and Yue (2014) (see also Kochurin et al., 2020 for collinear capillary wave turbulence). However, surface waves are subject to numerical instabilities that make it difficult to obtain a large inertial range (less than a decade has been obtained). Moreover, these simulations with an external force take particularly long to reach a steady state for which turbulence is fully developed. Finally, the discretization of the Fourier space induced

---

[3] A similar experiment has been done on board the International Space Station (ISS), but the regime obtained is that of strong capillary wave turbulence (Berhanu et al., 2019).

by the numerical box also has potential consequences for turbulence, whose analytical properties are obtained in a continuous medium (Nazarenko, 2011; Pan and Yue, 2017). Paradoxically, the simulation of surface wave turbulence, by nature two-dimensional, seems to be more difficult to do than that in three dimensions, for which nonlinear interactions are more numerous and numerical instabilities better controlled.

These direct numerical simulations complement the numerical simulations of Falkovich et al. (1995) made directly from the wave turbulence equations: the aim was to study the nonstationary phase in a freely decaying regime in order to highlight, in particular, the explosive properties of the spectrum. We will come back to this property in inertial wave turbulence (Chapter 6). Note that capillary wave turbulence has also been numerically simulated at the surface of a ferrofluid in the presence of a strong horizontal magnetic field that reduces the wave interactions in the direction of the magnetic field (Kochurin, 2020).

5.9 Direct Numerical Simulation

In order to illustrate the capillary wave turbulence regime, we present in this section the results of a direct numerical simulation. For numerical stability reasons, dissipation terms have been added to the primitive equations (5.4a)–(5.4b). The simulated equations are therefore:

$$\frac{\partial \eta}{\partial t} = -\nabla_\perp \phi|_0 \cdot \nabla_\perp \eta + \frac{\partial \phi}{\partial z}|_0 + \eta \frac{\partial^2 \phi}{\partial z^2}|_0 - \nu \Delta^2 \eta, \quad (5.69a)$$

$$\frac{\partial \phi}{\partial t}|_0 + \eta \frac{\partial^2 \phi}{\partial t \partial z}|_0 = -\frac{1}{2}(\nabla \phi)^2|_0 + \sigma \Delta \eta - \nu \Delta^2 \phi + f, \quad (5.69b)$$

$$\Delta \phi = 0, \quad (5.69c)$$

with $\nu = 3 \times 10^{-8}$ a hyperviscosity. The use of a hyperviscosity allows us to confine the effect of dissipation to the smallest scales and thus widen the inertial range. An external force f is applied only to the velocity potential, assuming initially the absence of deformation (i.e. $\eta(x, y, t = 0) = 0$). The force f is characterized by a large-scale spectrum (excitation at wavenumbers $k = 1, 2, 3$ at each time step) with a random fluctuation of the amplitudes around a mean value and a random phase. Classically, a pseudospectral numerical code is used with periodic boundary conditions in (x,y). The spatial resolution used is 512 points in each direction.

In Figure 5.3, we see the spatial variation of the deformation $\eta(x, y)$ of the fluid surface and the velocity potential $\phi(x, y)$ in fully developed turbulence. These fields are characterized by random fluctuations. This first figure is interesting, but it does not allow us to fully appreciate the excited scale range. In particular, the smallest scales are impossible to see. In Figure 5.8, we see how these fields evolve

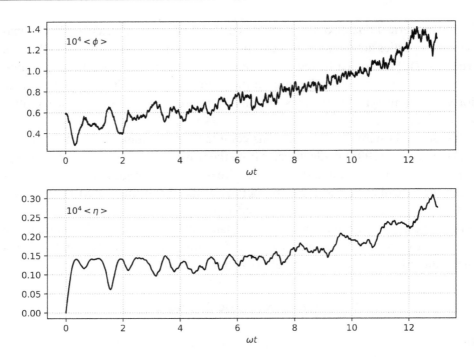

Figure 5.8 Temporal variations of the root mean square (in space) of the potential $\langle \phi(x,y) \rangle$ (top) and of the deformation $\langle \eta(x,y) \rangle$ (bottom) during the initial phase of development of capillary wave turbulence. The time is normalized with the angular frequency associated with the forcing.

over time from the initial instant for which $\langle \eta(x, y, t = 0) \rangle = 0$. More precisely, it is the temporal variations of the quadratic means that are shown. We find that the response of the deformation to the force applied to the velocity potential is relatively smooth. For both fields, the amplitude increases over time; in this window of time the stationary state has not yet been reached. This figure shows a qualitative behavior of the two fields: we can see that an increase in one field corresponds to a decrease in the other field.

The isotropic spectra of the surface potential energy $E^\eta(k, t)$ are shown in Figure 5.9 when the turbulence is fully developed. An inertial range clearly emerges, with a power law compatible with the prediction in $k^{-15/4}$, as we can see in the insert figure, when these spectra are compensated by the expected solution: a plateau over nearly a decade is indeed obtained.

The wave character of capillary wave turbulence can be revealed by tracing the spectrum $E^\eta(k, \omega)$. This spectrum is constructed from the field $\hat{\eta}(k_x, k_y, t)$. The values of the real part and the imaginary part are recorded in the course of time for $k_x = 1$ and $k_y \in [0, 45]$. A Fourier transform in time is then applied to these signals, weighted with a Hamming function in order to make them periodic (and thus reduce the noise produced by the nonperiodicity). The result is shown in

5.9 Direct Numerical Simulation

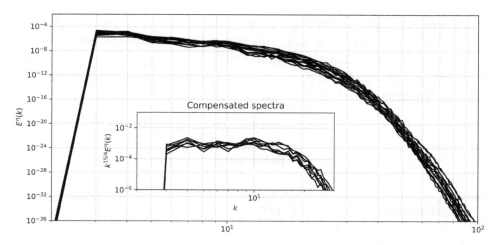

Figure 5.9 Spectra at different times of the surface potential energy $E^\eta(k,t)$ and (insert) compensated by the expected solution in $k^{-15/4}$.

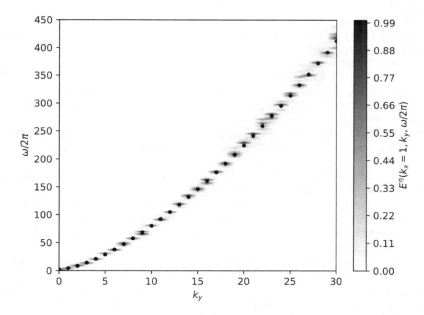

Figure 5.10 Surface potential energy spectrum $E^\eta(k_x, k_y, \omega)$ constructed from $\hat{\eta}_k(t)$ with $k_x = 1$ and $k_y \in [0, 45]$. Each signal in the vertical direction is normalized by its maximum. The black dots correspond to the dispersion relation $\omega = k^{3/2}$.

Figure 5.10: we plot the sum of the squared modulus of the Fourier coefficients. We see that the information is essentially concentrated along the curve $\omega = k^{3/2}$, which reveals the wave nature of this turbulence. The nonlinearities have the effect of widening the dispersion relationship. The ω–k spectrum is often used as a diagnostic tool to determine the type of turbulence. Often wave turbulence becomes strong at smaller scales. This is the situation encountered in magnetohydrodynamics (see Chapter 7): in this case, the ω–k spectrum shows a clear transition between the two regimes, with a signal that fills a large region of ω–k space in the strong wave turbulence regime (Meyrand et al., 2016).

References

Aubourg, Q., and Mordant, N. 2015. Nonlocal resonances in weak turbulence of gravity-capillary waves. *Phys. Rev. Lett.*, **114**(14), 144501.
Aubourg, Q., and Mordant, N. 2016. Investigation of resonances in gravity-capillary wave turbulence. *Phys. Rev. Fluids*, **1**(2), 023701.
Balk, A. M. 2000. On the Kolmogorov–Zakharov spectra of weak turbulence. *Physica D: Nonlinear Phenomena*, **139**(1–2), 137–157.
Benney, D. J. 1962. Non-linear gravity wave interactions. *J. Fluid Mech.*, **14**, 577–584.
Berhanu, M., and Falcon, E. 2013. Space-time-resolved capillary wave turbulence. *Phys. Rev. E*, **87**(3), 033003.
Berhanu, M., Falcon, E., and Deike, L. 2018. Turbulence of capillary waves forced by steep gravity waves. *J. Fluid Mech.*, **850**, 803–843.
Berhanu, M., Falcon, E., Michel, G., Gissinger, C., and Fauve, S. 2019. Capillary wave turbulence experiments in microgravity. *EPL (Europhysics Letters)*, **128**(3), 34001.
Brazhnikov, M. Yu., Kolmakov, G. V., Levchenko, A. A., and Mezhov-Deglin, L. P. 2001. Capillary turbulence at the surface of liquid hydrogen. *Sov. J. Exp. Theo. Phys. Lett.*, **73**(8), 398–400.
Brazhnikov, M. Yu., Kolmakov, G. V., and Levchenko, A. A. 2002. The turbulence of capillary waves on the surface of liquid hydrogen. *Sov. J. Exp. Theo. Phys.*, **95**(3), 447–454.
Case, K. M., and Chiu, S. C. 1977. Three-wave resonant interactions of gravity-capillary waves. *Phys. Fluids*, **20**(5), 742–745.
Cazaubiel, A., Haudin, F., Falcon, E., and Berhanu, M. 2019. Forced three-wave interactions of capillary-gravity surface waves. *Phys. Rev. Fluids*, **4**(7), 074803.
Deike, L., Berhanu, M., and Falcon, E. 2012. Decay of capillary wave turbulence. *Phys. Rev. Fluids*, **85**(6), 066311.
Deike, L., Fuster, D., Berhanu, M., and Falcon, E. 2014. Direct numerical simulations of capillary wave turbulence. *Phys. Rev. Lett.*, **112**(23), 234501.
Falcón, C., Falcon, E., Bortolozzo, U., and Fauve, S. 2009. Capillary wave turbulence on a spherical fluid surface in low gravity. *Europhys. Lett.*, **86**(1), 14002.
Falcon, É., Laroche, C., and Fauve, S. 2007. Observation of gravity-capillary wave turbulence. *Phys. Rev. Lett.*, **98**(9), 094503.
Falkovich, G. E., Shapiro, I. Ya, and Shtilman, L. 1995. Decay turbulence of capillary waves. *Europhys. Lett.*, **29**(1), 1–6.
Galtier, S. 2021. Wave turbulence: the case of capillary waves (a review). *Geophys. Astro. Fluid. Dyn.* **115**(3), 234–257.
Hassaini, R., and Mordant, N. 2018. Confinement effects on gravity-capillary wave turbulence. *Phys. Rev. Fluids*, **3**(9), 094805.
Henry, E., Alstrøm, P., and Levinsen, M. T. 2000. Prevalence of weak turbulence in strongly driven surface ripples. *Europhys. Lett.*, **52**(1), 27–32.
Holt, R. G., and Trinh, E. H. 1996. Faraday wave turbulence on a spherical liquid shell. *Phys. Rev. Lett.*, **77**(7), 1274–1277.

References

Kochurin, E., Ricard, G., Zubarev, N., and Falcon, E. 2020. Numerical simulation of collinear capillary-wave turbulence. *JETP Lett.*, **112**(12), 757–763.

Kochurin, E. A. 2020. Numerical simulation of the wave turbulence on the surface of a ferrofluid in a horizontal magnetic field. *J. Magn. Magn. Mater.*, **503**, 166607.

Lommer, M., and Levinsen, M. T. 2002. Using laser-induced fluorescence in the study of surface wave turbulence. *J. Floresc.*, **12**(1), 45–50.

Longuet-Higgins, M. S. 1963. The generation of capillary waves by steep gravity waves. *J. Fluid Mech.*, **16**, 138–159.

McGoldrick, L. F. 1965. Resonant interactions among capillary-gravity waves. *J. Fluid Mech.*, **21**, 305–331.

McGoldrick, L. F. 1970. An experiment on second-order capillary-gravity resonant wave interactions. *J. Fluid Mech.*, **40**, 251–271.

Meyrand, R., Galtier, S., and Kiyani, K. H. 2016. Direct evidence of the transition from weak to strong magnetohydrodynamic turbulence. *Phys. Rev. Lett.*, **116**(10), 105002.

Nazarenko, S. 2011. *Wave Turbulence*. Lecture Notes in Physics, vol. 825. Springer Verlag.

Pan, Y., and Yue, D. K. P. 2014. Direct numerical investigation of turbulence of capillary waves. *Phys. Rev. Lett.*, **113**(9), 094501.

Pan, Y., and Yue, D. K. P. 2017. Understanding discrete capillary-wave turbulence using a quasi-resonant kinetic equation. *J. Fluid Mech.*, **816**, R1 (11 pages).

Pushkarev, A. N., and Zakharov, V. E. 1996. Turbulence of capillary waves. *Phys. Rev. Lett.*, **76**(18), 3320–3323.

Pushkarev, A. N., and Zakharov, V. E. 2000. Turbulence of capillary waves – theory and numerical simulation. *Physica D Nonlinear Phenomena*, **135**(1), 98–116.

Wright, W. B., Budakian, R., and Putterman, S. J. 1996. Diffusing light photography of fully developed isotropic ripple turbulence. *Phys. Rev. Lett.*, **76**(24), 4528–4531.

Zakharov, V. E., and Filonenko, N. N. 1967. Weak turbulence of capillary waves. *J. Appl. Mech. Tech. Phys.*, **8**(5), 37–40.

6

Inertial Wave Turbulence

Inertial wave turbulence is the regime reached by an incompressible fluid subjected to a rapid and uniform rotation within the limit of a large Reynolds number. At first glance this situation seems to be close to the one discussed in Part I to introduce the main concepts of turbulence – eddy turbulence – since it is simply a matter of adding a well-known force to the Navier–Stokes equations, that is, the Coriolis force. However, there are two major differences to the standard case: inertial waves exist because of the rotation, and the spherical symmetry is broken by the presence of a privileged axis, that of rotation. This anisotropy makes the analytical development of wave turbulence more complex than in the case of capillary waves. As we will see here, it is nevertheless possible to derive the kinetic equations and find the main properties of such a system.

6.1 Introduction

Rotating turbulent flows are of interest to a wide variety of fields, such as industry, meteorology, and geophysics (oceans, atmosphere). In the latter case the effect of the earth's rotation is felt, for example, in large-scale atmospheric motions. The comparison between the earth and Venus illustrates this point well. The two planets, of similar size, rotate on themselves at very different speeds: by definition, it takes one day for an earth revolution to occur while Venus needs about 243 (Earth) days.[1] The effects of this difference are noticeable in the atmospheric motions of the two planets. The earth's atmosphere has a complex dynamics whose signature is revealed, in part, by the shape of the cloud structures (see Figure 6.1), while the Venusian atmosphere presents a more uniform appearance. Note that rotation also plays an important role in the geodynamo (we restrict ourselves here to the case of the earth): the earth's magnetic field owes its existence to a dynamo process operating in the earth's outer liquid core, where the Rossby number is about 10^{-6},

---

[1] The very slow rotation of Venus on itself makes the Venusian year (225 days) shorter than its day.

Figure 6.1 An example of a low Rossby number flow: large-scale terrestrial atmospheric motions. Image of the earth taken on September 4, 2019. Credits: NASA Earth Observatory/Joshua Stevens; NOAA National Environmental Satellite, Data, and Information Service.

and where inertial wave signatures emanating from this core have been detected (Aldridge and Lumb, 1987).

A precise description of geophysical flows requires the consideration of many parameters (rotation, stratification, geometry, etc.). The methodology often followed in research consists in isolating one parameter to study its effects in detail. Much work has been done on the effect of rotation on geophysical turbulent flows, for which Navier–Stokes equations are the simplest representation. These are written in the incompressible case (it is assumed that the mass density $\rho_0 = 1$):

$$\frac{\partial \mathbf{w}}{\partial t} - 2(\mathbf{\Omega} \cdot \nabla)\mathbf{u} = (\mathbf{w} \cdot \nabla)\mathbf{u} - (\mathbf{u} \cdot \nabla)\mathbf{w} + \nu \nabla^2 \mathbf{w}, \qquad (6.1)$$

with $\mathbf{w} \equiv \nabla \times \mathbf{u}$ the vorticity, $\mathbf{\Omega}$ the rotation rate associated with the Coriolis force, and ν the kinematic viscosity. Note that with this writing, the pressure term (including the effects of rotation) disappears. The intensity of the Coriolis force can be measured by the ratio between the advection term and the Coriolis force. This dimensionless quantity,

$$R_o = \frac{U}{L\Omega},\qquad(6.2)$$

is the Rossby number. A low Rossby number tells us that the effects of rotation are dominant. To fix these ideas, let us take the example of a wind flowing at $U = 0.1\,\text{ms}^{-1}$ and a length scale $L = 10\,\text{km}$ (the size of a large cloud structure). With the angular velocity of the earth, $\Omega_T \simeq 7.3 \times 10^{-5}\,\text{rad s}^{-1}$, we get the Rossby number $R_o \simeq 0.1$, which is the typical value of large-scale geophysical flows. With a kinematic viscosity $\nu \approx 0.1\,\text{cm}^2\text{s}^{-1}$, we arrive for this flow at the Reynolds number $R_e = UL/\nu \simeq 10^8$. The modeling of large-scale atmospheric flows therefore requires us to consider, among other things, the properties of rotating turbulence. The limit of interest in this chapter is the small Rossby number ($R_o \ll 1$). This number offers us a small parameter on which the theory of inertial wave turbulence can be built.

6.2 What Do We Know About Rotating Turbulence?

Laboratory Experiment

Numerous laboratory experiments have been dedicated to the study of rotating turbulent flows (see Figure 6.2). From an experimental point of view, it is not difficult to obtain a small Rossby number ($R_o < 1$) in a fast rotating tank; on the other hand, it is more difficult to reach a Reynolds number higher than 10^5. However, this number is high enough for the flow to be in a fully developed turbulent regime. These experiments (e.g. in a wind tunnel with a rotating section) have shown that rotation has the effect, when $R_o < 1$, of bi-dimensionalizing an initially isotropic turbulence (Jacquin et al., 1990; Lamriben et al., 2011). This results in the presence of vorticity filaments having their axes approximately parallel to the rotation axis (Ω) and a strong correlation of velocity along Ω (Hopfinger et al., 1982). An asymmetry is also measured in the distribution of cyclones and anticyclones (the former dominating the latter) as well as a slowing down (compared to the nonrotating case) of the temporal energy decay when the turbulence is free, that is, in absence of external forcing (Moisy et al., 2011).

Thanks to the PIV (particle image velocimetry) technique, which consists in illuminating a fluid in which small spherical particles of a few μm in diameter are introduced, it is possible to accurately measure the velocity of these particles and thus that of the fluid, whose dynamics is supposed not to be influenced by the presence of these intruders. A restriction exists, however: until now, the measurement has been made in a plane. It is often the plane perpendicular to the axis of rotation that is chosen. The energy spectra measured as a function of the perpendicular wavenumber $k_\perp$ show a stiffening of the power law passing from an index close to $-5/3$ (case without rotation at $R_o = +\infty$) to a value close to -2.2 for the fastest rotation (Morize et al., 2005).

Figure 6.2 Rotating platform dedicated to the experimental study of rotating turbulence and inertial waves. This mechanical platform of 2 m in diameter, rotating at up to 30 revolutions per minute, allows for the loading of one tonne of material, including the weight of the tank and the water (700 liters here) where flows are generated. The platform allows experimenters to remotely control an embedded PIV system composed of a pulsed laser and several high-sensitivity cameras. In the experiment presented here, an assembly of inertial wave beams is forced by 32 oscillating cylinders organized regularly around a virtual sphere of 80 cm diameter. This rotating platform and the experiments (Monsalve et al., 2020) have been developed by P.-P. Cortet & F. Moisy at the Fluids, Automatic and Thermal Systems (FAST) Laboratory (CNRS & University of Paris-Saclay) in Orsay.

The experiments also reveal that the p-order structure functions follow a self-similar law with $\xi_p = p/2$ (Baroud et al., 2002) or $\xi_p = 3p/4$ (van Bokhoven et al., 2009). Thus, rapidly rotating turbulence seems to be characterized by an absence of intermittency, in the sense that the curvature of $\xi_p(p)$ measured in hydrodynamics at $R_o = +\infty$ is absent (see Chapter 3).[2] Even if the origin of this self-similar behavior can be attributed to inhomogeneities, measurements, or forcing effects, it is interesting to mention that the $\xi_p = 3p/4$ law is dimensionally compatible with the exact solution ($p=2$) of inertial wave turbulence (Galtier, 2003) as well as some direct numerical simulations (Pouquet and Mininni, 2010). These experiments also show that the probability distribution functions of (increments of) velocity have non-Gaussian wings (Baroud et al., 2002; van Bokhoven et al., 2009).

The PIV technique has also made it possible to highlight the inertial wave turbulence regime by measuring the frequency spectrum. Like capillary waves,

---

[2] Sometimes, the simple deviation from Kolmogorov's law $\xi_p = p/3$ is considered as a signature of intermittency. However, this definition is restrictive because it includes any deviation – even self-similar – from the spectrum in $-5/3$.

this spectrum is characterized by a signal essentially concentrated along the inertial wave dispersion relation (Yarom and Sharon, 2014). This property is the one expected when nonlinearities are weak (Nazarenko, 2011). The experiments have shown the presence, in the perpendicular direction only, of a large-scale inverse cascade (Campagne et al., 2014) whose properties seem, however, to be different from those of a purely two-dimensional turbulence (Kraichnan, 1967).

To conclude, note that a quantitative experimental observation of weak inertial wave turbulence has been reported recently by Monsalve et al. (2020). In this experiment where inertial waves are directly excited and the geostrophic mode strongly damped, the spectra display the expected power-law behavior both in scaling and amplitude.

Numerical Simulations

The effects of rotation on hydrodynamic turbulence are also studied via numerical simulation. A reduction of the nonlinear transfer in the direction of Ω has been observed, as well as a slowing down of the temporal decay of energy (Bardina et al., 1985; Cambon et al., 1997). A stiffening of the power law followed by the energy spectrum was measured, as well as signatures of an inverse cascade (Hossain, 1994; Smith et al., 1996). In particular, Smith and Waleffe (1999) have shown with direct numerical simulations that when the flow is forced three-dimensionally at an intermediate wavenumber k_f, one observes a direct cascade of energy for $k > k_f$, with a one-dimensional isotropic spectrum close to k^{-2}, and an apparently inverse cascade for $k < k_f$, with a one-dimensional isotropic spectrum close to k^{-3}. Their analysis shows that large-scale energy is mainly contained in the two-dimensional state which is defined as the fluctuations in the mode $k_\parallel = 0$ (with $\mathbf{k} \cdot \mathbf{\Omega} = k_\parallel \Omega$; we also speak about slow mode), whereas at small scales the energy is mainly contained in the three-dimensional modes (fluctuations at modes $|k_\parallel| > 0$). The spectrum of the slow mode could be the result of nonlocal interactions between the two- and three-dimensional modes, rather than the consequence of a two-dimensional inverse cascade (Bourouiba et al., 2012). We will come back to this problem at the end of the chapter. These simulations also show that the behavior at small and large scales is strongly influenced by the aspect ratio between the vertical resolution, along Ω, and the horizontal resolution: a small aspect ratio, with a low resolution in the vertical direction, leads to a reduction in the number of resonant triads and a significant alteration of the energy spectrum. Their simulations reveal a global energy spectrum in $k^{-5/3}$ for a sufficiently small aspect ratio. This result suggests that the resonant triads have a fundamental role to play in turbulence with rotation.

The domination of cyclones (rotating in the same direction as the global rotation) over anticyclones (rotating in the opposite direction; see Figure 6.3) was also highlighted by numerical simulation in the case of a fast rotation (Buzzicotti et al., 2018). One can show that this asymmetry is linked to the nonstability of the anticyclones. The formation of larger and larger structures is studied through the notion of condensate, that is, the ability of a turbulent system to excite the smallest wavenumbers (Seshasayanan and Alexakis, 2018). Within the limit of a fast rotation, the use of frequency and wavenumber spectra has made it possible to highlight the wave character of turbulence (Clark di Leoni et al., 2014). Note finally that the most recent three-dimensional direct numerical simulations by Le Reun et al. (2020) show that the anisotropic Kolmogorov–Zakharov spectrum predicted by the theory is well reproduced (see also Sharma et al., 2018; Yokoyama and Takaoka, 2021) and that the energy transfer is mainly local. More details on this topic, and many other issues, are discussed in the reviews of Godeferd and Moisy (2015) and Alexakis and Biferale (2018).

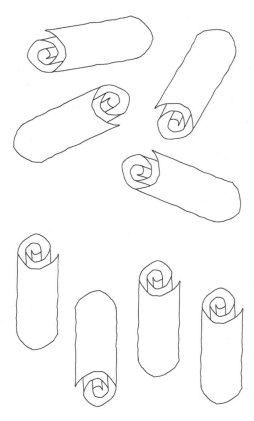

Figure 6.3 Schematic view of vorticity tubes in the case of hydrodynamic turbulence without rotation (top) and with rotation (bottom). The rotation polarizes the tubes along the axis of rotation, here in the vertical direction, and the direction of rotation of the tubes is mostly that of the rotation of the fluid (domination of cyclones over anticyclones).

Theories

Rotating turbulence can be analyzed phenomenologically. Early works on the subject (Zeman, 1994; Zhou, 1995) show that a steep spectrum in k^{-2} is expected at large scales, a region where the Coriolis force is nonnegligible, that is, for wavenumbers smaller than a critical wavenumber k_Ω. We can notice, however, that this phenomenological prediction does not introduce anisotropy, which is a fundamental property of this turbulence.

In spectral theory, the most striking results were obtained using the EDQNM closure model (Orszag, 1970) that was introduced in Chapter 3. Cambon and Jacquin (1989) have developed a formalism based on an eigenmode decomposition. The ad hoc closure used leads to dynamic equations for the invariants of the system. The simulation of these equations allows us to understand more precisely certain observed properties such as the anisotropic transfer mechanism (see also Cambon et al., 1997). Inhomogeneity effects have been introduced to take into account the existence of infinitely large boundaries in the direction perpendicular to the axis of rotation (Scott, 2014). It has been shown that the rapidly rotating confined fluid behaves differently from the free one, with essentially a dissipation effect, due to the walls, which dynamically emerges before the classical volume dissipation.

One can adapt Kolmogorov's exact law to this axisymmetric problem (Galtier, 2009). This law can be used in its nonintegrated form (Lamriben et al., 2011), whose expression is, at the origin, identical to that of hydrodynamics without rotation (exact law (2.25)) because the Coriolis force does not work. It can also be integrated by means of a critical balance hypothesis (see Chapter 7). The nonintegrated Kolmogorov law is also valid in wave turbulence, since we do not need for its derivation to make assumptions about the type of turbulence; the universality of Kolmogorov's law is thus reinforced. To conclude, we note that a recent theoretical study shows an unexpected proximity between rotating turbulence and solar wind turbulence at subionic scales (Galtier and David, 2020). This proximity opens interesting perspectives that will be discussed in final section of this chapter also Chapter 8).

6.3 Helical Inertial Waves

To obtain the inertial wave dispersion relation, we start from the incompressible equation (6.1), with by definition $\boldsymbol{\Omega} \equiv \Omega \hat{e}_{||}$ and $\hat{e}_{||}$ a unit vector such that $|\hat{e}_{||}| = 1$. Anticipating the helical character of these waves, we will introduce a complex helical decomposition (Craya, 1954):

$$\mathbf{h}_{\mathbf{k}}^{s} \equiv \mathbf{h}^{s}(\mathbf{k}) = (\hat{e}_k \times \hat{e}_{||}) \times \hat{e}_k + is(\hat{e}_k \times \hat{e}_{||}), \qquad (6.3)$$

with $\mathbf{k} = k\hat{\mathbf{e}}_k = \mathbf{k}_\perp + k_\parallel \hat{\mathbf{e}}$ ($k = |\mathbf{k}|$, $k_\perp = |\mathbf{k}_\perp|$, $|\hat{\mathbf{e}}_k| = 1$) and $s = \pm$ the directional polarity. This basis satisfies the following properties:

$$\mathbf{h}_\mathbf{k}^{-s} = \mathbf{h}_{-\mathbf{k}}^s, \tag{6.4a}$$

$$is(\hat{\mathbf{e}}_k \times \mathbf{h}_\mathbf{k}^s) = \mathbf{h}_\mathbf{k}^s, \tag{6.4b}$$

$$\mathbf{k} \cdot \mathbf{h}_\mathbf{k}^s = 0, \tag{6.4c}$$

$$\mathbf{h}_\mathbf{k}^s \cdot \mathbf{h}_\mathbf{k}^{s'} = \frac{2k_\perp^2}{k^2}\delta(s+s'). \tag{6.4d}$$

In particular, the relation (6.4c) means that the incompressibility condition will be implicitly verified by the helical fields. This property will allow us to lighten the wave turbulence formalism that we will briefly present later. The projection of the velocity on the helical basis leads us to the following definition for A_k^s:

$$\hat{\mathbf{u}}_k \equiv \hat{\mathbf{u}}(\mathbf{k}) = \sum_s A^s(\mathbf{k})\,\mathbf{h}_\mathbf{k}^s \equiv \sum_s A_k^s\,\mathbf{h}_\mathbf{k}^s, \tag{6.5}$$

with $\hat{\mathbf{u}}_k$ the Fourier transform of the velocity $\mathbf{u}$. One deduces from this:

$$\hat{\mathbf{w}}_k \equiv \hat{\mathbf{w}}(\mathbf{k}) = i\mathbf{k} \times \hat{\mathbf{u}}_k = k \sum_s s A_k^s\,\mathbf{h}_\mathbf{k}^s. \tag{6.6}$$

In the inviscid case ($\nu = 0$), the Fourier transform of equation (6.1) gives us:

$$\frac{\partial \hat{\mathbf{w}}_k}{\partial t} - 2i\Omega k_\parallel \hat{\mathbf{u}}_k = [(\mathbf{w}\cdot\nabla)\mathbf{u} - (\mathbf{u}\cdot\nabla)\mathbf{w}]_\mathbf{k}, \tag{6.7}$$

where the index $\mathbf{k}$ in the right-hand term means the Fourier transform. If we forget the nonlinear contribution and introduce the helical fields (6.5) and (6.6), we obtain, after projection and use of relation (6.4d):

$$sk\frac{\partial A_k^s}{\partial t} = -2i\Omega k_\parallel A_k^s, \tag{6.8}$$

which leads to the following dispersion relationship:

$$\boxed{\omega_k = \frac{2\Omega k_\parallel}{k}}. \tag{6.9}$$

Inertial waves are transverse, dispersive, and helical with a left polarization.

6.4 Phenomenological Predictions

Following the classical phenomenology presented in the previous chapter on capillary waves, we will define the nonlinear time from the Navier–Stokes equations; we obtain trivially:

$$\tau_{NL} \sim \frac{1}{ku_\ell}. \tag{6.10}$$

6.4 Phenomenological Predictions

The transfer (or cascade) time for three-wave interactions (nonlinearity is quadratic) is written in the case of fast rotation (see Chapter 4):

$$\tau_{tr} \sim \omega \tau_{NL}^2 \sim \frac{\Omega k_\parallel}{k^3 u_\ell^2}. \tag{6.11}$$

The difficulty here is to take into account the statistical anisotropy induced by rotation, whose theoretical justification will be given in Section 6.5 the resonance condition. To distinguish between perpendicular and parallel directions, we introduce the following two-dimensional axisymmetric energy spectrum:

$$u_\ell^2 \sim E(k_\perp, k_\parallel) k_\perp k_\parallel. \tag{6.12}$$

However, numerous experiments and numerical simulations show that the cascade is preferentially done in the direction perpendicular to the axis of rotation: this means that the energy is preferentially transferred in a region of Fourier space where $k_\perp \gg k_\parallel$. We will place ourselves in this region and thus substitute $k_\perp$ for k. We then introduce the mean rate of kinetic energy transfer ε in the inertial range:

$$\varepsilon \sim \frac{E(k_\perp, k_\parallel) k_\perp k_\parallel}{\tau_{tr}} \sim \frac{E^2(k_\perp, k_\parallel) k_\perp^5 k_\parallel}{\Omega}. \tag{6.13}$$

We finally get the following spectral prediction:

$$E(k_\perp, k_\parallel) \sim \sqrt{\varepsilon \Omega} \, k_\perp^{-5/2} k_\parallel^{-1/2}. \tag{6.14}$$

The effect of rotation is therefore to stiffen the energy spectrum compared to the standard case (at $R_o = +\infty$) while orienting the cascade in the transverse direction. The dependency in $k_\parallel$ that appears here is therefore not very relevant from the dynamic point of view. We will see in Section 6.5 that the spectrum (6.14) is the exact solution of the kinetic equations of inertial wave turbulence.

The ratio χ of linear to nonlinear times gives us an additional indication of the range of validity of the wave turbulence regime. We get, always with $k_\perp \gg k_\parallel$:

$$\chi = \frac{1/\omega}{\tau_{NL}} \sim \frac{k_\perp^2 u_\ell}{k_\parallel \Omega} \sim \frac{k_\perp^2 \sqrt{E(k_\perp, k_\parallel) k_\perp k_\parallel}}{k_\parallel \Omega} \sim \frac{k_\perp^{5/4}}{\Omega k_\parallel^{3/4}}. \tag{6.15}$$

As the energy is essentially transferred to larger and larger $k_\perp$, if the wave turbulence condition ($\chi \ll 1$) is met at a given $k_\perp$ scale, it will be less and less as the cascade develops in the perpendicular direction, and beyond a critical scale, $k_{\perp c}$, turbulence becomes strong. We see here a difference to the case of capillary wave turbulence, which becomes weaker and weaker as the cascade develops. Another limit of validity of wave turbulence exists at small wavenumbers. Indeed, the inequality $\chi \ll 1$ also means that the wavenumber $k_\parallel$ cannot be too small: in particular, it is necessary that $k_\parallel > 0$. This condition means that the slow mode $k_\parallel = 0$ cannot be described by weak wave turbulence.

6.5 Inertial Wave Turbulence Theory

Let us now take again the nonlinear expression (6.7) and introduce the helical fields (6.5) and (6.6); after projection and exchange of the dummy variables $\mathbf{p}$ and $\mathbf{q}$, s_p and s_q, we obtain:

$$\frac{\partial A_k^s}{\partial t} + is\omega_k A_k^s = i \sum_{s_p s_q} \int_{\mathbb{R}^6} L_{-\mathbf{k}\,\mathbf{p}\,\mathbf{q}}^{ss_p s_q} A_p^{s_p} A_q^{s_q} \delta(\mathbf{k} - \mathbf{p} - \mathbf{q}) \, d\mathbf{p} d\mathbf{q}, \qquad (6.16)$$

with the interaction coefficient:

$$L_{\mathbf{k}\,\mathbf{p}\,\mathbf{q}}^{ss_p s_q} = \frac{sk}{4k_\perp^2}(s_p p - s_q q)[(\mathbf{q} \cdot \mathbf{h}_\mathbf{p}^{s_p})(\mathbf{h}_\mathbf{q}^{s_q} \cdot \mathbf{h}_\mathbf{k}^{s}) - (\mathbf{p} \cdot \mathbf{h}_\mathbf{q}^{s_q})(\mathbf{h}_\mathbf{p}^{s_p} \cdot \mathbf{h}_\mathbf{k}^{s})]. \qquad (6.17)$$

Since the amplitude of the waves is assumed to be weak, the dynamics over short times – of the order of the wave period $1/\omega$ – will be dictated by linear terms. At longer times, such that $\tau \gg 1/\omega$, nonlinear terms will come into play by changing the amplitude of the waves. Therefore, the amplitude is separated from the phase (see Chapter 5 for a justification):

$$A_k^s \equiv \epsilon a_k^s e^{-is\omega_k t}, \qquad (6.18)$$

with ϵ a small parameter ($0 < \epsilon \ll 1$); hence the equation for the amplitude of inertial waves:

$$\frac{\partial a_k^s}{\partial t} = i\epsilon \sum_{s_p s_q} \int_{\mathbb{R}^6} L_{-\mathbf{k}\,\mathbf{p}\,\mathbf{q}}^{ss_p s_q} a_p^{s_p} a_q^{s_q} e^{i(s\omega_k - s_p\omega_p - s_q\omega_q)t} \delta(\mathbf{k} - \mathbf{p} - \mathbf{q}) d\mathbf{p} d\mathbf{q}. \qquad (6.19)$$

We find a classical form for three-wave interactions with a term in the right-hand side of weak amplitude (proportional to ϵ), a quadratic nonlinearity, and an exponential which, over long times, will give a nonzero contribution only when its coefficient cancels out.

After a few manipulations, we can write the resonance condition as follows:

$$\frac{s_q q - s_p p}{s\omega_k} = \frac{sk - s_q q}{s_p \omega_p} = \frac{s_p p - sk}{s_q \omega_q}. \qquad (6.20)$$

It is interesting to discuss the particular case of strongly local interactions that generally give a dominant contribution to turbulent dynamics. In this case, we have $k \simeq p \simeq q$ and the previous expression simplifies, at leading order, to:

$$\frac{s_q - s_p}{sk_\parallel} \simeq \frac{s - s_q}{s_p p_\parallel} \simeq \frac{s_p - s}{s_q q_\parallel}. \qquad (6.21)$$

If $k_\parallel$ is nonzero, the term on the left will give a nonnegligible contribution only when $s_p = -s_q$. We do not consider the case $s_p = s_q$, which is not relevant to the leading order in the case of local interactions, as we can see from the expression of the interaction coefficient (6.17), which then becomes negligible. The immediate consequence is that either the middle term or the right-hand term has a numerator which cancels itself out (at main order), which implies that the associated denominator must also cancel (at main order) to satisfy the equality: for example, if $s = s_p$ then $q_\parallel \simeq 0$. This condition means that the transfer in the parallel direction is negligible: indeed, the integration of equation (6.19) in the parallel direction is then reduced to a few modes (since $p_\parallel \simeq k_\parallel$), which strongly limits the transfer between parallel modes. The cascade in the parallel direction is thus possible, but relatively weak compared to that in the perpendicular direction. This situation is to be compared to that of incompressible MHD, which will be presented in the chapter 7: in this case, the uniform magnetic field $\mathbf{b}_0$ plays the role of the (uniform) rotation by introducing a privileged axis that breaks the spherical symmetry of the turbulence. In the MHD wave turbulence regime, it can be shown that no transfer in the $\mathbf{b}_0$ direction is possible; therefore, the direct cascade of the total (kinetic + magnetic) energy is exclusively in the perpendicular direction.

The kinetic equations of inertial wave turbulence are obtained in a classical way after a (long) systematic development.[3] They are partially simplified when the $k_\perp \gg k_\parallel$ limit is taken: this limit is fully justified by the arguments developed previously about the weak cascade in the parallel direction. Under these conditions, we finally get:

$$\partial_t \begin{Bmatrix} E_k \\ H_k \end{Bmatrix} = \frac{\epsilon^2 \Omega^2}{4} \sum_{ss_p s_q} \int_{\Delta_\perp} \frac{sk_\parallel s_p p_\parallel}{k_\perp^2 p_\perp^2 q_\perp^2} \quad (6.22)$$

$$\times \left(\frac{s_q q_\perp - s_p p_\perp}{\omega_k} \right)^2 (sk_\perp + s_p p_\perp + s_q q_\perp)^2 \sin\theta$$

$$\times \begin{Bmatrix} E_q(p_\perp E_k - k_\perp E_p) + (p_\perp s H_k/k_\perp - k_\perp s_p H_p/p_\perp)s_q H_q/q_\perp \\ sk_\perp \left[E_q(p_\perp s H_k/k_\perp - k_\perp s_p H_p/p_\perp) + (p_\perp E_k - k_\perp E_p)s_q H_q/q_\perp \right] \end{Bmatrix}$$

$$\times \delta(s\omega_k + s_p \omega_p + s_q \omega_q) \delta(k_\parallel + p_\parallel + q_\parallel) dp_\perp dq_\perp dp_\parallel dq_\parallel ,$$

with $E_k \equiv E(k_\perp, k_\parallel) = 2\pi \mathbf{k}_\perp E(\mathbf{k}_\perp, k_\parallel)$ and $H_k \equiv H(k_\perp, k_\parallel) = 2\pi \mathbf{k}_\perp H(\mathbf{k}_\perp, k_\parallel)$ the axisymmetric energy and kinetic helicity spectra, respectively. In this expression, θ is the angle between the wavevectors $\mathbf{k}_\perp$ and $\mathbf{p}_\perp$ in the triangle $\mathbf{k}_\perp + \mathbf{p}_\perp + \mathbf{q}_\perp = \mathbf{0}$ and $\Delta_\perp$ is the integration domain (infinitely extended band) corresponding to this triangle (see Figure 5.5). The anisotropic kinetic equations of inertial wave turbulence (6.22) have been obtained and analyzed for the first time by Galtier (2003). They were then studied numerically by Bellet et al. (2006), and then found analytically with a Hamiltonian method by Gelash et al. (2017).

---

[3] However, an additional step must be considered: it consists in introducing a local orthonormal vectorial basis for each triad (Galtier, 2014) in order to make all symmetries appear. It is only after that the statistical development can take place.

After applying the generalized Zakharov transformation to the axisymmetric case (Kuznetsov, 1972), one obtains the exact solutions with a non-zero constant energy flux (Kolmogorov–Zakharov spectrum):[4]

$$\boxed{E(k_\perp, k_\parallel) \sim k_\perp^{-5/2} k_\parallel^{-1/2}}, \tag{6.23}$$

and $H(k_\perp, k_\parallel) \sim k_\perp^{-3/2} k_\parallel^{-1/2}$. These solutions correspond to a direct cascade of energy. In particular, this means that the spectrum of kinetic helicity is not the consequence of a dynamics specific to helicity, but the trace of a dynamics induced by energy. These solutions are in the domain of convergence of integrals, for which a subtle calculation is necessary (which mixes the perpendicular and parallel modes). With the following definition $E \sim k_\perp^x k_\parallel^y$, we find the conditions of locality (see Exercise II.4):

$$-4 < x + 2y < -3, \tag{6.24}$$
$$-4 < x + y < -2. \tag{6.25}$$

This property demonstrates that the Kolmogorov–Zakharov spectrum (6.23) is relevant in describing this regime. As can be seen, it is placed exactly in the middle of the two intervals. Numerical simulations of the kinetic equations of wave turbulence performed by Bellet et al. (2006) show a spectral behavior compatible with, in particular, the prediction (6.23). Note that an anisotropic spectrum compatible with the Kolmogorov–Zakharov solution has been obtained recently with three-dimensional direct numerical simulations (Le Reun et al., 2020; Yokoyama and Takaoka, 2021).

Several other properties of inertial wave turbulence can be deduced from expression (6.22). First, we see that a state with zero kinetic helicity will not produce helicity at any scale. Energy is therefore the main driver of turbulence. Second, we observe that there is no nonlinear coupling when the wavevectors $\mathbf{p}_\perp$ and $\mathbf{q}_\perp$ are collinear (because then $\sin\theta = 0$). Third, there is no nonlinear coupling when $p_\perp$ and $q_\perp$ are equal, if at the same time their polarity s_p and s_q are equal. This is a property that seems to be quite general, since it is common to other types of helical waves (Kraichnan, 1973; Waleffe, 1992; Turner, 2000; Galtier and Bhattacharjee, 2003). Finally, we recall that these kinetic equations cannot describe the slow mode ($k_\parallel = 0$) and cannot describe too large three-dimensional modes (especially in $k_\perp$) for which the wave turbulence becomes strong. Kinetic equations are thus limited to a finite size domain in Fourier space. Moreover, the exact solutions obtained are restricted to a part of this authorized domain for which $k_\perp \gg k_\parallel$.

---

[4] The expressions of the energy flux components can be deduced under the assumption of statistical axisymmetry. For this, we use the relation $\partial_t E(\mathbf{k}_\perp, k_\parallel) = -\nabla \cdot \mathbf{\Pi}(\mathbf{k}_\perp, k_\parallel)$ in cylindrical coordinates from which we can show that $\Pi_\parallel / \Pi_\perp \propto k_\parallel / k_\perp$. Therefore, the parallel energy flux is expected to be smaller than the perpendicular flux since the kinetic equation used is valid in the limit $k_\parallel / k_\perp \ll 1$.

6.6 Local Triadic Interactions

6.6.1 Nonlinear Diffusion Equation

The kinetic equations of inertial wave turbulence can be simulated numerically to study more precisely this regime. However, the work on the subject (Bellet et al., 2006) shows that the problem remains difficult because it requires, in particular, solving properly the resonance condition, which strongly limits the number of triads to be considered. One way to get around this difficulty is to take into account only local interactions, which leads to a significant simplification of the equations. In our case, this limit concerns above all perpendicular wavenumbers. We will see later that in fact it is not possible to consider this limit for parallel wavenumbers.

In this study, we will limit ourselves to the particular case of energy, and we assume that the kinetic helicity is null. We have seen that within the highly anisotropic limit ($k_\perp \gg k_\parallel$), the kinetic equation of inertial wave turbulence takes the following form:

$$\frac{\partial E_k}{\partial t} = \sum_{ss_ps_q} \int T^{ss_ps_q}_{\mathbf{kpq}} dp_\perp dq_\perp dp_\parallel dq_\parallel, \qquad (6.26)$$

with by definition $T^{ss_ps_q}_{\mathbf{kpq}}$ the transfer function per mode.

$$T^{ss_ps_q}_{\mathbf{kpq}} = \frac{\Omega^2}{4} \frac{sk_\parallel s_p p_\parallel}{k_\perp^2 p_\perp^2 q_\perp^2} \left(\frac{s_q q_\perp - s_p p_\perp}{\omega_k}\right)^2 (sk_\perp + s_p p_\perp + s_q q_\perp)^2 \sin\theta$$
$$E_q(p_\perp E_k - k_\perp E_p)\delta(s\omega_k + s_p\omega_p + s_q\omega_q)\,\delta(k_\parallel + p_\parallel + q_\parallel). \qquad (6.27)$$

It can be noted that the small parameter ϵ has been absorbed in the time derivative and therefore no longer appears explicitly: this means that we are focusing on the long times of wave turbulence. The transfer function verifies the following symmetry property, which will be used later:

$$T^{s_p s s_q}_{\mathbf{pkq}} = -T^{ss_ps_q}_{\mathbf{kpq}}. \qquad (6.28)$$

Within the limit of strongly local interactions, we can write:

$$p_\perp = k_\perp(1 + \epsilon_p) \quad \text{and} \quad q_\perp = k_\perp(1 + \epsilon_q), \qquad (6.29)$$

with $0 < \epsilon_p \ll 1$ and $0 < \epsilon_q \ll 1$. We can then introduce an arbitrary function $f(k_\perp, k_\parallel)$ and integrate the kinetic equation; we get:

$$\frac{\partial}{\partial t}\left(\int f(k_\perp, k_\parallel)E_k dk_\perp dk_\parallel\right) = \sum_{ss_ps_q}\int f(k_\perp, k_\parallel)T^{ss_ps_q}_{\mathbf{kpq}} dk_\perp dk_\parallel dp_\perp dq_\perp dp_\parallel dq_\parallel$$
$$= \frac{1}{2}\sum_{ss_ps_q}\int [f(k_\perp, k_\parallel) - f(p_\perp, p_\parallel)]$$
$$\times T^{ss_ps_q}_{\mathbf{kpq}} dk_\perp dk_\parallel dp_\perp dq_\perp dp_\parallel dq_\parallel. \qquad (6.30)$$

For local interactions, we have at the main order (we neglect the contribution of the parallel wavenumber):

$$f(p_\perp, p_\|) = f(k_\perp, k_\|) + (p_\perp - k_\perp)\frac{\partial f(k_\perp, k_\|)}{\partial k_\perp}$$
$$= f(k_\perp, k_\|) + \epsilon_p k_\perp \frac{\partial f(k_\perp, k_\|)}{\partial k_\perp}. \quad (6.31)$$

One obtains:

$$\frac{\partial}{\partial t}\left(\int f(k_\perp, k_\|)E_k dk_\perp dk_\|\right) = -\frac{1}{2}\sum_{s s_p s_q}\int \epsilon_p k_\perp \frac{\partial f(k_\perp, k_\|)}{\partial k_\perp} \quad (6.32)$$
$$\times T_{\mathbf{kpq}}^{s s_p s_q} dk_\perp dk_\| dp_\perp dq_\perp dp_\| dq_\|.$$

An integration by part of the right-hand term allows us to write (after simplification):

$$\frac{\partial E_k}{\partial t} = \frac{1}{2}\frac{\partial}{\partial k_\perp}\left(\sum_{s s_p s_q}\int \epsilon_p k_\perp T_{\mathbf{kpq}}^{s s_p s_q} dp_\perp dq_\perp dp_\| dq_\|\right). \quad (6.33)$$

The local form of the transfer function $T_{\mathbf{kpq}}^{s s_p s_q}$ can be deduced by using the locality in the perpendicular direction. In particular, we have:

$$k_\perp^2 p_\perp^2 q_\perp^2 = k_\perp^6, \quad (6.34)$$

$$\left(\frac{s_q q_\perp - s_p p_\perp}{\omega_k}\right)^2 = \left(\frac{s_q - s_p + s_q \epsilon_q - s_p \epsilon_p}{2\Omega k_\|}\right)^2 k_\perp^4, \quad (6.35)$$

$$(sk_\perp + s_p p_\perp + s_q q_\perp)^2 = (s + s_p + s_q)^2 k_\perp^2, \quad (6.36)$$

$$E_q(p_\perp E_k - k_\perp E_p) = -\epsilon_p k_\perp^3 E_k \frac{\partial(E_k/k_\perp)}{\partial k_\perp}, \quad (6.37)$$

$$\sin\theta = \sin(\pi/3) = \frac{\sqrt{3}}{2}, \quad (6.38)$$

$$\delta(g_{kpq}) = \frac{k_\perp}{2\Omega}\delta(sk_\| + s_p p_\| + s_q q_\|). \quad (6.39)$$

After simplification, we arrive at:

$$T_{\mathbf{kpq}}^{s s_p s_q} = -\frac{\sqrt{3}}{64\Omega}\frac{s s_p p_\|}{k_\|}(s_q - s_p + s_q \epsilon_q - s_p \epsilon_p)^2 (s + s_p + s_q)^2 \epsilon_p k_\perp^4$$
$$E_k \frac{\partial(E_k/k_\perp)}{\partial k_\perp}\delta(sk_\| + s_p p_\| + s_q q_\|)\delta(k_\| + p_\| + q_\|). \quad (6.40)$$

The previous expression shows us that the transfer will be dominant when $s_p s_q = -1$. Therefore, we will only consider this type of interaction. The expression is then reduced to:

$$T_{\mathbf{kpq}}^{s s_p s_q} = -\frac{\sqrt{3}}{16\Omega}\frac{s s_p p_\|}{k_\|}\epsilon_p k_\perp^4 E_k \frac{\partial(E_k/k_\perp)}{\partial k_\perp}\delta(sk_\| + s_p p_\| - s_p q_\|)\delta(k_\| + p_\| + q_\|). \quad (6.41)$$

The resonance condition leads us to two possible combinations for parallel wavenumbers:

$$k_\| + p_\| - q_\| = 0 \quad \text{and} \quad k_\| + p_\| + q_\| = 0, \tag{6.42}$$

or

$$k_\| - p_\| + q_\| = 0 \quad \text{and} \quad k_\| + p_\| + q_\| = 0. \tag{6.43}$$

The solution is either $q_\| = 0$ or $p_\| = 0$, which means that the strong locality assumption does not apply for the parallel direction. The second solution cancels the transfer function, so we will consider only the first solution, for which we obtain:

$$\frac{\partial E_k}{\partial t} = \frac{\sqrt{3}}{16\Omega} \frac{\partial}{\partial k_\perp} \left(k_\perp^7 E_k \frac{\partial (E_k/k_\perp)}{\partial k_\perp} \right) \int_{-\epsilon}^{+\epsilon} \epsilon_p^2 d\epsilon_p \int_{-\epsilon}^{+\epsilon} d\epsilon_q. \tag{6.44}$$

After integration, we finally arrive at the following nonlinear diffusion equation for energy:

$$\boxed{\frac{\partial E_k}{\partial t} = C \frac{\partial}{\partial k_\perp} \left(k_\perp^7 E_k \frac{\partial (E_k/k_\perp)}{\partial k_\perp} \right)}, \tag{6.45}$$

with $C = \epsilon^4/(4\sqrt{3}\Omega)$. This nonlinear diffusion equation describes inertial wave turbulence in the limit of strongly local interactions in the perpendicular direction (Galtier and David, 2020). It is an equation that has been rigorously deduced from the kinetic equation. Note that it is also possible to obtain this diffusion equation by phenomenological arguments. The calculation is then simpler, but it does not allow us to obtain the exact expression of the constant C. Consequently, we will renormalize the time and take $C = 1$.

6.6.2 Stationary and Self-Similar Solutions

We will verify that the nonlinear energy diffusion equation (6.45) has the same exact solutions as the kinetic equation (6.22). To do this, we introduce into equation (6.45) the energy flux $\Pi(k_\perp)$, which is defined by the relation:

$$\frac{\partial E(k_\perp)}{\partial t} = -\frac{\partial \Pi(k_\perp)}{\partial k_\perp}, \tag{6.46}$$

as well as the kinetic energy spectrum $E(k_\perp) = A k_\perp^x$. By definition, A is a positive constant because the energy spectrum is a positive definite quantity. We obtain:

$$\Pi(k_\perp) = A^2 (1-x) k_\perp^{5+2x}. \tag{6.47}$$

Constant energy flux solutions are therefore $x = 1$ and $x = -5/2$. The first value cancels the flux and corresponds to the thermodynamic solution. The second value corresponds to the spectrum discussed in Section 6.5: it is the non-zero constant flux solution (Kolmogorov–Zakharov spectrum). In this case, we can also show that:

$$\lim_{x \to -5/2} \Pi(k_\perp) = \varepsilon = \frac{7}{2}A^2, \qquad (6.48)$$

which is positive, as expected for a direct cascade.

The stationary solution – known as the Kolmogorov–Zakharov spectrum – is sometimes preceded by a nonstationary transient solution of a different nature. We will see, in the numerical study in the Section 6.6.3, that this is the case for inertial wave turbulence. It is a self-similar solution whose form is:

$$E_k = \frac{1}{\tau^a} E_0 \left(\frac{k_\perp}{\tau^b} \right), \qquad (6.49)$$

with $\tau = t_* - t$ and t_* the time it takes for the spectrum to reach the largest available wavenumber. This time is finite, which means that in principle the front of the energy spectrum can reach infinity in a finite amount of time. This property reminds us of the discussion in Chapter 2 on the emergence of a singularity in finite time. By introducing the top expression into equation (6.45), one obtains the condition:

$$\boxed{a = 4b + 1}. \qquad (6.50)$$

A second condition can appear, assuming that $E_0(\xi) \sim \xi^m$ behind the front. The stationarity condition then gives us the relation:

$$\boxed{a + mb = 0}. \qquad (6.51)$$

The combination of the two relationships finally gives us:

$$m = -\frac{a}{b} = -4 - \frac{1}{b}. \qquad (6.52)$$

This last expression means that we have a direct relation between the exponent m of the power-law spectrum and the propagation law of the front $k_f \sim \tau^b$. For example, if we assume that the stationary solution is established immediately when the front propagates, then $m = -5/2$ and thus $b = -2/3$ (and $a = -5/3$). In this case, the prediction for the front propagation is:

$$k_f \sim (t_* - t)^{-2/3}. \qquad (6.53)$$

In practice, we will see that the nonstationary solution is different: it verifies well the conditions (6.51)–(6.52), but the values of a, b, m are not predictable. We then speak of a self-similar solution of the second kind (only solutions of the first kind are predictable).

6.6.3 Numerical Study

The numerical study of equation (6.45) does not pose any particular problem in terms of implementation; however, it is necessary to consider a viscous term to dissipate energy at small scales in order to avoid the emergence of numerical instability. We make the choice of a hyperviscosity to extend the inertial range. Therefore, the normalized equation with $C = 1$ that we are going to simulate is:

$$\frac{\partial E_k}{\partial t} = \frac{\partial}{\partial k_\perp} \left(k_\perp^7 E_k \frac{\partial (E_k/k_\perp)}{\partial k_\perp} \right) - \nu k_\perp^6 E_k. \quad (6.54)$$

One can use a logarithmic discretization for the $k_\perp$ axis, with $k_{\perp i} = 2^{i/8}$ and i an integer between 0 and 160. The Crank–Nicolson and Adams–Bashforth numerical schemes are used for the linear and nonlinear terms, respectively. The initial condition ($t = 0$) corresponds to a spectrum localized at large-scale such that:

$$E_k \sim k_\perp^3 \exp(-(k_\perp/k_0)^2), \quad (6.55)$$

with $k_0 = 5$. No forcing is introduced in this simulation and the time step is $\Delta T = 2 \times 10^{-13}$.

Figure 6.4 shows (top) the evolution of the energy spectrum from $t = 0$ to t_*. During this nonstationary phase a power law in $k_\perp^{-8/3}$ appears progressively over about three decades. This nonstationary phase is characterized by a nonconstant energy flux (bottom of Figure 6.4). Therefore, the spectrum does not correspond to the constant flux solution obtained analytically. On the other hand, it is compatible with the solution $\sim k_\perp^{-1/3}$ when we take $x = -8/3$ in equation (6.47).

In order to check if these results correspond to the self-similar solution introduced previously, we show on Figure 6.5 the temporal evolution of the front $k_f(t)$. This spectral front is defined by taking as a reference $E(k_\perp) = 10^{-15}$ in Figure 6.4. We then follow the point of intersection between this threshold and the spectral tail. From this figure, we can define the singular time t_* that the front takes to reach the maximum wavenumber (in principle $k_\perp = +\infty$). We get approximately $t_* = 6.7537 \times 10^{-7}$. In Figure 6.5 (insert) we see k_f as a function of $t_* - t$: a very clear power law appears over three decades, with an exponent close to -0.750. The negative exponent value illustrates the explosive nature of the direct energy cascade. These values are well compatible with:

$$\boxed{a = -2, \quad b = -3/4, \quad \text{and} \quad m = -8/3}, \quad (6.56)$$

which demonstrates the self-similar nature of the nonstationary solution. Note that the power law in $-8/3$ was apparently obtained with a numerical simulation of the kinetic equations of inertial wave turbulence when the spectrum is in a phase of development and before it is affected by inhomogeneities (Eremin, 2019).

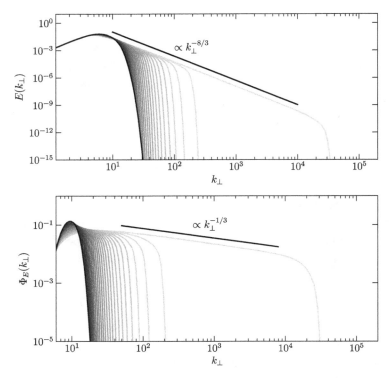

Figure 6.4 Top: Temporal evolution (every $4 \times 10^3 \Delta T$) of the energy spectrum between $t = 0$ (large-scale localized spectrum) and t_*; a spectrum in $k_\perp^{-8/3}$ emerges over about three decades. Bottom: The spectrum of the energy flux (at the same time) is compatible with $k_\perp^{-1/3}$. Simulation made by V. David.

In Figure 6.6 we show the temporal evolutions of the energy and flux (insertion) spectra for $t > t_*$, with t close to t_* (top) and $t \gg t_*$ (bottom). The stationary solution is finally obtained with a spectrum in $k_\perp^{-5/2}$ and a constant flux. We can notice that the stationary solution is established following a bounce of the spectrum when it reaches the smallest scales. We then have a relatively slow self-similar decay of the stationary solutions.

6.6.4 Universality of the Anomalous Scaling

The behavior that we have just described in detail for inertial wave turbulence is qualitatively universal for the solutions with finite capacity. We recall that these solutions correspond to spectra in power law whose integral (from the initial excitation mode k_i, to infinity for a direct cascade, or zero for an inverse cascade) is convergent. In the case of a direct cascade of energy, this means that the system can redistribute the initial energy in the neighborhood of k_i, over the interval $k_i < k < +\infty$ in a finite time (we assume that there is no external force).

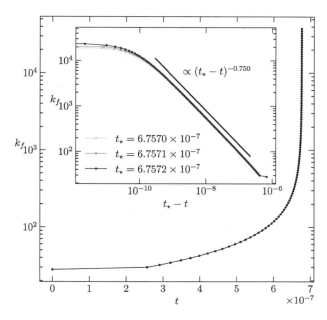

Figure 6.5 Temporal evolution of the front k_f of the spectrum for $t \leq t_*$ (semilogarithmic scale). A sudden increase of k_f is observed from which one can precisely define the time $t_* = 6.7537 \times 10^{-7}$. Insert: temporal evolution of k_f as a function of $t_* - t$ (logarithmic scale). The dashed line corresponds to $(t_* - t)^{-0.750}$. For comparison two other values of $t_* - t$ are taken. Simulation made by V. David.

The discovery of this anomalous scaling is relatively recent in the history of turbulence. The first detection was carried out within the framework of the study of a Bose condensate process (here an inverse cascade) with an exponent around -1.24 instead of $-7/6$ for the stationary solution (Semikoz and Tkachev, 1995). A second detection was performed in magnetohydrodynamics (see Chapter 7) with an exponent for the energy spectrum in $-7/3$ instead of -2 (Galtier et al., 2000). In the case of gravitational waves (see Chapter 9), the difference between these exponents for the wave action is tiny: -0.6517 instead of $-2/3$. Spectral anomalous scalings are also detected in strong turbulence. For example, using a nonlinear diffusion model to describe three-dimensional incompressible hydrodynamic turbulence (Leith, 1967), it is possible to show numerically that the Kolmogorov spectrum in $-5/3$ is preceded by a nonstationary spectrum around -1.856 (Connaughton and Nazarenko, 2004).

To date, an exact prediction for the anomalous exponent seems out of reach (Grebenev et al., 2014). Only the numerical simulation can help us to estimate it, for example, by framing the value (Thalabard et al., 2015). As we have seen, this nonstationary solution is followed by a bounce of the spectrum to finally form the stationary solution. We can show that this bounce is described by a self-similar solution of the third kind (Nazarenko and Grebenev, 2017). We can ask whether

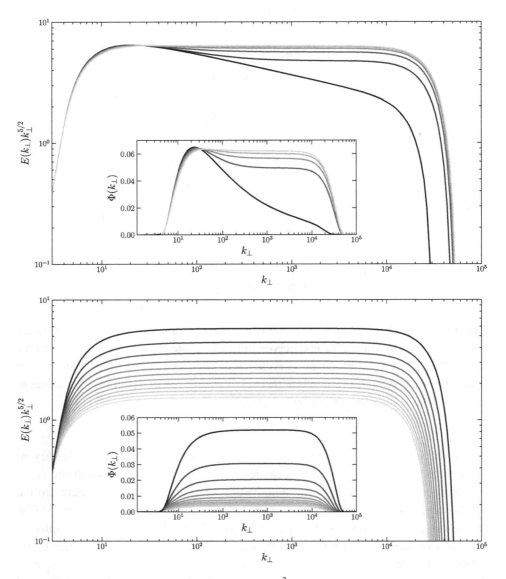

Figure 6.6 Top: Temporal evolution (every $4 \times 10^3 \Delta T$) for $t > t_*$ of the energy spectrum compensated by $k_\perp^{5/2}$, and (insert) the temporal evolution of the energy flux at the same time (semilogarithmic scale). Bottom: Final temporal evolution of these quantities for $t \gg t_*$. Simulation made by V. David.

there is a quantitative relationship between these anomalous exponents and the possible appearance of singularities in a finite time (see Chapter 2). The answer to this question offers an interesting perspective.

6.7 Perspectives

Recently, several experimental studies have been devoted to rotating turbulence (see, e.g., Le Reun et al., 2017). An interesting perspective is the experimental reproduction of the inertial wave turbulence regime (Yarom and Sharon, 2014; Monsalve et al., 2020) and an understanding of, in particular, the role of the slow mode. Experiments show a first difficulty: that of avoiding the too fast excitation of the slow mode which corresponds to geostrophic structures. In many laboratory experiments dedicated to rotating turbulence, the slow mode is often excited by the forcing. Now we know that this mode can become dominant and prevent the detection of three-dimensional modes. One possible approach is to use an inertia wave attractor device from which the desired regime could develop by instability (Brunet et al., 2020). The underlying fundamental question is to understand by which mechanism the slow mode can be excited and how this can, in turn, modify the dynamics of wave turbulence. From a theoretical point of view, the study of quasi-resonant interactions as an efficient source of slow mode excitation is an interesting perspective (Clark di Leoni and Mininni, 2016).

A new question concerns the proximity between inertial wave turbulence and a seemingly very different problem, that of kinetic Alfvén wave turbulence, a relevant regime in understanding the solar wind at subionic scales (see Chapter 8). Indeed, in the approximation of local interactions we can show analytically that the two problems are described by the same nonlinear diffusion equation (Galtier and David, 2020). It is then perhaps not so surprising to note that experimental measurements of intermittency in the case of rotation (van Bokhoven et al., 2009) and in the solar wind (Kiyani et al., 2009) give very close self-similar exponents. Is this a pure coincidence or is there really a common physics? We have here, potentially, an example of close interdisciplinarity.

Another interesting project would be to highlight the transition between the weak and strong wave turbulence regimes that is expected to occur at small scales. As predicted by relation (6.15), the time ratio increases with the mode $k_\perp$. According to the phenomenology of Nazarenko and Schekochihin (2011), one can expect to fall into the regime of strong wave turbulence (described by the critical balance – see Chapter 7), then, on an even smaller scale, to no longer feel the rotation effect and find the fluid in the classic regime of strong eddy isotropic turbulence (Chapter 2). As shown by the numerical study of Mininni et al. (2012) (see also the work of Meyrand et al., 2016 in magnetohydrodynamics), the measurement of a such a transition requires very high spatial resolution, at the limits of current computing facilities. It also requires measuring precisely the energy flux vector in anisotropic turbulence (Yokoyama and Takaoka, 2021).

References

Aldridge, K. D., and Lumb, L. I. 1987. Inertial waves identified in the earth's fluid outer core. *Nature*, **325**(6103), 421–423.

Alexakis, A., and Biferale, L. 2018. Cascades and transitions in turbulent flows. *Phys. Rep.*, **767**, 1–101.

Bardina, J., Ferziger, J. H., and Rogallo, R. S. 1985. Effect of rotation on isotropic turbulence: Computation and modelling. *J. Fluid Mech.*, **154**, 321–336.

Baroud, C. N., Plapp, B. B., She, Z. S., and Swinney, H. L. 2002. Anomalous self-similarity in a turbulent rapidly rotating fluid. *Phys. Rev. Lett.*, **88**(11), 114501.

Bellet, F., Godeferd, F. S., Scott, J. F., and Cambon, C. 2006. Wave turbulence in rapidly rotating flows. *J. Fluid Mech.*, **562**, 83–121.

Bourouiba, L., Straub, D. N., and Waite, M. L. 2012. Non-local energy transfers in rotating turbulence at intermediate Rossby number. *J. Fluid Mech.*, **690**, 129–147.

Brunet, M., Gallet, B., and Cortet, P.-P. 2020. Shortcut to geostrophy in wave-driven rotating turbulence: The quartetic instability. *Phys. Rev. Lett.*, **124**(12), 124501.

Buzzicotti, M., Aluie, H., Biferale, L., and Linkmann, M. 2018. Energy transfer in turbulence under rotation. *Phys. Rev. Fluids*, **3**(3), 034802.

Cambon, C., and Jacquin, L. 1989. Spectral approach to non-isotropic turbulence subjected to rotation. *J. Fluid Mech.*, **202**, 295–317.

Cambon, C., Mansour, N. N., and Godeferd, F. S. 1997. Energy transfer in rotating turbulence. *J. Fluid Mech.*, **337**(1), 303–332.

Campagne, A., Gallet, B., Moisy, F., and Cortet, P.-P. 2014. Direct and inverse energy cascades in a forced rotating turbulence experiment. *Phys. Fluids*, **26**(12), 125112.

Clark di Leoni, P., Cobelli, P. J., Mininni, P. D., Dmitruk, P., and Matthaeus, W. H. 2014. Quantification of the strength of inertial waves in a rotating turbulent flow. *Phys. Fluids*, **26**(3), 035106.

Clark di Leoni, P., and Mininni, P. D. 2016. Quantifying resonant and near-resonant interactions in rotating turbulence. *J. Fluid Mech.*, **809**, 821–842.

Connaughton, C., and Nazarenko, S. 2004. Warm cascades and anomalous scaling in a diffusion model of turbulence. *Phys. Rev. Lett.*, **92**(4), 044501.

Craya, A. 1954. Contribution à l'analyse de la turbulence associée à des vitesses moyennes. *PST Ministère de l'Air*, **345**.

Eremin, A. 2019. *Implémentation numérique de la fermeture de turbulence d'ondes dans un canal rotatif*. Thesis, University of Lyon.

Galtier, S. 2003. Weak inertial-wave turbulence theory. *Phys. Rev. E*, **68**(1), 015301.

Galtier, S. 2009. Exact vectorial law for homogeneous rotating turbulence. *Phys. Rev. E*, **80**(4), 046301.

Galtier, S. 2014. Weak turbulence theory for rotating magnetohydrodynamics and planetary flows. *J. Fluid Mech.*, **757**, 114–154.

Galtier, S., and Bhattacharjee, A. 2003. Anisotropic weak whistler wave turbulence in electron magnetohydrodynamics. *Phys. Plasmas*, **10**(8), 3065–3076.

Galtier, S., and David, V. 2020. Inertial/kinetic-Alfvén wave turbulence: A twin problem in the limit of local interactions. *Phys. Rev. Fluids*, **5**(4), 044603.

Galtier, S., Nazarenko, S. V., Newell, A. C., and Pouquet, A. 2000. A weak turbulence theory for incompressible magnetohydrodynamics. *J. Plasma Phys.*, **63**, 447–488.

Gelash, A. A., L'vov, V. S., and Zakharov, V. E. 2017. Complete Hamiltonian formalism for inertial waves in rotating fluids. *J. Fluid Mech.*, **831**, 128–150.

Godeferd, F. S., and Moisy, F. 2015. Structure and dynamics of rotating turbulence: A review of recent experimental and numerical Results. *Appl. Mech. Rev.*, **67**(3), 030802.

Grebenev, V. N., Nazarenko, S. V., Medvedev, S. B., Schwab, I. V., and Chirkunov, Y. A. 2014. Self-similar solution in the Leith model of turbulence: Anomalous power law and asymptotic analysis. *J. Physics A Math. General*, **47**(2), 025501.

Hopfinger, E. J., Gagne, Y., and Browand, F. K. 1982. Turbulence and waves in a rotating tank. *J. Fluid Mech.*, **125**, 505–534.

Hossain, M. 1994. Reduction in the dimensionality of turbulence due to a strong rotation. *Phys. Fluids*, **6**(3), 1077–1080.

Jacquin, L., Leuchter, O., Cambon, C., and Mathieu, J. 1990. Homogeneous turbulence in the presence of rotation. *J. Fluid Mech.*, **220**, 1–52.

Kiyani, K. H., Chapman, S. C., Khotyaintsev, Y. V., Dunlop, M. W., and Sahraoui, F. 2009. Global scale-invariant dissipation in collisionless plasma turbulence. *Phys. Rev. Lett.*, **103**, 075006.

Kraichnan, R. H. 1967. Inertial ranges in two-dimensional turbulence. *Phys. Fluids*, **10**(7), 1417–1423.

Kraichnan, R. H. 1973. Helical turbulence and absolute equilibrium. *J. Fluid Mech.*, **59**, 745–752.

Kuznetsov, E. A. 1972. Turbulence of ion sound in a plasma located in a magnetic field. *J. Exp. Theor. Phys.*, **35**, 310–314.

Lamriben, C., Cortet, P.-P., and Moisy, F. 2011. Direct measurements of anisotropic energy transfers in a rotating turbulence experiment. *Phys. Rev. Lett.*, **107**(2), 024503.

Le Reun, T., Favier, B., Barker, A. J., and Le Bars, M. 2017. Inertial wave turbulence driven by elliptical instability. *Phys. Rev. Lett.*, **119**(3), 034502.

Le Reun, T., Favier, B., and Le Bars, M. 2020. Evidence of the Zakharov–Kolmogorov spectrum in numerical simulations of inertial wave turbulence. *EPL (Europhysics Letters)*, **132**(6), 64002.

Leith, C. E. 1967. Diffusion approximation to inertial energy transfer in isotropic turbulence. *Phys. Fluids*, **10**(7), 1409–1416.

Meyrand, R., Galtier, S., and Kiyani, K. H. 2016. Direct evidence of the transition from weak to strong magnetohydrodynamic turbulence. *Phys. Rev. Lett.*, **116**(10), 105002.

Mininni, P. D., Rosenberg, D., and Pouquet, A. 2012. Isotropization at small scales of rotating helically driven turbulence. *J. Fluid Mech.*, **699**, 263–279.

Moisy, F., Morize, C., Rabaud, M., and Sommeria, J. 2011. Decay laws, anisotropy and cyclone–anticyclone asymmetry in decaying rotating turbulence. *J. Fluid Mechanics*, **666**, 5–35.

Monsalve, E., Brunet, M., Gallet, B., and Cortet, P.-P. 2020. Quantitative experimental observation of weak inertial-wave turbulence. *Phys. Rev. Lett.*, **125**, 254502.

Morize, C., Moisy, F., and Rabaud, M. 2005. Decaying grid-generated turbulence in a rotating tank. *Phys. Fluids*, **17**(9), 095105–095105–11.

Nazarenko, S. 2011. *Wave Turbulence*. Lecture Notes in Physics, vol. 825. Springer Verlag.

Nazarenko, S. V., and Grebenev, V. N. 2017. Self-similar formation of the Kolmogorov spectrum in the Leith model of turbulence. *J. Physics A Math. General*, **50**(3), 035501.

Nazarenko, S. V., and Schekochihin, A. A. 2011. Critical balance in magnetohydrodynamic, rotating and stratified turbulence: Towards a universal scaling conjecture. *J. Fluid Mech.*, **677**, 134–153.

Orszag, S. A. 1970. Analytical theories of turbulence. *J. Fluid Mech.*, **41**, 363–386.

Pouquet, A., and Mininni, P. D. 2010. The interplay between helicity and rotation in turbulence: Implications for scaling laws and small-scale dynamics. *Phil. Trans. Royal Soc. London Series A*, **368**(1916), 1635–1662.

Scott, J. F. 2014. Wave turbulence in a rotating channel. *J. Fluid Mech.*, **741**, 316–349.

Semikoz, D. V., and Tkachev, I. I. 1995. Kinetics of Bose condensation. *Phys. Rev. Lett.*, **74**(16), 3093–3097.

Seshasayanan, K., and Alexakis, A. 2018. Condensates in rotating turbulent flows. *J. Fluid Mech.*, **841**, 434–462.

Sharma, M. K., Verma, M. K., and Chakraborty, S. 2018. On the energy spectrum of rapidly rotating forced turbulence. *Phys. Fluids*, **30**(11), 115102.

Smith, L. M., and Waleffe, F. 1999. Transfer of energy to two-dimensional large scales in forced, rotating three-dimensional turbulence. *Phys. Fluids*, **11**(6), 1608–1622.

Smith, L. M., Chasnov, J. R., and Waleffe, F. 1996. Crossover from two- to three-dimensional turbulence. *Phys. Rev. Lett.*, **77**(12), 2467–2470.

Thalabard, S., Nazarenko, S., Galtier, S., and Medvedev, S. 2015. Anomalous spectral laws in differential models of turbulence. *J. Physics A Math. General*, **48**(28), 285501.

Turner, L. 2000. Using helicity to characterize homogeneous and inhomogeneous turbulent dynamics. *J. Fluid Mech.*, **408**(1), 205–238.

van Bokhoven, L. J. A., Clercx, H. J. H., van Heijst, G. J. F., and Trieling, R. R. 2009. Experiments on rapidly rotating turbulent flows. *Phys. Fluids*, **21**(9), 096601.

Waleffe, F. 1992. The nature of triad interactions in homogeneous turbulence. *Phys. Fluids A*, **4**(2), 350–363.

Yarom, E., and Sharon, E. 2014. Experimental observation of steady inertial wave turbulence in deep rotating flows. *Nature Physics*, **10**(7), 510–514.

Yokoyama, N., and Takaoka, M. 2021. Energy-flux vector in anisotropic turbulence: Application to rotating turbulence. *J. Fluid Mech.*, **908**, A17.

Zeman, O. 1994. A note on the spectra and decay of rotating homogeneous turbulence. *Phys. Fluids*, **6**(10), 3221–3223.

Zhou, Ye. 1995. A phenomenological treatment of rotating turbulence. *Phys. Fluids*, **7**(8), 2092–2094.

7

Alfvén Wave Turbulence

Magnetohydrodynamics (MHD) is the study of the magnetic properties and behavior of electrically conducting fluids. It is an extension of hydrodynamics that couples the Navier–Stokes equations and Maxwell electrodynamics. MHD fluids are essentially of two kinds: on the one hand, there are liquid metals such as sodium, used, for example, in laboratory experiments to study the mechanism of generation of the magnetic field – the dynamo effect – which occurs naturally in the heart of our planet via the turbulent movements of liquid iron; on the other hand, there are partially ionized gases – called plasmas – whose properties differ deeply from those of neutral gases.

MHD is primarily used in astrophysics because about 99 percent of the visible matter in the Universe consists of plasma. There are many examples of applications: planetary magnetospheres, the Sun, stars, solar and stellar winds, interstellar clouds, accretion disks, and galaxies. Most of these media are very turbulent, with, in the case of the interstellar medium, a turbulent Mach number often of several dozen (see Chapter 2). Controlled thermonuclear fusion is the second-best-known field of application of MHD, with the famous ITER (International Thermonuclear Experimental Reactor) reactor in Cadarache. Indeed, the control of a magnetically confined plasma requires an understanding of large-scale equilibria and the solution of stability problems whose theoretical framework is basically MHD. Here, turbulence is also detected, but by nature it is inhomogeneous and non-MHD. Tokamak turbulence is, however, considered harmful because it is an obstacle to plasma confinement (Fujisawa, 2021).

Incompressible MHD waves – called Alfvén waves – play an important role in many processes because they are carried by the large-scale magnetic field that structures the medium. Figure 7.1 displays solar coronal loops: these structures are interpreted as magnetic loops along which Alfvén waves propagate and where the regime of Alfvén wave turbulence is expected. The objective of this chapter is to give the main properties of Alfvén wave turbulence, which bears a certain resemblance to inertial wave turbulence (Chapter 6) with a strongly anisotropic

Figure 7.1 Coronal loops along which Alfvén waves propagate. Observation made on June 6, 1999 at the wavelength 17.1 nm by the TRACE/NASA space telescope. Credits: M. Aschwanden et al. (LMSAL), TRACE, NASA.

dynamics. Some properties of incompressible MHD will be exposed, including the critical balance phenomenology that describes the strong wave turbulence regime. We will assume the reader is familiar with the MHD equations; otherwise, we recommend reading Galtier (2016), where the same notations are used.

7.1 Incompressible MHD

We will consider the incompressible MHD equations. In this case, the mass density ρ_0 is constant and the magnetic field $\mathbf{B}$ can be expressed dimensionally as a velocity field $\mathbf{b}$ with the normalization $\mathbf{b} = \mathbf{B}/\sqrt{\mu_0 \rho_0}$ (μ_0 is the magnetic permeability of vacuum). The MHD equations are then written as (Alfvén, 1942):[1]

$$\nabla \cdot \mathbf{u} = 0, \tag{7.1a}$$

$$\frac{\partial \mathbf{u}}{\partial t} + \mathbf{u} \cdot \nabla \mathbf{u} = -\nabla P_* + \mathbf{b} \cdot \nabla \mathbf{b} + \nu \Delta \mathbf{u}, \tag{7.1b}$$

$$\frac{\partial \mathbf{b}}{\partial t} + \mathbf{u} \cdot \nabla \mathbf{b} = \mathbf{b} \cdot \nabla \mathbf{u} + \eta \Delta \mathbf{b}, \tag{7.1c}$$

---

[1] The MHD theory was proposed by the Swedish astrophysicist H. Alfvén, who received the Nobel Prize in Physics in 1970 "for fundamental works and discoveries in magnetohydrodynamics with fruitful applications in different parts of plasma physics."

$$\nabla \cdot \mathbf{b} = 0, \quad (7.1d)$$

with **u** the velocity, $P_* = P/\rho_0 + b^2/2$ the total pressure, ν the kinematic viscosity, and η the magnetic diffusivity. The induction equation (7.1c) is obtained from Maxwell's equations and Ohm's law; it reflects the dynamics of the magnetic field when the plasma behaves like a monofluid within a nonrelativistic limit. This MHD approximation implies, in particular, plasma electroneutrality. We can refine the description by adding, for example, the Hall term in expression (7.1c) in order to take into account the decoupling between the ions and the electrons: we then obtain Hall MHD (see Chapter 8).

MHD turbulence is characterized by two dimensionless numbers, the classical (kinetic) Reynolds number, R_e, and the magnetic Reynolds number, R_m, such that:

$$R_e = \frac{UL}{\nu}, \quad R_m = \frac{UL}{\eta}, \quad (7.2)$$

where U and L are characteristic velocity and macroscopic lengths of the fluid. These numbers are naturally very high in astrophysical media, as we can see in Figure 7.2.

7.2 Strong Alfvén Wave Turbulence

7.2.1 Alfvén Wave Packets

Most astrophysical plasmas evolve in media structured on a large scale by a magnetic field. The solar wind plasma illustrates this situation well, since on

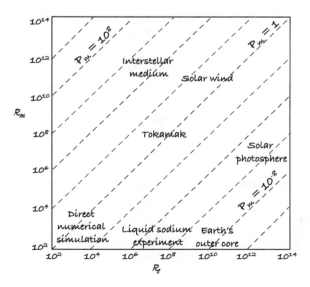

Figure 7.2 Kinetic (R_e) and magnetic (R_m) Reynolds numbers for different media. Diagonals correspond to constant magnetic Prandtl numbers $P_m = \nu/\eta$.

average the interplanetary magnetic field forms a spiral (called Parker's spiral) with an amplitude of the order of fluctuations (see Chapter 8). The Sun with its numerous coronal loops (see Figure 7.1), is another example: in this case, the large-scale magnetic field along the loops is significantly more intense than the fluctuations.

It is legitimate to ask whether MHD turbulence behaves in the same way parallel and perpendicular to a uniform magnetic field $\mathbf{b}_0$. Theoretical, numerical, and observational analyses show that this is not the case: MHD turbulence becomes anisotropic with a reduction of the cascade along the direction of $\mathbf{b}_0$ (see Section 7.4.2). It is therefore fundamental – even phenomenological – to make this distinction in our analysis of MHD turbulence. In this section, we present a model that accounts for this dynamics when $\mathbf{b}_0$ is moderate: it is the critical balance phenomenology proposed by Goldreich and Sridhar (1995) on the basis of a previous study by Higdon (1984) (see Oughton and Matthaeus, 2020, for a critical review on the subject). Basically, it is a phenomenology of strong MHD wave turbulence. Let us note that Phillips (1958) had already proposed a similar conjecture (the author speaks of "equilibrium interval") in the framework of strong gravity wave turbulence.

To introduce this phenomenology, we will write the incompressible MHD equations using Elsässer's variables $\mathbf{z}^\pm \equiv \mathbf{u} \pm \mathbf{b}$ (Elsässer, 1950) and introduce a uniform magnetic field $\mathbf{b}_0$. One obtains (with $\nu = \eta = 0$):

$$\nabla \cdot \mathbf{z}^\pm = 0, \qquad (7.3a)$$

$$\frac{\partial \mathbf{z}^\pm}{\partial t} \mp \mathbf{b}_0 \cdot \nabla \mathbf{z}^\pm = -\nabla P_* - \mathbf{z}^\mp \cdot \nabla \mathbf{z}^\pm . \qquad (7.3b)$$

In this form, MHD equations appear compact and symmetrical. We can immediately see that the linear resolution of these incompressible MHD equations leads to the dispersion relationship for Alfvén waves:

$$\boxed{\omega_k^2 = (\mathbf{k} \cdot \mathbf{b}_0)^2} . \qquad (7.4)$$

We notice that if the field $\mathbf{z}^+ = 0$, the field $\mathbf{z}^-$ is an exact nonlinear solution of the MHD equations (at $\nu = \eta = 0$), since in this case:

$$\frac{\partial \mathbf{z}^-}{\partial t} + \mathbf{b}_0 \cdot \nabla \mathbf{z}^- = 0 . \qquad (7.5)$$

Note that the total pressure is absent because the application of the null divergence condition on the MHD equations shows that P_* is proportional to the nonlinear term. Equation (7.5) is interpreted as an Alfvén wave packet $\mathbf{z}^-$ propagating along the uniform magnetic field $\mathbf{b}_0$, at a speed b_0, without deformation. This reasoning is also valid for $\mathbf{z}^- = 0$ with, in this case, an Alfvén wave packet $\mathbf{z}^+$ propagating in the opposite direction. Therefore, the $\mathbf{z}^\pm$ fields can be interpreted as Alfvén wave packets propagating in opposite directions (see Figure 7.3) and which are

Figure 7.3 Propagation of two Alfvén wave packets along a quasi-uniform magnetic field that defines the parallel direction.

deformed nonlinearly when they collide. Here we touch a fundamental physical property of MHD turbulence: it is the collisions between Alfvén wave packets z^+ and z^- and their successive deformations that produce a cascade. This collision property is intrinsically linked to the nondispersive nature of Alfvén waves: two waves that follow each other will never catch up and will therefore not suffer any nonlinear deformation. Note that the nonlinear interaction between two counter-propagating Alfvén waves has been measured experimentally by Drake et al. (2013), while the first clear experimental detection of an Alfvén wave was realized in 1959 (Allen et al., 1959). We will come back in Section 7.3.1 to this aspect in Alfvén wave turbulence where successive collisions play a fundamental role in the cascade process.

7.2.2 Critical Balance Phenomenology

With the linear term in the left-hand side of equation (7.3b), a new characteristic time appears; it is the Alfvén time:

$$\tau_A \sim \frac{\ell_\parallel}{b_0}, \tag{7.6}$$

which can be interpreted as the collision time of two wave packets z^+ and z^- of length $\ell_\parallel$ (see Figure 7.3). Hereafter, we will assume, to simplify, that MHD turbulence is balanced and that $z^+ \sim z^- \sim z$. The original idea of critical balance is that in the inertial range a natural equilibrium is established between the linear term that carries the Alfvén wave packet and the nonlinear term which corresponds to its deformation by collision, whatever the scale ℓ considered. In other words, it means that the Alfvén time τ_A balances with the nonlinear time τ_{NL}:

$$\tau_A \sim \tau_{NL}, \tag{7.7}$$

and that a significant transfer is achieved after a single collision. With the introduction of anisotropy, the nonlinear time becomes:

$$\tau_{NL} \sim \frac{\ell_\perp}{z_\ell}. \tag{7.8}$$

This means that we consider that the cascade is preferentially in the perpendicular direction. From this new definition and relation (7.7), we get:

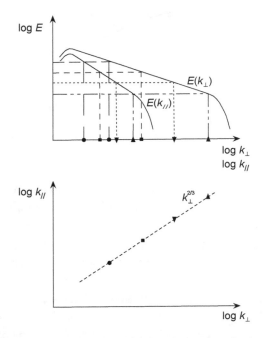

Figure 7.4 The critical balance relationship (7.12) can be obtained from the perpendicular and parallel energy spectra (top). There are two wavenumbers that correspond to each value of energy. Relation (7.12) emerges when these pairs of points are plotted (bottom) – see Bigot et al. (2008b). Reprinted with permission (Galtier, 2016).

$$z_\ell \sim b_0 \frac{\ell_\perp}{\ell_\parallel}. \tag{7.9}$$

Insofar as there is a balance between the only two times available, the phenomenology is reduced to that of Kolmogorov, and one obtains classically:

$$\varepsilon \sim \frac{E_\ell}{\tau_{NL}} \sim \frac{z_\ell^3}{\ell_\perp}. \tag{7.10}$$

The spectrum of critical balance can be deduced directly; it can be written:

$$\boxed{E(k_\perp) \sim k_\perp^{-5/3}}. \tag{7.11}$$

The combination of relations (7.9) and (7.10) finally gives the scaling relation for the critical balance (Goldreich and Sridhar, 1995):

$$\boxed{\ell_\parallel \sim \left(\frac{b_0}{\varepsilon^{1/3}}\right) \ell_\perp^{2/3} \sim \ell_0^{1/3} \ell_\perp^{2/3}}, \tag{7.12}$$

with ℓ_0 the (isotropic) injection scale of energy.

Relation (7.12) expresses the fact that the MHD cascade develops preferentially – but not exclusively – in the transverse direction to the magnetic field $\mathbf{b_0}$

and that this turbulence is more anisotropic at small scales, as shown schematically in Figure 7.4. In practice, the difficulty lies in determining the parallel direction, which may vary slightly depending on the scale on which one is standing. However, we can show by direct numerical simulations that when $\mathbf{b}_0$ is at least three times greater than the fluctuations, this variation is negligible (Bigot et al., 2008c).

The conjecture of critical balance was verified by direct numerical simulations (Cho and Vishniac, 2000; Bigot et al., 2008c; Beresnyak, 2011). From the observational point of view, its verification is more difficult for technical reasons: it requires measurement of the plasma in several directions at the same time, which is only possible if we have several satellites at our disposal (Osman et al., 2011; see, however, Luo and Wu, 2010). The critical balance spectrum is in principle easier to check since one can concentrate on a single transverse direction. In practice, the measurements of the solar wind magnetic field at 1 AU show good agreement with this prediction (see Figure 8.2); however, the velocity fluctuations follow a $-3/2$ exponent power law (Podesta et al., 2007).

To conclude, one can note that the critical balance conjecture has been extended to sub-MHD scales within the electron MHD approximation. It leads to the relation $\ell_\parallel \sim \ell_\perp^{1/3}$ (see Chapter 8). Note that the critical balance is also mentioned in rotating turbulence (Chapter 6) and for gravitational wave turbulence (Chapter 9). In the first case, the critical balance regime is supposed to appear at the smallest scales of the system after a weak turbulence direct cascade of energy, while in the second case it is expected to appear at the largest scales of the system after a weak turbulence inverse cascade of wave action.

7.3 Phenomenology of Wave Turbulence

7.3.1 Iroshnikov–Kraichnan Isotropic Spectrum

The discussion on Alfvén wave packets leads us to introduce one of the major ideas of MHD turbulence: the (IK) phenomenology proposed independently by Iroshnikov (1964) and Kraichnan (1965). This approach is interesting for this part of the book, since the ideas that emerged from this phenomenology date from the incubation period of the wave turbulence theory (see Chapter 4). They have contributed to a better understanding of the physics of wave turbulence in the case of triadic interactions.

The major difference between hydrodynamics and incompressible MHD is the presence of waves in the latter. To these waves, we can associate an Alfvén time, τ_A, which characterizes the interaction time – that is, the collision time – between two wave packets. If these two packets have a typical length ℓ (isotropic turbulence is assumed), then:

$$\tau_A \sim \frac{\ell}{b_0} \sim \frac{1}{\omega_k}, \tag{7.13}$$

where b_0 is the Alfvén (group) velocity. Even if this hypothesis of isotropy is not physically justified, it will allow us in a first step to simplify the analysis. The main idea of IK's phenomenology is that at each scale of the inertial range the structures see locally a uniform magnetic field, even if there is no uniform magnetic field at large scales. In other words, even if the medium is isotropic, at the turbulent scales of the inertial range we will consider that the physics is governed by the process described in Figure 7.3. Using this approach, the collision between Alfvén wave packets becomes the central element in the analysis of MHD turbulence: it is the multiplicity of collisions that deforms the wave packets and finally transfers the energy from the large scales to smaller ones.

We will place ourselves in the balanced situation where $z^+ \sim z^- \sim z$, that is, when the waves propagating in opposite directions are approximately balanced. Therefore:

$$\varepsilon \sim \frac{z_\ell^2}{\tau_{tr}}, \tag{7.14}$$

where ε is the mean rate of energy transfer in the inertial range (at scale ℓ) and τ_{tr} is the associated transfer time. In the process we describe, turbulence is supposed to develop as the result of a large number of stochastic collisions between wave packets. Moreover, at a scale ℓ in the inertial range, the locally uniform magnetic field is assumed to be significantly larger than the magnetic fluctuations. The more intense the field, the shorter the duration of the collision, and the weaker the associated deformation will be. Therefore, the number of collisions required to produce a cascade must increase with the intensity of the field $\mathbf{b}_0$. So in strong field, we have $\tau_{tr} \gg \tau_A$. To estimate the transfer time, let us first evaluate the deformation of a wave packet at scale ℓ produced by a single collision. By placing ourselves on the wave packet, we have:

$$z_\ell(t + \tau_A) \sim z_\ell(t) + \tau_A \frac{\partial z_\ell}{\partial t} \sim z_\ell(t) + \tau_A \frac{z_\ell^2}{\ell}. \tag{7.15}$$

The deformation of the wave packet for one collision is therefore estimated at:

$$\Delta_1 z_\ell \sim \tau_A \frac{z_\ell^2}{\ell}. \tag{7.16}$$

This deformation will increase with time and for N stochastic collisions the cumulative effect can be evaluated in the same way as a random walk:

$$\sum_{i=1}^{N} \Delta_i z_\ell \sim \tau_A \frac{z_\ell^2}{\ell} \sqrt{\frac{t}{\tau_A}}. \tag{7.17}$$

The transfer time we are looking for is the one from which the cumulative deformation is of order one, that is, of the wave packet itself:

$$z_\ell \sim \sum_{1}^{N} \Delta_i z_\ell \sim \tau_A \frac{z_\ell^2}{\ell} \sqrt{\frac{\tau_{tr}}{\tau_A}}. \tag{7.18}$$

Then, we obtain:
$$\tau_{tr} \sim \frac{1}{\tau_A} \frac{\ell^2}{z_\ell^2} \sim \frac{\tau_{NL}^2}{\tau_A} \sim \omega \tau_{NL}^2, \qquad (7.19)$$

where τ_{NL} is the nonlinear time that emerges from the MHD equations (which is similar to that of hydrodynamics). We see here for the first time a phenomenological justification of the cascade time expression used for turbulence of capillary and inertial waves: it is the angular frequency of the wave weighted by the square of nonlinear time. Note that if we take as reference the collision time τ_A and introduce the small parameter $\epsilon \sim \tau_A/\tau_{NL}$, we come to the expression:

$$\boxed{\tau_{tr} \sim \epsilon^{-2} \tau_A}. \qquad (7.20)$$

This characteristic time now shows the small parameter on which the development of the wave turbulence is realized. By adapting the writing, this expression is generalized to all triadic problems (see Chapter 4).

All that remains is to complete the calculation using expression (7.14); assuming $E(k) \sim E^{\pm}(k)$, we get:

$$\varepsilon \sim \frac{z_\ell^2}{\omega \tau_{NL}^2} \sim \frac{z_\ell^4}{\ell b_0} \sim \frac{E^2(k) k^3}{b_0}, \qquad (7.21)$$

hence the one-dimensional isotropic Iroshnikov–Kraichnan (IK) spectrum of wave turbulence:

$$\boxed{E(k) = C_{IK} \sqrt{\varepsilon b_0} \, k^{-3/2}}, \qquad (7.22)$$

with C_{IK} a constant of the order of unity. The isotropic IK prediction thus differs from that of Kolmogorov, with a slightly less steep spectrum. It is interesting to note that in this approach the transfer time is longer than in hydrodynamics (we assume $\tau_A \ll \tau_{NL}$). Physically, it is the fact of having sporadic collisions between wave packets – and not continuous interactions between eddies – that explains phenomenologically the slowdown of energy transfer to small scales. This slowing down is observed, for example, in direct numerical simulations where the free decay of energy is clearly slower in MHD than in hydrodynamics (Biskamp and Welter, 1989; Kinney et al., 1995; Galtier et al., 1997; Bigot et al., 2008c). This decay is of course related to the dissipation, whose efficiency weakens when the cascade process slows down.

Today, the IK spectrum is still under discussion. For example, in situ measurements made in the solar wind (see Figure 8.2) at one astronomical unit show a power-law spectrum in $-5/3$ for the magnetic field fluctuations while it is in $-3/2$ for the velocity.[2] However, these properties may depend, on the one hand, on the heliocentric distance because of the expansion of the wind (Verdini and Grappin,

---

[2] Note that turbulence in liquid metals is characterized by kinetic energy spectra close to $k^{-5/3}$ over more than three decades (Crémer and Alemany, 1981).

2015) and, on the other hand, on the wave character of the turbulence, which is probably more pronounced near the Sun where the magnetic field is stronger. In general, it is the critical balance of strong wave turbulence that is used to describe the solar wind at MHD scales for which a $k_\perp^{-5/3}$ is predicted. As far as direct numerical simulation is concerned, the question of the spectral exponent is not yet decided, because the small difference between the IK and Kolmogorov predictions requires a wide inertial range (at least two decades) to conclude (Beresnyak, 2011). Note that the situation is different at subionic scales where the prediction of wave turbulence is in relatively good agreement with the solar wind data (see Chapter 8).

7.3.2 Anisotropic Spectrum

We have just seen that IK's phenomenology leads to an isotropic spectrum in $k^{-3/2}$. In this approach, we have assumed the existence, locally, of a uniform magnetic field to justify the notion of Alfvén wave packet. As shown, for example, in direct numerical simulations (Oughton et al., 1994; Müller et al., 2003; Bigot et al., 2008b; Teaca et al., 2009), the presence of a uniform magnetic field is a source of anisotropy with a cascade limited to the perpendicular direction, which contradicts the hypothesis of isotropy.[3] We are here in a situation similar to that of hydrodynamics under rotation (Chapter 6). Let us take up again this IK phenomenology by incorporating anisotropy through the wavenumbers $k_\perp$ and $k_\parallel$. In the simple case where the Alfvén wave packets are balanced ($z^+ \sim z^- \sim z$), we obtain for the transfer time:

$$\tau_{tr} \sim \omega \tau_{NL}^2 \sim \frac{(\ell_\perp/z_\ell)^2}{\ell_\parallel/b_0} \sim \frac{k_\parallel b_0}{k_\perp^2 z_\ell^2}. \tag{7.23}$$

We deduce from this:

$$\varepsilon \sim \frac{z_\ell^2}{\tau_{tr}} \sim \frac{k_\perp^2 z_\ell^4}{k_\parallel b_0} \sim \frac{k_\perp^2 (E(k_\perp, k_\parallel) k_\perp k_\parallel)^2}{k_\parallel b_0} \sim \frac{k_\perp^4 k_\parallel E^2(k_\perp, k_\parallel)}{b_0}, \tag{7.24}$$

hence the two-dimensional anisotropic (axisymmetric) spectrum:

$$\boxed{E(k_\perp, k_\parallel) \sim \sqrt{\varepsilon b_0}\, k_\perp^{-2} k_\parallel^{-1/2}}. \tag{7.25}$$

This anisotropic MHD spectrum is therefore the result of the cumulative effect of collisions of Alfvén wave packets propagating in opposite directions. It was first proposed by Ng and Bhattacharjee (1997). Phenomenology finds here its limits insofar as the dependence in $k_\parallel$ is only apparent: indeed, we can demonstrate analytically that the nonlinear transfer (the cascade) is totally frozen along the $\mathbf{b_0}$ magnetic field. We will see in Section 7.4.3 that the spectral prediction on $k_\perp^{-2}$ is

---

[3] MHD equations are invariant by Galilean transformation on a uniform velocity field, but not on a uniform magnetic field $\mathbf{b_0} = b_0 \mathbf{e}_\parallel$ (with $|\mathbf{e}_\parallel| = 1$). The observed anisotropy can be seen as a consequence of this property.

in fact an exact solution of Alfvén wave turbulence. These two results are part of the theory published by Galtier et al. (2000).

7.4 Theory of Alfvén Wave Turbulence

7.4.1 Canonical Variables

Let us go back to the MHD equations (with $\nu = \eta = 0$), and use the directional polarization $s = \pm$; we get:

$$\frac{\partial \mathbf{z}^s}{\partial t} - s\mathbf{b_0} \cdot \nabla \mathbf{z}^s = -\mathbf{z}^{-s} \cdot \nabla \mathbf{z}^s - \nabla P_*, \quad (7.26a)$$

$$\nabla \cdot \mathbf{z}^s = 0. \quad (7.26b)$$

The direction of the uniform magnetic field $\mathbf{b_0}$ defines the parallel direction. The Fourier transform (by component) of the Elsässer field is introduced:

$$z_j^s(\mathbf{x}, t) \equiv \int_{\mathbb{R}^3} A_j^s(\mathbf{k}, t) e^{i\mathbf{k} \cdot \mathbf{x}} d\mathbf{k}, \quad (7.27)$$

as well as the decomposition:

$$A_j^s(\mathbf{k}, t) \equiv \epsilon a_j^s(\mathbf{k}, t) e^{-is\omega_k t}, \quad (7.28)$$

with $\omega_k \equiv k_\parallel b_0$ and $0 < \epsilon \ll 1$. The introduction of the variable a_j^s allows us to place ourselves directly on an Alfvén wave packet and to follow its slow evolution over time whose origin is purely nonlinear. By applying the Fourier transform on expression (7.26a), we obtain the exact relation, whatever the value of ϵ:

$$\boxed{\frac{\partial a_j^s(\mathbf{k})}{\partial t} = -i\epsilon k_m P_{jn} \int_{\mathbb{R}^6} a_m^{-s}(\mathbf{q}) a_n^s(\mathbf{p}) e^{is(\omega_k - \omega_p + \omega_q)t} \delta(\mathbf{k} - \mathbf{p} - \mathbf{q}) d\mathbf{p} d\mathbf{q}}, \quad (7.29)$$

where $P_{jn}(k) \equiv \delta_{jn} - k_j k_n / k^2$ is the projection operator which ensures that the incompressibility condition (7.26b) is satisfied. Note that obtaining expression (7.29) requires writing the total pressure as a function of Elsässer's variables beforehand. This is possible by applying the divergence operator to equation (7.26a); one obtains the relation:

$$\Delta P_* = -\nabla \cdot (\mathbf{z}^{-s} \cdot \nabla \mathbf{z}^s). \quad (7.30)$$

Expression (7.29) represents the MHD equations in Fourier space in the presence of a uniform magnetic field. It is the starting point for the analytical development of wave turbulence. We see that the Elsässer variables are the canonical variables of the problem, since they directly lead to the classical form with a relatively simple interaction coefficient which is, here, essentially made up of the projection operator. Since only the counter-propagating Alfvén wave packets give a nonlinear contribution, the sum on the directional polarities is reduced to a unique combination involving only the s polarity.

7.4.2 Resonance Condition

Equation (7.29) shows the resonance condition:

$$\begin{cases} \omega_k = \omega_p - \omega_q, \\ \mathbf{k} = \mathbf{p} + \mathbf{q}, \end{cases} \tag{7.31}$$

which can be rewritten for its parallel part:

$$k_\parallel = p_\parallel \pm q_\parallel. \tag{7.32}$$

Therefore, the only possible solution is:

$$q_\parallel = 0 \quad \text{and} \quad k_\parallel = p_\parallel. \tag{7.33}$$

This condition implies the absence of coupling between wavenumbers in the parallel direction when equation (7.29) is solved (a mode coupling is only possible if the sum implies several modes). Therefore, no transfer along the $\mathbf{b_0}$ field is expected in this problem and the cascade is only in the perpendicular direction. Alfvén wave turbulence is thus even more anisotropic than that of inertial waves, a problem in which the transfer parallel to the axis of rotation is weak, but not totally inhibited.

The anisotropic character of MHD turbulence in a strong magnetic field was discussed in detail for the first time by Montgomery and Turner (1981). This theoretical study was based, in particular, on measurements made in magnetic confinement experiments (z-pinch or tokamak type) (Robinson and Rusbridge, 1971; Zweben et al., 1979), where it was possible to highlight an important difference between the correlation length along the mean field and those in the transverse directions, the former being greater than the latter two, fluctuations of the magnetic field in the transverse direction being always dominant. In the theoretical work of Montgomery and Turner (1981) the resonance condition is, evoked but it is really with Shebalin et al. (1983) that the analysis presented above, based on wave turbulence, was proposed for the first time to explain the results of direct numerical simulation of two-dimensional MHD turbulence in a strong magnetic field.

It is interesting to mention that at the end of the 1980s there was some confusion about the nature of MHD turbulence, because the Alfvén-based IK phenomenology had a fault: it did not take anisotropy into account. Wave turbulence theory could provide a rigorous answer, but the first attempt by Sridhar and Goldreich (1994) proved to be erroneous. The error comes from the fact that the authors considered that the fluctuations associated with the (slow) mode $q_\parallel = 0$ solution of the resonance condition had no energy, since they were not wavelike in nature. The authors therefore proposed a four-wave theory; at the same time they questioned the IK phenomenology based on triadic interactions. After strong criticisms (Montgomery and Matthaeus, 1995; Ng and Bhattacharjee, 1996; Chen and Kraichnan, 1997) explaining why, in particular, the slow mode could contain energy (such as in rotating hydrodynamics; see Chapter 6), the analytical theory of Alfvén wave turbulence was finally published by Galtier et al. (2000). The importance of this paper goes far beyond the resolution of a classical wave turbulence problem: the theory is demonstrated for three-wave interactions, which consequently validates IK's triadic approach and clarifies the nature of MHD turbulence (see also Lithwick and Goldreich, 2003).

7.4.3 Kinetic Equations

The formalism of wave turbulence applies in a classical way to equation (7.29). Note that the simplified derivation of Galtier et al. (2002) allows us to arrive at the result more quickly. The kinetic equations take the following form:

$$\frac{\partial e^s(\mathbf{k})}{\partial t} = \frac{\pi \epsilon^2}{b_0} \int_{\mathbb{R}^6} \frac{(\mathbf{k}_\perp \cdot \mathbf{p}_\perp)^2 (\mathbf{k} \times \mathbf{q})_\parallel^2}{k_\perp^2 p_\perp^2 q_\perp^2} e^{-s}(\mathbf{q}) \qquad (7.34)$$
$$\times [e^s(\mathbf{p}) - e^s(\mathbf{k})] \, \delta(q_\parallel) \delta(\mathbf{k} - \mathbf{p} - \mathbf{q}) \, d\mathbf{p} d\mathbf{q},$$

where $e^s(\mathbf{k})$ is the energy spectrum associated with shear-Alfvén waves. These waves correspond to the transverse part of the Elsässer fields. The waves associated with the parallel part are called pseudo-Alfvén waves: their dynamics is enslaved to shear-Alfvén waves, and they can therefore be forgotten, to lighten the calculations.[4]

We introduce the reduced axisymmetric spectrum $E^s(k_\perp)$ defined by the relation:

$$2\pi k_\perp e^s(\mathbf{k}) \equiv E^s(k_\perp) f(k_\parallel), \qquad (7.35)$$

with $f(k_\parallel)$ a function that arbitrarily depends on the parallel wavenumbers. The introduction of such a dependency in $k_\parallel$ is justified by the absence of transfer in the $\mathbf{b}_0$ direction which is mathematically translated by the presence of $\delta(q_\parallel)$ in the kinetic equations. In the same way as for capillary wave turbulence, we can then return to an expression involving only wavenumbers. After simplification, we get:

$$\frac{\partial E^s(k_\perp)}{\partial t} = \frac{\epsilon^2 f(0)}{2b_0} \int_{\Delta_\perp} \frac{k_\perp}{q_\perp} (\cos \theta_q)^2 \sin \theta_p \, E^{-s}(q_\perp) \qquad (7.36)$$
$$\times [k_\perp E^s(p_\perp) - p_\perp E^s(k_\perp)] \, dp_\perp dq_\perp,$$

---

[4] Pseudo-Alfvén waves are the incompressible limit of the slow magneto-acoustic waves, whereas shear-Alfvén waves remain unchanged in the compressible case: these are the "true" Alfvén waves.

with by definition $\theta_p \equiv \widehat{(\mathbf{k}_\perp, \mathbf{q}_\perp)}$ and $\theta_q \equiv \widehat{(\mathbf{k}_\perp, \mathbf{p}_\perp)}$ the angles in the triangle $\mathbf{k}_\perp = \mathbf{p}_\perp + \mathbf{q}_\perp$, and $\Delta_\perp$ the integration band corresponding to this triangle (see Figure 5.5). The presence of $f(0)$ corresponds to the contribution $q_\parallel = 0$. For the development of wave turbulence to remain valid, the contribution of $f(0)$ must remain finite: in other words, in the case of a strong condensation the theory becomes invalid because the term on the right is no longer in $O(\epsilon^2)$ (see discussion in Section 7.4.4).

The simplest exact solution of the kinetic equations is the one corresponding to zero energy flux. This is the thermodynamic solution (see Chapter 5) for which:

$$E^s(k_\perp) \sim k_\perp . \tag{7.37}$$

Numerical simulations show that over long times compared to the timescale of the direct cascade, this solution can appear at the largest scales when the system is forced at an intermediate scale (Galtier and Nazarenko, 2008). In this case, the thermodynamic solution can contribute to the slow regeneration of a very large-scale magnetic field (dynamo effect). One application envisaged by the authors is the regeneration of the galactic magnetic field.

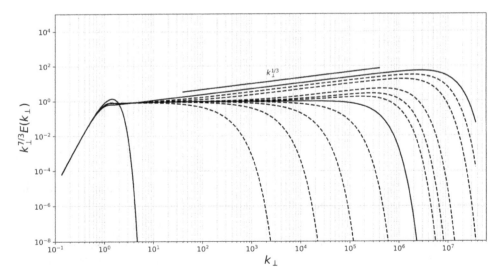

Figure 7.5 Numerical simulation of the kinetic equations (7.36) in which a dissipative term corresponding to a Laplacian is introduced. Balanced Alfvén wave turbulence is considered, with an energy spectrum $E(k_\perp)$ initially excited at large scales. The compensation by the nonstationary solution $\propto k_\perp^{-7/3}$ reveals the transition to the Kolmogorov–Zakharov spectrum $\propto k_\perp^{-2}$ after a spectral bounce (which is less localized at small scales than for rotating turbulence (see Chapter 6) where a hyperviscosity is used).

The exact constant-flux solution of equation (7.36), with nonzero ε, is:

$$\boxed{n_+ + n_- = -4}, \tag{7.38}$$

with by definition $E^\pm(k_\perp) \sim k_\perp^{n_\pm}$. In this case, the convergence condition implies that:[5]

$$-3 < n_\pm < -1. \tag{7.39}$$

For the particular case of a balanced Alfvén wave turbulence, the exact solution is written (see Figure 7.5):

$$E^+(k_\perp) = E^-(k_\perp) \equiv E(k_\perp) \sim k_\perp^{-2}. \tag{7.40}$$

We can see that the phenomenology (7.25) predicts well the right spectrum for its transverse part. However, the theory allows us to go further than the phenomenology and to demonstrate that the flux is positive – the cascade is therefore direct. One can also find the analytical form of the Kolmogorov constant which, in the balanced case, is:[6]

$$C_K \simeq 1.467. \tag{7.41}$$

This constant can also be evaluated directly (see Figure 7.6) from the simulation of the kinetic equation (7.36) and the form of the energy spectrum

$$\boxed{E(k_\perp) = C_K \sqrt{b_0 \varepsilon} k_\perp^{-2}}. \tag{7.42}$$

In this last expression, the small parameter ϵ^2 is absorbed by the time variable as well as the contribution $f(0)$. The formation of this spectrum from the kinetic equations is shown in Figure 7.5: a spectrum in $k_\perp^{-2}$ appears over approximately seven decades. Note that the first observational evidence of weak MHD turbulence, which includes this spectrum, was detected in the middle magnetosphere of Jupiter (Saur et al., 2002).

A last remark is to be made on the condition of validity of this Alfvén wave turbulence regime. Indeed, this regime is not uniformly valid in Fourier space because it is conditioned by the inequality:

$$\chi \equiv \frac{\tau_A}{\tau_{NL}} \ll 1, \tag{7.43}$$

underlying to the asymptotic development. Using the spectrum of balanced turbulence, this condition can be rewritten:

$$\chi \sim \frac{k_\perp z_\ell}{k_\| b_0} \sim \sqrt{\frac{k_\perp}{k_\|}} \ll 1. \tag{7.44}$$

In the absence of a parallel cascade, energy will populate the $k_\perp$ modes more and more. If the condition on χ is verified initially in a spectral domain, it will

---

[5] Unlike the three-dimensional case discussed here, two-dimensional Alfvén wave turbulence is nonlocal because the solutions obtained are out of the convergence domain (Tronko et al., 2013).
[6] There is a difference (actually a correction) of $\sqrt{2\pi}$ with the prediction published by Galtier et al. (2000).

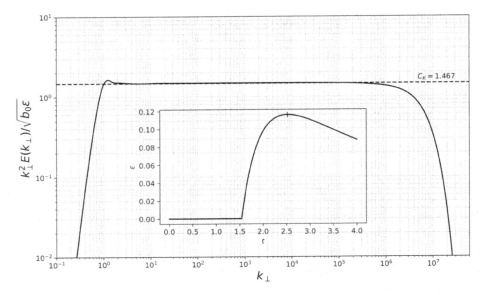

Figure 7.6 Numerical estimation of the Kolmogorov constant C_K from a simulation of the kinetic equation (7.36). The analytical prediction is obtained with a good approximation. Insert: time evolution of the mean rate of energy dissipation ε. Its value used to estimate C_K is taken at time $t = 2.5$.

no longer be in a finite time, since this ratio will grow $\propto k_\perp^{1/2}$ with the cascade. The turbulence will become strong at small scales when the two characteristic times become of the same order of magnitude. Then, the regime can be described by the critical balance phenomenology (Goldreich and Sridhar, 1995), in which it is assumed that the relation $\tau_A \sim \tau_{NL}$ remains true in the inertial range under consideration. This transition was highlighted clearly for the first time by Meyrand et al. (2016).

7.4.4 Treatment of the 2D Mode ($k_\parallel = 0$)

The kinetic equations (7.34) show a term $e^{-s}(\mathbf{q})$ whose contribution is reduced to $q_\parallel = 0$ because of the resonance condition. Since this is the slow mode, these fluctuations are not associated with an Alfvén wave and the inequality (7.43) cannot be verified. How then can this different term be treated?

First of all, there is no reason to think that $e^{-s}(\mathbf{q}_\perp, q_\parallel = 0) = 0$, because low-frequency fluctuations usually have energy (this is also the case in rotating hydrodynamics). This means that the dominant interactions in this problem are indeed triadic. Can we have a strong condensation in $k_\parallel = 0$ that would invalidate the analytical approach? This situation cannot be totally excluded, but in practice it requires very specific conditions. For example, forcing around the slow

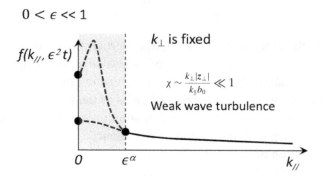

Figure 7.7 The asymptotic theory of Alfvén wave turbulence is valid in a domain of k-space where the inequality (7.43) is satisfied; it corresponds to the region $k_\| > \epsilon^\alpha$ (with $\alpha > 0$). However, the three-wave interactions induce the energy spectrum (schematically represented by the function f) of the slow mode ($k_\| = 0$). The implicit assumption in the theory of Alfvén wave turbulence is the smooth behavior of f in the narrow shaded domain such that $f(k_\| = \epsilon^\alpha) \sim f(k_\| = 0)$. The opposite situation corresponds to a singular behavior, which leads to a much higher value of $f(k_\| = 0)$ (strong condensation) that could break the asymptotic (at time $\epsilon^2 t$) and make the theory irrelevant.

mode could perhaps cause such condensation. According to the direct numerical simulations of Bigot and Galtier (2011), in which a forcing of the type $E^s(k_\perp, k_\|) = F(k_\|)k_\perp^3$ was used for the perpendicular and parallel modes 1 and 2, an increase in energy at $k_\| = 0$ was clearly observed. This increase was stronger for larger b_0, however, no strong condensation, that is, a singular tendency to transfer energy to the slow mode, has been observed. Indeed, after a phase of rapid increase, a saturation was measured. Considering a continuous variation of $e^{-s}(\mathbf{q})$ around the slow mode – without condensation – therefore seems a good hypothesis. From a theoretical point of view, this means that we consider fields whose correlation decreases with distance in any direction: in this case, $e^{-s}(\mathbf{q}_\perp, 0)$ can be seen as the limit of $e^{-s}(\mathbf{q}_\perp, k_\|)$ when $q_\| \to 0$ and the solutions described given in Section 7.4.3 (see Figure 7.7). If the fluctuations of the slow mode have their own dynamics – a possible situation if the associated energy is higher than that of the waves – kinetic equations can still inform us about solutions. Indeed, equation (7.38) can always be used with n_- the exponent of the slow mode energy spectrum. For example, if we assume a Kolmogorovian two-dimensional dynamics with $n_- = -5/3$, then for the wave energy spectrum we get $n_+ = -7/3$.[7] These questions around condensation have also been discussed in phenomenological and numerical studies (Boldyrev and Perez, 2009; Wang et al., 2011; Schekochihin et al., 2012).

---

[7] In two-dimensional MHD, strong turbulence behaves as in the three-dimensional case, with a direct energy cascade for which a spectrum in $-5/3$ is a realistic solution.

7.4.5 Other Results

We conclude with several remarks. First, like inertial waves, the kinetic equations (7.36) can be studied in the limited case of local triadic interactions (see Exercise II.1). We can then show that the nonlinear diffusion equations that emerge are structurally simpler and have the same solutions as the kinetic equations (Galtier and Buchlin, 2010). Second, a nonstationary solution was highlighted by a numerical simulation of the kinetic equation in the balanced case (see Figure 7.5). It is a self-similar solution of the second kind compatible with $E(k_\perp) \sim k_\perp^{-7/3}$ (Galtier et al., 2000). The nature of this solution is identical to that discussed in Chapter 6 on inertial wave turbulence. Third, when the magnetic field b_0 is too strong, Alfvén wave turbulence becomes too weak and a problem of discretization appears which can modify the dynamics (Nazarenko, 2007).

7.5 Direct Numerical Simulation

Direct numerical simulation has recently made it possible to reproduce the main properties of Alfvén wave turbulence, including its transition to a small-scale regime of strong turbulence (Meyrand et al., 2016). A result of such a simulation is shown in Figure 7.8: the modulus of the magnetic field fluctuations is presented on a slice perpendicular to $\mathbf{b_0}$. Contrary to the case of strong turbulence, we do not see a hierarchy of coherent structures. This is explained by the phase mixing induced by $\mathbf{b_0}$, which tends to oppose the emergence of structures.

From such a simulation, it is possible to construct the ω–$k_\parallel$ spectrum to highlight the wave character of such a turbulence. The result is shown in Figure 7.9: we see the two branches of the dispersion relation corresponding to Alfvén waves. These have a certain thickness, whose origin is attributed to nonlinear effects. We can observe also the presence of a thick band at low ω: we find there the contribution of the slow mode.

These numerical studies have highlighted the intermittency nature of Alfvén wave turbulence (Meyrand et al., 2015). The structure functions used to take into account the coupling between Alfvénic fluctuations of different polarities are:

$$S_p = \langle (\delta z^+)^{p/2} \rangle \langle (\delta z^-)^{p/2} \rangle = C_p \ell_\perp^{\zeta_p}, \tag{7.45}$$

with C_p a constant. Due to the absence of cascade in the parallel direction, only the transverse increments ($\ell_\perp$) are considered. The data verify with a great accuracy the following law:

$$\zeta_p = \frac{p}{8} + 1 - \left(\frac{1}{4}\right)^{p/2}. \tag{7.46}$$

This law has been constructed by adapting the approach of She and Leveque (1994) to this problem (see Chapter 2) where structures oriented along the external

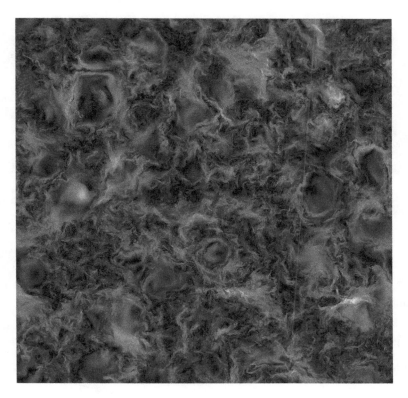

Figure 7.8 Three-dimensional direct numerical simulation of Alfvén wave turbulence. We see here the modulus of magnetic field fluctuations on a slice of spatial resolution 3072×3072 perpendicular to $\mathbf{b}_0$. Simulation made by R. Meyrand. Reprinted with permission (Galtier, 2016).

magnetic field in the form of sheets are present. Figure 7.10 shows these measurements and the correspondence with the theoretical model. The origin of the intermittency is partly attributed to the slow mode: indeed, the artificial reduction of the interactions of this mode with Alfvén's modes reveals a significant reduction of intermittency (at the PDF level). At the same time, a modification of the Elsässer spectrum is observed with a power law going from $k_\perp^{-2}$ (interaction with the slow mode) to $k_\perp^{-3/2}$ (no interaction) (Meyrand et al., 2015).

7.6 Application: The Solar Corona

The Sun has a strong magnetic activity that can be detected through the appearance of sunspots in the photosphere or flares in the corona. These active regions are made up of a network of magnetic loops in perpetual reorganization and whose activity can be measured by emissions in the ultraviolet or X-ray range (Reale,

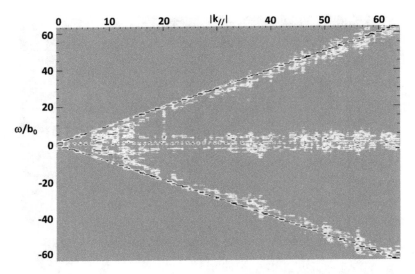

Figure 7.9 Normalized ω–$k_\parallel$ spectrum built from magnetic energy fluctuations at $k_\perp = 4$. Simulation made by R. Meyrand.

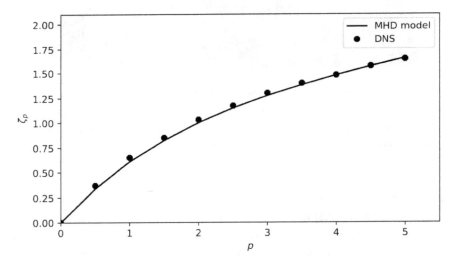

Figure 7.10 Measurements of the ζ_p coefficients of the structure functions (7.45) realized from a direct numerical simulation show a good agreement with the law (7.46) in solid line. Adapted from Meyrand et al. (2015).

2014). Space telescopes or imagers such as SDO/NASA allow us to follow this activity, but the current instruments are not yet able to resolve coronal loops at more than one million degrees and whose thickness is probably less than one kilometer (see Figure 7.1). The coronal temperature is very inhomogeneous: between the quiet and active regions there is nearly a factor of ten. A major problem in solar physics is to understand why such a high temperature is measured in the corona even though the photosphere, at a lower altitude, is at $\simeq 6400$ K. The energy available at the surface of the Sun seems to be largely sufficient to compensate for the loss of coronal energy, estimated at about $10^4 \mathrm{Jm}^{-2}\mathrm{s}^{-1}$ in active regions, and about one or even two orders of magnitude lower for the quiet corona or coronal holes, which are located essentially at the poles. The main problem is to understand how this energy available on the surface of the Sun is deposited at the corona and then finally released to heat the plasma (Priest, 2014).

Spectrometric analyses of coronal loops in active regions reveal plasma movements at several tens of km/s (Brekke et al., 1997), with nonthermal velocities up to 30 km/s (Chae et al., 1998). These observations, made with the SOHO/ESA satellite, show, moreover, that the widening of the spectral lines is due to motions at very small spatial and temporal scales that are not resolved by the instruments (of the order of a second for the temporal resolution). The ubiquity of Alfvén waves was also highlighted by measurements made from the Hinode/JAXA satellite (De Pontieu et al., 2007). These various observations constitute a bundle of clues that pleads in favor of a turbulent description of the coronal plasma in which Alfvén waves play an important role (Matthaeus and Velli, 2011; Pontin and Hornig, 2020).

At lower altitudes, at the level of the photosphere, it is possible to accurately measure (Zeeman effect) the magnetic field component transverse to the surface of the Sun. Based on two-dimensional magnetic maps of active regions, structure functions and spectra can be calculated. In this case, it is the correlation in the perpendicular direction ($k_\perp$ for the spectrum) that is available. Studies show spectra compatible with wave turbulence predictions, especially during a flare, that is, in a phase of high light emission caused by an increase in heating. The spectrum, which is close to $-5/3$ in a quiet phase, becomes much steeper during a flare with exponents generally between -2 and -2.3 (Abramenko, 2005; Hewett et al., 2008; Mandage and McAteer, 2016).

In the scenario of a coronal heating by MHD turbulence, the energy transport to the corona is done via Alfvén waves excited at the photospheric level.[8] This forcing is in fact produced by the convective movements of the plasma that can be observed at the surface of the Sun in the form of granulation. As shown in Figure 7.11, the energy transported by Alfvén wave packets cascades to small scales due to the numerous collisions between contra-propagating wave packets. In the end, an intermittent heating is produced. Thermodynamics then allows us to make the link between heating and cooling, by conduction and radiation.

Many models have been published on the subject. The first ones hypothesized an isotropic turbulence (Heyvaerts and Priest, 1992), then anisotropy was introduced with a numerical modeling in different dimensions (Einaudi et al., 1996; Hendrix and van Hoven, 1996; Galtier and Pouquet, 1998; Walsh and Galtier,

---

[8] The problem of solar coronal heating is sufficiently complex to make room for several other scenarios, such as the phase mixing mechanism proposed by Heyvaerts and Priest (1983).

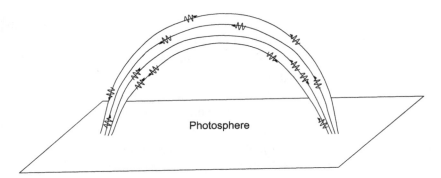

Figure 7.11 Heating of a magnetic loop by Alfvén wave turbulence. The coronal loop can act as a resonant cavity for the Alfvén wave packets generated in the photosphere. The direct cascade produced by the numerous collisions between contra-propagating wave packets eventually leads to a plasma heating. Reprinted with permission (Galtier, 2016).

2000; Buchlin et al., 2003; Gudiksen and Nordlund, 2005; Buchlin and Velli, 2007; Cranmer et al., 2007). The Alfvén wave turbulence regime was then studied by Rappazzo et al. (2007), and on the basis of the isotropic analytical model of Heyvaerts and Priest (1992), an anisotropic version has been proposed by Bigot et al. (2008a) in which the coronal heating is estimated from the turbulent viscosities deduced from the Alfvén wave turbulence kinetic equations (7.36). In this case, the heating rate obtained proves to be high enough to explain the coronal observations. This heating rate (integrated along the loop) is written (in IS units):

$$\varepsilon_\perp \sim 10^3 \left(\frac{L}{10^6}\right)^{2/3} \left(\frac{u}{10^3}\right)^{4/3} \left(\frac{B_0}{10^{-2}}\right)^{5/3} \left(\frac{\rho_0}{10^{-12}}\right)^{1/6} \text{Jm}^{-2}\text{s}^{-1}, \quad (7.47)$$

where L is the diameter of the coronal loop, u the photospheric velocity of the footpoints of the loop (forcing velocity), B_0 the coronal magnetic field and ρ_0 the mass density in the corona. This prediction is comparable to the one made by Heyvaerts and Priest (1992) for the strong turbulence regime. Therefore, the reduction of the cascade to two directions does not fundamentally change the heating produced.

7.7 Perspectives

An important question concerns the physics of the solar wind near the Sun. The two new probes Parker Solar Probe (NASA) and Solar Orbiter (ESA), launched in 2018 and 2020 respectively, aim, among other things, to measure the turbulent properties of the solar wind at the intersection with the low solar corona (as close as 10 solar radii). We can expect that the wave aspect of the wind will be more pronounced near the Sun since the magnetic field is more intense there, and that the properties differ from those measured essentially at one astronomical unit (Bale et al., 2019). Interesting new questions will have to be tackled, such as, for example, on physics at electron scales (so beyond the MHD approximation but still in the incompressible limit), taking into account electron inertia (Roytershteyn et al., 2019). Wave turbulence is a solid approach to address this issue (see Chapter 8).

A second interesting perspective concerns MHD turbulence subject to uniform rotation. This regime is relevant in the context of the geodynamo, where the Coriolis force plays a central role: the Rossby number is indeed of the order of 10^{-6} (see Chapter 6). The liquid metal of the Earth's outer core must therefore be described by incompressible MHD under rapid rotation. This turbulence has been little studied so far (Favier et al., 2012), because in general the emphasis is on convection. Rotating MHD in the presence of a uniform magnetic field generally has only one invariant, the total energy: indeed, cross-helicity and magnetic helicity, the two other invariants of standard MHD, lose their invariant status because of the rotation and the uniform magnetic field, respectively. However, in the particular case where the axis of rotation is parallel to the uniform magnetic field, a second invariant appears: hybrid helicity, a combination of cross-helicity and magnetic helicity. The strong wave turbulence regime has been recently studied using direct numerical simulation (Menu et al., 2019). It is interesting to note that the inverse cascade of the magnetic field is relatively large precisely when the axis of rotation is parallel (or almost) to the uniform field. This is the situation observed for the Earth: the angle between the two axes is nonzero but small – about 10°. Understanding this regime requires further studies according to the available parameters (e.g. the magnetic Prandtl number). Note that this theory of wave turbulence has been published by Galtier (2014) and numerically simulated by Bell and Nazarenko (2019). As is very often the case, these results can be used as a guide to better understand the regime of strong wave turbulence.

References

Abramenko, V. I. 2005. Relationship between magnetic power spectrum and flare productivity in solar active regions. *Astrophys. J.*, **629**(2), 1141–1149.

Alfvén, H. 1942. Existence of electromagnetic-hydrodynamic waves. *Nature*, **150**, 405–406.

Allen, T. K., Baker, W. R., Pyle, R. V., and Wilcox, J. M. 1959. Experimental generation of plasma Alfvén waves. *Phys. Rev. Lett.*, **2**, 383–384.

Bale, S. D., Badman, S. T., Bonnell, J. W. et al. 2019. Highly structured slow solar wind emerging from an equatorial coronal hole. *Nature*, **576**(7786), 237–242.

Bell, N. K., and Nazarenko, S. V. 2019. Rotating magnetohydrodynamic turbulence. *J. Phys. A: Math. Theo.*, **52**(44), 445501.

Beresnyak, A. 2011. Spectral slope and Kolmogorov constant of MHD turbulence. *Phys. Rev. Lett.*, **106**, 075001.

Bigot, B., and Galtier, S. 2011. Two-dimensional state in driven magnetohydrodynamic turbulence. *Phys. Rev. E*, **83**(2), 026405.

Bigot, B., Galtier, S., and Politano, H. 2008a. An anisotropic turbulent model for solar coronal heating. *Astron. Astrophys.*, **490**, 325–337.

Bigot, B., Galtier, S., and Politano, H. 2008b. Development of anisotropy in incompressible magnetohydrodynamic turbulence. *Phys. Rev. E*, **78**(6), 066301.

Bigot, B., Galtier, S., and Politano, H. 2008c. Energy decay laws in strongly anisotropic magnetohydrodynamic turbulence. *Phys. Rev. Lett.*, **100**(7), 074502.

Biskamp, D., and Welter, H. 1989. Dynamics of decaying two-dimensional magnetohydrodynamic turbulence. *Phys. of Fluids B*, **1**(10), 1964–1979.

Boldyrev, S., and Perez, J. C. 2009. Spectrum of weak magnetohydrodynamic turbulence. *Phys. Rev. Lett.*, **103**(22), 225001.

Brekke, P., Kjeldseth-Moe, O., and Harrison, R. A. 1997. High-velocity flows in an active region loop system observed with the coronal diagnostic spectrometer on SOHO. *Sol. Phys.*, **175**, 511–521.

Buchlin, E., and Velli, M. 2007. Shell models of RMHD turbulence and the heating of solar coronal loops. *Astrophys. J.*, **662**, 701–714.

Buchlin, E., Aletti, V., Galtier, S. et al. 2003. A simplified numerical model of coronal energy dissipation based on reduced MHD. *Astron. Astrophys.*, **406**, 1061–1070.

Chae, J., Schühle, U., and Lemaire, P. 1998. SUMER measurements of nonthermal motions: Constraints on coronal heating mechanisms. *Astrophys. J.*, **505**, 957–973.

Chen, S., and Kraichnan, R. H. 1997. Inhibition of turbulent cascade by sweep. *J. Plasma Phys.*, **57**(1), 187–193.

Cho, J., and Vishniac, E. T. 2000. The anisotropy of magnetohydrodynamic Alfvénic turbulence. *Astrophys. J.*, **539**, 273–282.

Cranmer, S. R., van Ballegooijen, A. A., and Edgar, R. J. 2007. Self-consistent coronal heating and solar wind acceleration from anisotropic magnetohydrodynamic turbulence. *Astrophys. J. Supp. Series*, **171**, 520–551.

Crémer, P., and Alemany, A. 1981. Aspects expérimentaux du brassage électromagnétique en creuset. *J. Mécanique Appl.*, **5**, 37–50.

De Pontieu, B., McIntosh, S. W., Carlsson, M. et al. 2007. Chromospheric Alfvénic waves strong enough to power the solar wind. *Science*, **318**, 1574–1577.

Drake, D. J., Schroeder, J. W. R., Howes, G. G. et al. 2013. Alfvén wave collisions, the fundamental building block of plasma turbulence. IV. Laboratory experiment. *Phys. Plasmas*, **20**(7), 072901.

Einaudi, G., Velli, M., Politano, H., and Pouquet, A. 1996. Energy release in a turbulent corona. *Astrophys. J. Lett.*, **457**, L113–L116.

Elsässer, W. M. 1950. The hydromagnetic equations. *Phys. Rev.*, **79**, 183–183.

Favier, B. F. N., Godeferd, F. S., and Cambon, C. 2012. On the effect of rotation on magnetohydrodynamic turbulence at high magnetic Reynolds number. *Geophys. Astrophys. Fluid Dynamics*, **106**, 89–111.

Fujisawa, A. 2021. Review of plasma turbulence experiments. *Proc. Japan Acad., Series B*, **97**(3), 103–119.

Galtier, S. 2014. Weak turbulence theory for rotating magnetohydrodynamics and planetary flows. *J. Fluid Mech.*, **757**, 114–154.

Galtier, S. 2016. *Introduction to Modern Magnetohydrodynamics*. Cambridge University Press.

Galtier, S., and Buchlin, E. 2010. Nonlinear diffusion equations for anisotropic magnetohydrodynamic turbulence with cross-helicity. *Astrophys. J.*, **722**(2), 1977–1983.

Galtier, S., and Nazarenko, S. V. 2008. Large-scale magnetic field sustainment by forced MHD wave turbulence. *J. Turbulence*, **9**(40), 1–10.

Galtier, S., and Pouquet, A. 1998. Solar flare statistics with a one-dimensional MHD model. *Sol. Phys.*, **179**, 141–165.

Galtier, S., Politano, H., and Pouquet, A. 1997. Self-similar energy decay in magnetohydrodynamic turbulence. *Phys. Rev. Lett.*, **79**(15), 2807–2810.

Galtier, S., Nazarenko, S. V., Newell, A. C., and Pouquet, A. 2000. A weak turbulence theory for incompressible magnetohydrodynamics. *J. Plasma Phys.*, **63**, 447–488.

Galtier, S., Nazarenko, S. V., Newell, A. C., and Pouquet, A. 2002. Anisotropic turbulence of shear-Alfvén waves. *Astrophys. J.*, **564**(1), L49–L52.

Goldreich, P., and Sridhar, S. 1995. Toward a theory of interstellar turbulence. 2: Strong alfvenic turbulence. *Astrophys. J.*, **438**, 763–775.

Gudiksen, B. V., and Nordlund, A. A. 2005. An ab initio approach to the solar coronal heating problem. *Astrophys. J.*, **618**, 1020–1030.

Hendrix, D. L., and van Hoven, G. 1996. Magnetohydrodynamic turbulence and implications for solar coronal heating. *Astrophys. J.*, **467**, 887–893.

Hewett, R. J., Gallagher, P. T., McAteer, R. T. J. et al. 2008. Multiscale analysis of active region evolution. *Solar Phys.*, **248**(2), 311–322.

Heyvaerts, J., and Priest, E. R. 1983. Coronal heating by phase-mixed shear Alfven waves. *Astron. Astrophys.*, **117**, 220–234.

Heyvaerts, J., and Priest, E. R. 1992. A self-consistent turbulent model for solar coronal heating. *Astrophys. J.*, **390**, 297–308.

Higdon, J. C. 1984. Density fluctuations in the interstellar medium: Evidence for anisotropic magnetogasdynamic turbulence. I – Model and astrophysical sites. *Astrophys. J.*, **285**, 109–123.

Iroshnikov, P. S. 1964. Turbulence of a conducting fluid in a strong magnetic field. *Soviet Astron.*, **7**, 566–571.

Kinney, R., McWilliams, J. C., and Tajima, T. 1995. Coherent structures and turbulent cascades in two-dimensional incompressible magnetohydrodynamic turbulence. *Phys. Plasmas*, **2**(10), 3623–3639.

Kraichnan, R. H. 1965. Inertial-range spectrum of hydromagnetic turbulence. *Phys. Fluids*, **8**, 1385–1387.

Lithwick, Y., and Goldreich, P. 2003. Imbalanced weak magnetohydrodynamic turbulence. *Astrophys. J.*, **582**(2), 1220–1240.

Luo, Q. Y., and Wu, D. J. 2010. Observations of anisotropic scaling of solar wind turbulence. *Astrophys. J. Lett.*, **714**, L138–L141.

Mandage, R. S., and McAteer, R. T. J. 2016. On the non-Kolmogorov nature of flare-productive solar active regions. *Astrophys. J.*, **833**(2), 237 (7 pages).

Matthaeus, W. H., and Velli, M. 2011. Who needs turbulence? A review of turbulence effects in the heliosphere and on the fundamental process of reconnection. *Space Sci. Rev.*, **160**(1–4), 145–168.

Menu, M. D., Galtier, S., and Petitdemange, L. 2019. Inverse cascade of hybrid helicity in **B**–Ω MHD turbulence. *Phys. Rev. Fluids*, **4**(7), 073701.

Meyrand, R., Kiyani, K. H., and Galtier, S. 2015. Weak magnetohydrodynamic turbulence and intermittency. *J. Fluid Mech.*, **770**, R1.

Meyrand, R., Galtier, S., and Kiyani, K. H. 2016. Direct evidence of the transition from weak to strong magnetohydrodynamic turbulence. *Phys. Rev. Lett.*, **116**(10), 105002.

Montgomery, D., and Matthaeus, W. H. 1995. Anisotropic modal energy transfer in interstellar turbulence. *Astrophys. J.*, **447**, 706.

Montgomery, D., and Turner, L. 1981. Anisotropic magnetohydrodynamic turbulence in a strong external magnetic field. *Phys. Fluids*, **24**(5), 825–831.

Müller, W.-C., Biskamp, D., and Grappin, R. 2003. Statistical anisotropy of magnetohydrodynamic turbulence. *Phys. Rev. E*, **67**(6), 066302.

Nazarenko, S. 2007. 2D enslaving of MHD turbulence. *New J. Physics*, **9**(8), 307.

Ng, C. S., and Bhattacharjee, A. 1996. Interaction of shear-Alfven wave packets: Implication for weak magnetohydrodynamic turbulence in astrophysical plasmas. *Astrophys. J.*, **465**, 845–854.

Ng, C. S., and Bhattacharjee, A. 1997. Scaling of anisotropic spectra due to the weak interaction of shear-Alfvén wave packets. *Phys. Plasmas*, **4**(3), 605–610.

Osman, K. T., Wan, M., Matthaeus, W. H., Weygand, J. M., and Dasso, S. 2011. Anisotropic third-moment estimates of the energy cascade in solar wind turbulence using multispacecraft data. *Phys. Rev. Lett.*, **107**(16), 165001.

Oughton, S., and Matthaeus, W. H. 2020. Critical balance and the physics of magnetohydrodynamic turbulence. *Astrophys. J.*, **897**(1), 37 (24 pages).

Oughton, S., Priest, E. R., and Matthaeus, W. H. 1994. The influence of a mean magnetic field on three-dimensional magnetohydrodynamic turbulence. *J. Fluid Mech.*, **280**, 95–117.

Phillips, O. M. 1958. The equilibrium range in the spectrum of wind-generated waves. *J. Fluid Mech.*, **4**, 426–434.

Podesta, J. J., Roberts, D. A., and Goldstein, M. L. 2007. Spectral exponents of kinetic and magnetic energy spectra in solar wind turbulence. *Astrophys. J.*, **664**, 543–548.

Pontin, D. I., and Hornig, G. 2020. The Parker problem: Existence of smooth force-free fields and coronal heating. *Living Rev. Solar Phys.*, **17**(1), 5.

Priest, E. 2014. *Magnetohydrodynamics of the Sun*. Cambridge University Press.

Rappazzo, A. F., Velli, M., Einaudi, G., and Dahlburg, R. B. 2007. Coronal heating, weak MHD turbulence, and scaling laws. *Astrophys. J. Lett.*, **657**, L47–L51.

Reale, F. 2014. Coronal loops: Observations and modeling of confined plasma. *Living Rev. Sol. Phys.*, **11**, 4.

Robinson, D. C., and Rusbridge, M. G. 1971. Structure of turbulence in the ZETA plasma. *Phys. Fluids*, **14**(11), 2499–2511.

Roytershteyn, V., Boldyrev, S., Delzanno, G.-L. et al. 2019. Numerical study of inertial kinetic-Alfvén turbulence. *Astrophys. J.*, **870**(2), 103.

Saur, J., Politano, H., Pouquet, A., and Matthaeus, W. H. 2002. Evidence for weak MHD turbulence in the middle magnetosphere of Jupiter. *Astrophys. Astron.*, **386**, 699–708.

Schekochihin, A. A., Nazarenko, S. V., and Yousef, T. A. 2012. Weak Alfvén-wave turbulence revisited. *Phys. Rev. E*, **85**(3), 036406.

She, Z.-S., and Leveque, E. 1994. Universal scaling laws in fully developed turbulence. *Phys. Rev. Lett.*, **72**(3), 336–339.

Shebalin, J. V., Matthaeus, W. H., and Montgomery, D. 1983. Anisotropy in MHD turbulence due to a mean magnetic field. *J. Plasma Phys.*, **29**, 525–547.

Sridhar, S., and Goldreich, P. 1994. Toward a theory of interstellar turbulence. I. Weak Alfvenic turbulence. *Astrophys. J.*, **432**, 612–621.

Teaca, B., Verma, M. K., Knaepen, B., and Carati, D. 2009. Energy transfer in anisotropic magnetohydrodynamic turbulence. *Phys. Rev. E*, **79**(4), 046312.

Tronko, N., Nazarenko, S. V., and Galtier, S. 2013. Weak turbulence in two-dimensional magnetohydrodynamics. *Phys. Rev. E*, **87**(3), 033103.

Verdini, A., and Grappin, R. 2015. Imprints of expansion on the local anisotropy of solar wind turbulence. *Astrophys. J.*, **808**(2), L34.

Walsh, R. W., and Galtier, S. 2000. Intermittent heating in a model of solar coronal loops. *Solar Phys.*, **197**(1), 57–73.

Wang, Y., Boldyrev, S., and Perez, J. C. 2011. Residual energy in magnetohydrodynamic turbulence. *Astrophys. J.*, **740**(2), L36 (4 pages).

Zweben, S. J., Menyuk, C. R., and Taylor, R. J. 1979. Small-scale magnetic fluctuations inside the Macrotor tokamak. *Phys. Rev. Lett.*, **42**(19), 1270–1274.

8

Wave Turbulence in a Compressible Plasma

Plasmas are naturally compressible. In astrophysics, most visible matter is in a state of plasma and a wide variety of turbulence is encountered: for example, in the interstellar medium, turbulence can be supersonic, with turbulent Mach numbers greater than 10 (see Chapter 2), while in the solar wind, it is subsonic and weakly compressible with a relative mass density variation of about 5 percent.[1] The Earth's magnetosphere is another example where the mass density variations can be strong, often with the presence of a bow shock at the interface with the solar wind (see Figure 8.1).

A major difference between incompressible hydrodynamics and plasmas is the plethora of waves supported by the latter. Therefore, the theory of wave turbulence applies well beyond the approximation of incompressible magnetohydrodynamics (MHD) introduced in the previous chapter. The most immediate extension is the weak compressible MHD case, for which the Hamiltonian (Kuznetsov, 2001) and Eulerian (Chandran, 2008) approaches have been used. In both cases, the thermodynamic pressure is neglected compared to the magnetic pressure. As for incompressible hydrodynamics, the turbulent dynamics is dominated by triadic interactions. However, the uniformity of the development has not been checked.

It is also necessary to go beyond the MHD approximation when we deal with sub-MHD scales. A simple model that can be used is incompressible Hall MHD which, within the limit of the small scales, becomes electron MHD. The associated theories (always for triadic interactions) use a complex helical basis, as for inertial waves (Galtier and Bhattacharjee, 2003; Galtier, 2006). The direct numerical simulations of Meyrand et al. (2018) show that this system is nontrivial, with the possibility of having at the same time a wave turbulence regime for right-polarized waves (carried by electrons) and a strong turbulence regime for left-polarized waves (carried by ions). Weakly compressible Hall MHD is a much more difficult case to treat and, to date, no complete theory has been developed (Sahraoui et al., 2003). Only one attempt has been made for the case of kinetic Alfvén waves:

---

[1] The solar wind has a velocity (400–800 km/s) higher than the speed of sound ($\sim$ 50 km/s); however, with velocity fluctuations $\sim$ 10 km/s, the turbulent Mach number M_t is less than one.

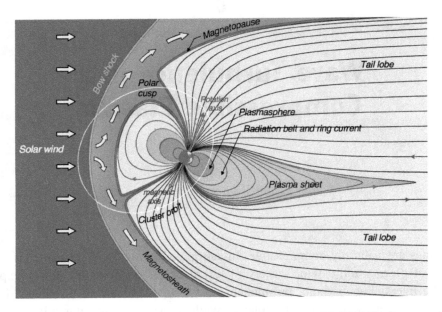

Figure 8.1 Earth's magnetoshere (image created by P. Robert, LPP/CNRS). Compressible turbulence is clearly observed in the magnetosheath. Reprinted with permission (Galtier, 2016).

in the Hall MHD approach, this corresponds to very oblique wavevectors, that is close to the direction perpendicular to the uniform magnetic field (Voitenko, 1998; Galtier and Meyrand, 2015).

At much smaller spatial and temporal scales, we encounter Langmuir wave turbulence, studied by Zakharov (1967) (see also Zakharov, 1972; Musher et al., 1995). The purpose here is to describe the behavior of a plasma when the electroneutrality is momentarily broken. It this case, the electron population oscillates and produces waves characterized by a wavenumber smaller than the Debye length (which is the characteristic distance over which electrostatic potentials are screened out by a redistribution of the charged particles). Contrary to the situations mentioned above, this turbulence involves four-wave interactions.

We will start this chapter by presenting some turbulent properties of the solar wind, which is at the heart of many questions related to strong and weak wave turbulence. Then we will derive a weakly compressible model that describes the dynamics of electrons on a quasi-static ocean of ions. The kinetic equations will be derived for this compressible system and their properties and exact solutions will be discussed. We will finally show the strong similarity between this problem and inertial wave turbulence.

8.1 Multiscale Solar Wind

The Sun is a wonderful gift for people working on turbulence, because it acts like a giant natural wind tunnel (Goldstein and Roberts, 1999; Bruno and Carbone,

8.1 Multiscale Solar Wind

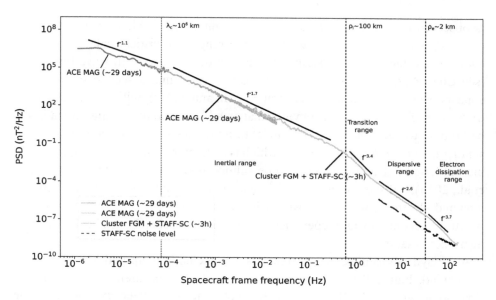

Figure 8.2 Magnetic spectrum measured in the solar wind by the ACE/NASA and Cluster/ESA spacecraft at one astronomical unit. The scales ρ_i and ρ_e are the ion and electron Larmor radius, respectively. There are several power laws, including that of MHD around $f^{-5/3}$ over nearly four decades. Figure created by L. Hadid and based on Sahraoui et al. (2020).

2013). The generated solar wind propagates in the interplanetary medium at several hundred km/s. There exists two solar winds: a slow equatorial wind with a velocity of the order of 400 km/s, and a fast polar wind at about 800 km/s. Nowadays, the properties of the interplanetary plasma are well known, thanks to the numerous in situ measurements carried out by the ESA and NASA probes, which can be seen as nonintrusive instruments. In Figure 8.2 we show a synthetic spectrum of the interplanetary magnetic field fluctuations carried by the solar wind. It is a frequency f spectrum as the in situ measurements are based on a time signal. Since the solar wind propagates relatively quickly, we can assume that the recorded time signal gives a fairly accurate picture of the plasma at a given time, because the plasma remains relatively fixed during the duration of the measurements (this is the Taylor hypothesis commonly used in wind tunnels). In other words, the frequency spectrum can be interpreted in a first approximation as a wavenumber spectrum ($f \sim k$). This spectrum shows several power laws over a total of eight decades, making it the widest spectrum ever measured in turbulence.

At very low frequencies ($f < 10^{-4}$Hz), we have a law close to f^{-1} whose origin is generally attributed to physical processes of the lower solar corona: this frequency range can therefore be interpreted as the domain of injection (the reservoir) of energy. At higher frequencies (up to 0.5Hz), a second power law is observed around $f^{-5/3}$. This frequency range, where Alfvén waves are also detected, is interpreted as an MHD inertial range; the critical balance phenomenology is often used to describe this regime of strong Alfvén wave turbulence (see Chapter 7). After a very steep transition spectrum, a second inertial zone appears over about a decade (between ~ 2 Hz and 40 Hz), with a magnetic spectrum often around $f^{-8/3}$. It is obvious that an additional physical ingredient has to be brought to the standard MHD model to reproduce such a law at spatial scales smaller than the ion Larmor radius (which is here approximatively equal to the ion inertial length). This ingredient can be simply modeled by the Hall term (Franci et al., 2015), a dispersive term, which takes into account the decoupling between ions and electrons, without, however, including kinetic effects. From 40 Hz, the spectrum becomes even steeper: electron scales are reached beyond which current instruments saturate.

As in hydrodynamics, intermittency is also a clearly identified property of the solar wind. Figure 8.3 shows a reproduction of the measurements made in the slow solar wind at MHD scales (main plot): the normalized exponents of the structure functions of the velocity and (parallel and perpendicular components

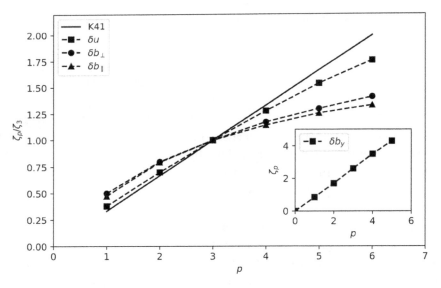

Figure 8.3 Main plot: Normalized (with respect to ζ_3) exponents of structure functions $\langle(\delta X)^p\rangle \sim \ell^{\zeta_p}$ of fluctuations in velocity ($X = u$) and magnetic field (components parallel $X = b_\parallel$ and perpendicular $X = b_\perp$ to the wind direction) measured in the slow solar wind (data from the German probe Helios; after Carbone et al. (2004). Kolmogorov's self-similar law in $p/3$ is indicated in solid line (K41). Inset: the plot of ζ_p for the y-component of the magnetic field at sub-MHD scales (data from Cluster/ESA) reveals a monoscaling behavior (with $\zeta_p \sim 0.85p$). After Kiyani et al. (2009).

to the wind direction) magnetic field reveal a stronger intermittency for the latter, with a stronger curvature. At sub-MHD scales (inset) a completely different behavior is observed for the three components of the magnetic field fluctuations (only the results for the y-component are reproduced here; note that the velocity fluctuations are currently not accessible for this high-frequency domain): a monoscaling behavior is observed. This result is similar to the experimental measurements made in fast-rotating hydrodynamics (van Bokhoven et al., 2009). This comparison is more than anecdotal, as will be explained in Section 8.7, and may be interpreted as a signature of the regime of weak wave turbulence.

The evolution of the solar wind with the heliocentric distance r offers an interesting question for turbulence. Indeed, the measurements (made by Voyager 1 and 2) of the ion temperature as a function of r reveal a decrease in $r^{-\alpha}$ with $\alpha < 1$ (and $r \in [0.3, 50 \text{AU}]$), which is significantly slower than an adiabatic cooling in $r^{-4/3}$ (Marsch et al., 1982; Matthaeus et al., 1999). Therefore, a local source of heating seems required. Currently, many studies are looking at this problem by considering a heating of turbulent origin (see the discussion in Section 8.2). The idea is that the turbulent cascade is an efficient process for transporting the energy supplied by the Sun at large scales towards sub-MHD scales, where it is dissipated by kinetic effects (Matthaeus, 2021). Without knowing the details of the kinetic processes that lead to the heating of the plasma, it is possible to estimate the heating rate by using the mean rate of energy transfer in the MHD inertial range. It can be shown that the measured values are close to that required to heat the solar wind (Sorriso-Valvo et al., 2007; Hadid et al., 2017).

8.2 Exact Law in Compressible Hall MHD

Kolmogorov's exact laws are valid for both eddy turbulence and wave turbulence. In plasma physics, exact laws were derived initially in the incompressible case, first for MHD (Politano and Pouquet, 1998) and later for Hall MHD (Galtier, 2008). Exact laws were then obtained for compressible plasmas (Banerjee and Galtier, 2013; Andrés et al., 2018; Ferrand et al., 2021a). As explained in Section 8.1, these different laws can be used to measure the mean rate of energy transfer ε in the solar wind (see Figure 8.4), which is a proxy for evaluating the mean rate of energy dissipation (Sorriso-Valvo et al., 2007; Vasquez et al., 2007; Osman et al., 2011; Banerjee et al., 2016; Hadid et al., 2017; Bandyopadhyay et al., 2020). The main conclusion of these observational studies is that weak compressibility has a stronger impact at sub-MHD scales than at MHD scales in the evaluation of ε.

In this section, we shall present such a law for compressible Hall MHD, a system described by the following set of equations (Galtier, 2016):

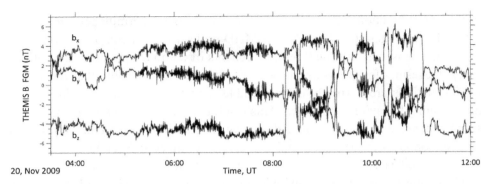

Figure 8.4 Variation of the magnetic field components (in nT) of the solar wind (November 2009; data from THEMIS/NASA spacecraft). A statistical study based on this interval shows that the heating rate ε is close to $10^{-16} \mathrm{Jm}^{-3}\mathrm{s}^{-1}$ (Hadid et al., 2017).

$$\frac{\partial \rho}{\partial t} + \nabla \cdot (\rho \mathbf{u}) = 0, \qquad (8.1\mathrm{a})$$

$$\rho \left(\frac{\partial \mathbf{u}}{\partial t} + \mathbf{u} \cdot \nabla \mathbf{u} \right) = -\nabla P + \frac{1}{\mu_0}(\nabla \times \mathbf{B}) \times \mathbf{B} + \mu \Delta \mathbf{u} + \frac{\mu}{3}\nabla(\nabla \cdot \mathbf{u}), \qquad (8.1\mathrm{b})$$

$$\frac{\partial \mathbf{B}}{\partial t} = \nabla \times (\mathbf{u} \times \mathbf{B}) - \nabla \times \left(\frac{(\nabla \times \mathbf{B}) \times \mathbf{B}}{\mu_0 n e} \right) + \eta \Delta \mathbf{B}, \qquad (8.1\mathrm{c})$$

$$\nabla \cdot \mathbf{B} = 0, \qquad (8.1\mathrm{d})$$

where $\mathbf{B}$ is the magnetic field, μ_0 the magnetic permeability of free space, n the plasma density (due to the electroneutrality, the electron and ion mass densities are equal), and e the absolute value of the electron charge (other variables are defined in Sections 2.9.2 and 7.1). From this system, using an isothermal closure, one can obtain the following compact exact law (Ferrand et al., 2021a):[2]

$$\begin{aligned}
-4\bar{\varepsilon} = \nabla_\ell \cdot &\left\langle \frac{\bar{\delta\rho}}{\rho_0} |\delta\mathbf{u}|^2 \delta\mathbf{u} + |\delta\mathbf{b}|^2 \delta\mathbf{u} - 2(\delta\mathbf{u} \cdot \delta\mathbf{b})\delta\mathbf{b} \right\rangle \\
&- \left\langle \frac{(\rho\theta' + \rho'\theta)}{2\rho_0} |\delta\mathbf{u}|^2 \right\rangle - 2\left\langle \frac{\delta\rho}{\rho_0} \delta\mathbf{u} \cdot \bar{\delta}\left(\mathbf{j_c} \times \mathbf{b} \right) \right\rangle \\
&+ d_i \nabla_\ell \cdot \left\langle 2(\delta\mathbf{b} \cdot \delta\mathbf{j_c})\delta\mathbf{b} - |\delta\mathbf{b}|^2 \delta\mathbf{j_c} \right\rangle - 2d_i \left\langle \delta(\mathbf{j} \times \mathbf{b}) \cdot \delta\mathbf{j_c} \right\rangle, \qquad (8.2)
\end{aligned}$$

where by definition $\bar{\varepsilon} \equiv \varepsilon/\rho_0$, $\bar{\delta}X \equiv (X + X')/2$, $\mathbf{b} \equiv \mathbf{B}/\sqrt{\mu_0 \rho_0}$ is the magnetic field normalized to a velocity, $\rho_0 \equiv \langle \rho \rangle$, $(\rho/\rho_0)\mathbf{j_c} \equiv \mathbf{j} = \nabla \times \mathbf{b}$, and $d_i \equiv m_i/\sqrt{\mu_0 \rho_0 e^2}$ is the ion inertial length which defines the characteristic scale

---

[2] Note that for a given system, there is no single expression for the exact law. The law given here is the most compact ever found.

below which the dynamics of ions and electrons decouples. In the first line of expression (8.2), we find the structure of the incompressible MHD law (Politano and Pouquet, 1998). The second line is purely compressible, with an HD term on the left and an MHD term on the right. In the third line, we finally find the contribution of the Hall effect, which is proportional to d_i; its expression can only be written as a flux in the incompressible limit. Overall, we see that the exact law for isothermal compressible Hall MHD has a similar expression to that of compressible hydrodynamics (which is a particular limit; see expression (2.97)) in that we have a flux $\mathbf{F}$ and a source S.

The exact law (8.2) is valid for eddy turbulence and wave turbulence; however, in the latter case, a relatively strong uniform magnetic field ($b_0 \gg b$) must be introduced. The weak turbulence regime has been studied in detail in the incompressible limit by Galtier (2006). This requires a tedious calculation, because one has to deal with two types of waves: left and right circularly polarized waves. Weakly compressible Hall MHD is a much more difficult case to treat and, to date, no complete theory has been proposed. In Section 8.6 we will present a theory of weak wave turbulence in a particular case where the dynamics is given by small-scale electrons evolving in a weakly compressible medium, and we will use from the beginning the anisotropy induced by the uniform magnetic field to simplify the problem. Under this limit, one can simplify expression (8.2). First, if we neglect ion velocity then $\mathbf{u} \to \mathbf{0}$. Second, we can write $\rho \equiv \rho_0 + \rho_1$ with $\rho_1/\rho_0 \ll 1$; then, $\mathbf{j_c} \simeq \mathbf{j} - \mathbf{j_{c_1}}$ with by definition $\mathbf{j_{c_1}} \equiv (\rho_1/\rho_0)\mathbf{j}$ is the (normalized) compressible electric current. With these assumptions, we obtain after some manipulation:

$$-4\bar{\varepsilon}/d_i = \nabla_\ell \cdot \left\langle (\delta\mathbf{b} \cdot \delta\mathbf{j})\delta\mathbf{b} - \frac{1}{2}|\delta\mathbf{b}|^2\delta\mathbf{j} \right\rangle + 2\langle \delta(\mathbf{j} \times \mathbf{b}_0) \cdot \delta\mathbf{j_{c_1}} \rangle \quad (8.3)$$
$$- \nabla_\ell \cdot \langle (2\delta\mathbf{b} \cdot \delta\mathbf{j_{c_1}})\delta\mathbf{b} - |\delta\mathbf{b}|^2\delta\mathbf{j_{c_1}} \rangle + 2\langle \delta(\mathbf{j} \times \mathbf{b}) \cdot \delta\mathbf{j_{c_1}} \rangle .$$

This exact law valid for wave turbulence reduces to the incompressible case when $\mathbf{j_{c_1}} = \mathbf{0}$. The contribution to the leading order comes from the first line: it includes the last term on the right because it is proportional to the strong uniform magnetic field $\mathbf{b}_0$. The second line corresponds to a correction in $\mathcal{O}(\rho_1/\rho_0)$ due to the small fluctuations in mass density; smaller corrections in $\mathcal{O}((\rho_1/\rho_0)^2)$ are neglected. As in the case of compressible hydrodynamics, the introduction of compressibility leads to a source which here consists of second-order mixed structure functions. Note that in the presence of $\mathbf{b}_0$, turbulence becomes anisotropic and the law cannot be integrated on a full sphere as it is usually done in isotropic hydrodynamics (see Chapter 2).

8.3 Weakly Compressible Electron MHD

When the Cluster/ESA spacecraft initially dedicated to the Earth's magnetosphere (see Figure 8.1) moved to the solar wind, a new physics became accessible in detail: the physics of turbulence at sub-MHD scales (Bale et al., 2005). This has contributed to the development of several new theoretical studies where plasma waves are the main ingredients. This includes the critical balance phenomenology for strong wave turbulence (see Chapter 7) in the context of electron MHD (Cho and Lazarian, 2004, 2009) or a gyrokinetic model (Howes et al., 2008, 2011), and the regime of weak wave turbulence in Hall MHD (Galtier and Bhattacharjee, 2003; Galtier, 2006). On the other hand, the exact laws, which are valid for both strong and weak wave turbulence, have also been widely used (see the Section 8.2).

The purpose of this section is to see how the weak compressibility assumption can be used in plasma physics to derive a new set of fluid equations, valid at sub-MHD scales, that can model in a simple way a part of the collisionless solar wind plasma, and from which wave turbulence may be used to obtain spectral predictions. The main assumption of the model called compressible electron MHD is that ions (mostly protons in the solar wind) are mainly at rest. It is not strictly speaking a kinetic model, which requires the use of velocity distribution functions to describe the particles in phase space and include kinetic effects, but a fluid model. However, the word kinetic is very often employed: for example, the weakly compressible waves associated with this system are generally called kinetic Alfvén waves.

8.3.1 From Bi-fluid to Compressible Electron MHD

We shall start with a bi-fluid system that describes separately the ions (for simplicity, we only consider protons) and the electrons. We have:

$$\frac{\partial \rho_i}{\partial t} + \nabla \cdot (\rho_i \mathbf{u}_i) = 0, \tag{8.4a}$$

$$\rho_i \left(\frac{\partial \mathbf{u}_i}{\partial t} + \mathbf{u}_i \cdot \nabla \mathbf{u}_i \right) = -\nabla P_i + n_i e \mathbf{E} + n_i e \mathbf{u}_i \times \mathbf{B}, \tag{8.4b}$$

$$\frac{\partial \rho_e}{\partial t} + \nabla \cdot (\rho_e \mathbf{u}_e) = 0, \tag{8.4c}$$

$$\rho_e \left(\frac{\partial \mathbf{u}_e}{\partial t} + \mathbf{u}_e \cdot \nabla \mathbf{u}_e \right) = -\nabla P_e - n_e e \mathbf{E} - n_e e \mathbf{u}_e \times \mathbf{B}, \tag{8.4d}$$

$$\frac{\partial \mathbf{B}}{\partial t} = -\nabla \times \mathbf{E}, \tag{8.4e}$$

$$\nabla \times \mathbf{B} = \mu_0 \mathbf{J} = \mu_0 e n_e (\mathbf{u}_i - \mathbf{u}_e), \tag{8.4f}$$

8.3 Weakly Compressible Electron MHD

where the subscript s denotes the species (i for ion and e for electron), ρ_s is the mass density, m_s the mass, n_s the particle density (due to the electroneutrality $n_e = n_i$), $\mathbf{u}_s$ the velocity, P_s the pressure, and $\mathbf{E}$ the electric field (other notations are introduced in Section 8.2). Note that the electroneutrality relation imposes a constant adjustment of the ion density to the variation of the electron density. This is possible if $\nabla \cdot \mathbf{u}_i \sim \nabla \cdot \mathbf{u}_e$, which means that the ion velocity is negligible compared to the electron velocity, but not its divergence (note that if $u_i/u_e \ll 1$, electrons must be more incompressible than ions to satisfy the ordering).

We limit ourselves to scales smaller than the typical scale at which ions and electrons decouple, which is the ion inertial length d_i (or the ion Larmor radius; note that they are roughly the same in the solar wind at 1 AU where $\beta_i \sim 1$; its definition will be given later in this section). We seek to model the solar wind where a strong large-scale magnetic field $\mathbf{B_0} = B_0 \mathbf{e}_\parallel$ (with $|\mathbf{e}_\parallel| = 1$) is present. This leads to an anisotropic dynamics (see Chapter 7) for which we have the inequality $\partial_\parallel \ll \partial_\perp$ or, equivalently, $k_\parallel \ll k_\perp$. (Note that at the leading order of the present derivation, no distinction will be made between the $\parallel$ direction along the *total* magnetic field line and the z direction; this approximation is correct as long as $b_0 \gg b$.) At these scales, the dynamics is relatively fast and we can assume that $u_i/u_e \sim \mathcal{O}(\epsilon)$, with $\epsilon \ll 1$. From the momentum equation (8.4b), we obtain at main order the relation valid for a perfect gas (with k_B a the Boltzmann constant and T_s the temperature):

$$n_{i0} e \mathbf{E} = \nabla P_i = \nabla(n_{i1} k_B T_i), \tag{8.5}$$

where n_{i0} is a uniform particle density. For a weakly compressible plasma, we have $n_s = n_{s0} + n_{s1}$ with $n_{s1}/n_{s0} \sim \mathcal{O}(\epsilon)$. Later, we will assume that both ions and electrons are isothermal.

We further simplify the problem by neglecting the inertia of the electrons and thus keeping only the right-hand-side terms of equation (8.4d). By introducing these terms into the Maxwell–Faraday equation (8.4e), we obtain:

$$\frac{\partial \mathbf{B}}{\partial t} = \frac{m_e}{e} \nabla \times \left(\frac{\nabla P_e}{\rho_e} \right) + \nabla \times (\mathbf{u}_e \times \mathbf{B}). \tag{8.6}$$

After introducing the relation $\nabla \cdot \mathbf{u}_e = -\partial_t \rho_e / \rho_e - (\mathbf{u}_e \cdot \nabla \rho_e)/\rho_e$ and the use of some vector identities, we find:

$$\frac{\partial \mathbf{B}}{\partial t} - \frac{\mathbf{B}}{\rho_e} \frac{\partial \rho_e}{\partial t} = \frac{m_e}{e\rho_e^2} \nabla P_e \times \nabla \rho_e + \mathbf{B} \cdot \nabla \mathbf{u}_e - \mathbf{u}_e \cdot \nabla \mathbf{B} + \frac{\mathbf{B}}{\rho_e}(\mathbf{u}_e \cdot \nabla)\rho_e. \tag{8.7}$$

We introduce the normalized magnetic field $\mathbf{b} \equiv \mathbf{B}/\sqrt{\mu_0 \rho_{e0}} = \nabla \times \mathbf{a}$ and the normalized vector potential $\mathbf{a} \equiv \mathbf{A}/\sqrt{\mu_0 \rho_{e0}}$, with ρ_{e0} a uniform mass density ($\rho_e = \rho_{e0} + \rho_{e1}$ with $\rho_{e1}/\rho_{e0} \sim \mathcal{O}(\epsilon)$).[3] Furthermore, the use of the Maxwell–Ampère equation (8.4f) leads to the relation $\mathbf{u}_e = d_e \Delta \mathbf{a}$, with $d_e \equiv m_e/\sqrt{\mu_0 \rho_{e0} e^2}$ the electron inertial length. With these new variables, one obtains (at main order):

---

[3] Note that the normalization is made with the electron mass density and not with the ion mass density, which leads to the introduction of the ion inertial length d_i.

$$\frac{\partial \mathbf{b}}{\partial t} - \frac{\mathbf{b}_0}{\rho_{e0}}\frac{\partial \rho_{e1}}{\partial t} = \frac{d_e}{\rho_{e0}^2}\nabla P_e \times \nabla \rho_{e1}$$
$$+ d_e\left[(\mathbf{b}\cdot\nabla)\Delta\mathbf{a} - (\Delta\mathbf{a}\cdot\nabla)\mathbf{b} + \frac{d_e}{\rho_{e0}}\mathbf{b}_0(\Delta\mathbf{a}\cdot\nabla)\rho_{e1}\right]. \quad (8.8)$$

The implicit assumption is that in terms of velocity, we have $u_e/b_0 \sim \mathcal{O}(\epsilon)$.

We introduce the following decomposition for the normalized magnetic field:

$$\mathbf{b} = b_0\mathbf{e}_\| - \nabla \times (\psi\mathbf{e}_\| + g\mathbf{e_x}) = \begin{pmatrix} -\partial_y\psi \\ \partial_x\psi - \partial_\| g \\ b_0 + b_\| \end{pmatrix}, \quad (8.9)$$

where ψ is the stream function and $b_\| = \partial_y g$. Then, we find at main order:

$$\Delta\mathbf{a} = \begin{pmatrix} -\partial_y b_\| \\ \partial_x b_\| \\ -\Delta_\perp \psi \end{pmatrix}, \quad (8.10)$$

which leads to the relation (at main order), $\mathbf{u}_e \times \mathbf{B}_0 = d_e B_0 \nabla_\perp b_\|$. Coming back to the momentum equation for electrons (8.4d) and neglecting the electron inertia, we find (at main order) the following relation:

$$\nabla P_e + n_{e0}e\mathbf{E} + n_{e0}e\mathbf{u}_e \times \mathbf{B}_0 = 0. \quad (8.11)$$

The combination of the latter expression with relations (8.5) gives:

$$\nabla_\perp(n_{e1}k_B(T_e + T_i)) = -en_{e0}d_e B_0 \nabla_\perp b_\|. \quad (8.12)$$

In the isothermal case, the latter expression leads to the closure relationship:

$$-\frac{2}{\beta_i(1+1/\tau)}\frac{b_\|}{b_0} = \frac{n_{e1}}{n_{e0}} = \frac{\rho_{e1}}{\rho_{e0}}, \quad (8.13)$$

with by definition the plasma parameters $\beta_i \equiv 2\mu_0 n_i k_B T_i/B_0^2$ and $\tau \equiv T_i/T_e$. The latter relation means that in a weakly compressible plasma, the density and the parallel magnetic field are anti-correlated. With the closure relation (8.13), the isothermal assumption, and the use of vector identities, equation (8.8) reduces to:

$$\frac{\partial \mathbf{b}}{\partial t} + \frac{2\mathbf{e}_\|}{\beta_i(1+1/\tau)}\frac{\partial b_\|}{\partial t} = \quad (8.14)$$
$$d_e\left[b_0\partial_\|\Delta_\perp\mathbf{a} + (\mathbf{b}_\perp\cdot\nabla_\perp)\Delta_\perp\mathbf{a} - (\Delta_\perp\mathbf{a}\cdot\nabla_\perp)\mathbf{b}\right].$$

Note that the only difference with incompressible electron MHD is the presence of the second term in the left-hand side. The last step consists in introducing the stream function and again using some vector identities; we finally get, after simplification:[4]

---

[4] In particular, the operator $\mathbf{e}_\| \times \nabla_\perp$ must appear in front of each term of the first equation.

$$\frac{\partial \psi}{\partial t} = d_e b_0 \partial_\parallel b_\parallel + d_e (\mathbf{e}_\parallel \times \nabla_\perp \psi) \cdot \nabla_\perp b_\parallel, \qquad (8.15)$$

$$\mathcal{K} \frac{\partial b_\parallel}{\partial t} = -d_e b_0 \partial_\parallel \Delta_\perp \psi - d_e (\mathbf{e}_\parallel \times \nabla_\perp \psi) \cdot \nabla_\perp (\Delta_\perp \psi), \qquad (8.16)$$

with $\mathcal{K} \equiv 1 + \frac{2}{\beta_i(1+1/\tau)}$. We obtain a set of self-consistent equations that describe the nonlinear evolution of magnetic field fluctuations in a weakly compressible isothermal plasma embedded in a strong uniform magnetic field (with $\partial_\parallel \ll \nabla_\perp$). This system is called *reduced compressible electron MHD*. Compressibility only affects the equation for $b_\parallel$, leaving the equation for ψ intact. The incompressible limit is obtained when $\beta_i \to +\infty$ (i.e. $\mathcal{K} \to 1$): this is the well-known electron MHD system introduced by Kingsep et al. (1990). Actually, through a simple change of variables ($\psi \to \psi/\sqrt{\mathcal{K}}$ and $b_0 \to b_0/\sqrt{\mathcal{K}}$; see also Galtier and Meyrand, 2015), one can recover the reduced incompressible electron MHD equations. The consequence is remarkable: the regime of (weak or strong) wave turbulence is the same for both systems, weakly compressible and incompressible electron MHD. Therefore, the nature of the waves (kinetic Alfvén or whistler; see Section 8.4 for the wave derivation) becomes a secondary point for the understanding of the turbulence regime.

8.3.2 Fluid versus Kinetic Models

Weakly compressible electron MHD has been derived from the bi-fluid system (8.4a)–(8.4f). A derivation from the gyrokinetic equations is also possible (Schekochihin et al., 2009); it is a more rigorous way of deriving equations (8.15)–(8.16). A basic assumption is, however, necessary concerning the distribution functions, which must be close to Maxwellian. We note that this situation is relatively far from the conditions encountered in the collisionless solar wind, where it has long been known that the velocity distribution functions are more complex, for example bi-Maxwellian or anisotropic (Montgomery et al., 1968; Marsch et al., 1982; Pilipp et al., 1987). We arrive here to an interesting debate in the plasma community. Fluid models are often criticized for their simplicity by plasma physicists, who sometimes consider that a fluid model is not relevant to describe reality. It is obvious that kinetic effects (e.g. Landau damping, finite Larmor radius effect) cannot be captured with a fluid model such as Hall MHD; however, it must be admitted that limiting oneself to a kinetic description that only considers a plasma with a velocity distribution function close to Maxwellian does not lead to the correct description of space plasmas either. When we work on turbulence in plasma physics, the goal for a theoretician is to find a prediction (very often for the energy spectrum), and in practice we generally come back to basic concepts that are fundamentally rooted in incompressible fluid turbulence. But we must remain modest in our quest for fundamental laws and accept fluid prediction that frequently gives a qualitatively correct answer. This comment does not mean that kinetic effects are unimportant. In the case of solar wind turbulence, their role in energy dissipation is of crucial importance: they can potentially modify the power-law predictions and lead to a nonconstant value (a decrease) of the mean rate of energy transfer (see, e.g., Ferrand et al., 2021b).

8.4 Kinetic Alfvén Waves (KAW)

The linearization of the Fourier transform of equations (8.15)–(8.16) leads to the system (hereafter, for simplicity the ∥ index for the parallel component of the magnetic field is not written in Fourier space and we use the index k for the **k**-dependence of the variables):

$$-\omega_{emhd}\hat{\psi}_k = d_e b_0 k_\parallel \hat{b}_k, \qquad (8.17)$$

$$-\mathcal{K}\omega_{emhd}\hat{b}_k = d_e b_0 k_\parallel k_\perp^2 \hat{\psi}_k, \qquad (8.18)$$

and eventually to the dispersion relation:

$$\boxed{\omega_{emhd} = d_e b_0 k_\parallel k_\perp / \sqrt{\mathcal{K}}}. \qquad (8.19)$$

The waves associated with a finite β_i are called kinetic Alfvén waves, while in the $\beta_i \to +\infty$ limit (i.e. $\mathcal{K} \to 1$) they are called (oblique) whistler waves. The latter are incompressible and the former weakly compressible. Note that whistler waves were first detected during World War I. They are audio frequency waves, often produced by lightning. Once produced in the magnetosphere, these waves travel along closed magnetic field lines from one hemisphere to the other. Their phase and group velocities are both proportional to k, implying that higher-frequency waves have higher group and phase velocities. Thus, the high-frequency part of the whistler wave packet will reach a detector earlier than its low-frequency part, and it will appear as a falling tone in a frequency–time sonogram (see Figure 8.5).

8.5 Spectral Phenomenology

8.5.1 Strong KAW Turbulence

The spectral prediction for the magnetic energy is straightforward. In strong kinetic Alfvén wave turbulence we have:

$$\varepsilon \sim \frac{E_\ell^b}{\tau_{NL}}, \qquad (8.20)$$

with

$$\tau_{NL} \sim \frac{1}{d_e k_\perp^2 b_\parallel} \sim \frac{\mathcal{K} b_\parallel}{d_e k_\perp^4 \psi^2}. \qquad (8.21)$$

8.5 Spectral Phenomenology

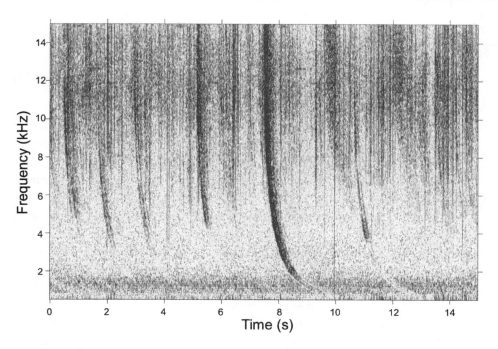

Figure 8.5 Spectrogram showing several whistler signals recorded by the Palmer Station (Antarctica) on August 24, 2005. These waves occur at audio frequencies. Credit: D. I. Golden, Stanford University. CC BY-SA 3.0.

By introducing the axisymmetric spectrum $E_\ell^b \sim E^b(k_\perp, k_\parallel) k_\perp k_\parallel$ and the balance relation $\sqrt{\mathcal{K}} b_\parallel \sim k_\perp \psi$, we find:

$$\boxed{E^b(k_\perp, k_\parallel) \sim \mathcal{K}^{1/3} \left(\frac{\varepsilon}{d_e}\right)^{2/3} k_\perp^{-7/3} k_\parallel^{-1}}, \quad (8.22)$$

which is the anisotropic generalization of the prediction made by Biskamp et al. (1996) for isotropic incompressible electron MHD. Note that this spectral prediction is compatible dimensionally with the small-scale limit of the exact law (8.3). Assuming a perpendicular dynamics, the exact law is written dimensionally as:

$$\bar{\varepsilon} \sim d_i b^2 j / \ell_\perp \sim d_i b^3 / \ell_\perp^2 \sim d_i k_\perp^2 (E^b(k_\perp, k_\parallel) k_\perp k_\parallel)^{3/2}, \quad (8.23)$$

hence the spectral prediction (8.22). Clearly, it is the exact law that gives solid support to the phenomenology, not the other way around.

On the other hand, we can use the critical balance conjecture:

$$\tau_{emhd} \sim \tau_{NL}, \quad (8.24)$$

with $\tau_{emhd} \sim 1/\omega_{emhd}$. This leads to the relation:

$$k_\| \sim \mathcal{K}^{1/6}\left(\frac{\varepsilon}{d_e b_0^3}\right)^{1/3} k_\perp^{1/3} \sim k_0^{2/3} k_\perp^{1/3}, \qquad (8.25)$$

with k_0 the characteristic wavenumber of magnetic energy injection. The latter expression shows that kinetic Alfvén wave turbulence is more anisotropic than (incompressible) MHD turbulence, for which the critical balance conjecture leads to the relation $k_\| \sim k_0^{1/3} k_\perp^{2/3}$, which was verified numerically by Cho and Lazarian (2009).

Strong kinetic Alfvén wave turbulence is often cited as a model for the solar wind at sub-MHD scales, however, the comparison with the data reveals a discrepancy, with a frequency spectrum characterized by a power-law index close to $-8/3$ instead of $-7/3$ (see Figure 8.2). Furthermore, a monoscaling behavior is measured for the magnetic field (see Figure 8.3) and the mass density fluctuations (Kiyani et al., 2009; Roberts et al., 2020), which are in strong contrast to the classical intermittency found at MHD scales and more generally in strong turbulence (see Chapter 2). Although no firm conclusion has been reached for the interpretation, these anomalous properties can find a natural explanation in the regime of weak wave turbulence.

8.5.2 Weak KAW Turbulence

The phenomenology of weak wave turbulence is based on three-wave interactions. Therefore, we have:

$$\varepsilon \sim \frac{E_\ell^b}{\tau_{tr}}, \qquad (8.26)$$

with the cascade time:

$$\tau_{tr} \sim \omega_{emhd} \tau_{NL}^2 \sim \frac{\sqrt{\mathcal{K}} b_0 k_\|}{d_e k_\perp^5 \psi^2}. \qquad (8.27)$$

As before, we introduce the axisymmetric spectrum $E_\ell^b \sim E^b(k_\perp, k_\|) k_\perp k_\|$ and get in this case:

$$E^b(k_\perp, k_\|) \sim \mathcal{K}^{1/4} \sqrt{\frac{b_0 \varepsilon}{d_e}} k_\perp^{-5/2} k_\|^{-1/2}, \qquad (8.28)$$

which is also the exact solution found by Galtier and Bhattacharjee (2003) in the framework of incompressible electron MHD ($\mathcal{K} = 1$). In Section 8.6.2, we will show that it is also an exact solution for kinetic Alfvén wave turbulence.

8.6 Theory of Weak KAW Turbulence

8.6.1 Canonical Variables

We start from the reduced compressible electron MHD equations (8.15)–(8.16). Their Fourier transforms give:

$$\frac{\partial \hat{\psi}_k}{\partial t} = i d_e b_0 k_\| \hat{b}_k - d_e \int_{\mathbb{R}^6} (\mathbf{e}_\| \times \mathbf{p}_\perp) \cdot \mathbf{q}_\perp \hat{\psi}_p \hat{b}_q \delta(\mathbf{k} - \mathbf{p} - \mathbf{q}) d\mathbf{p} d\mathbf{q}, \quad (8.29)$$

$$\mathcal{K}\frac{\partial \hat{b}_k}{\partial t} = i d_e b_0 k_\| k_\perp^2 \hat{\psi}_k - \int_{\mathbb{R}^6} d_e (\mathbf{e}_\| \times \mathbf{p}_\perp) \cdot \mathbf{q}_\perp q_\perp^2 \hat{\psi}_p \hat{\psi}_q \delta(\mathbf{k} - \mathbf{p} - \mathbf{q}) d\mathbf{p} d\mathbf{q}. \quad (8.30)$$

We introduce the canonical variables:

$$A_k^s = k_\perp \hat{\psi}_k - s\sqrt{\mathcal{K}} \hat{b}_k, \quad (8.31)$$

where the directional polarity $s = \pm$. This choice leads to a simple writing for the energy E. Indeed, for a compressible plasma we have:

$$E = \frac{1}{2}\rho_e(\mathbf{b}_\perp^2 + b_\|^2) + \rho_e U, \quad (8.32)$$

with U the internal energy. For an isothermal closure, we have $U = c_s^2 \ln(\rho_{e1}/\rho_{e0})$; this leads to:

$$\rho_e U \simeq c_s^2 \rho_{e0} \rho_{e1} + \frac{1}{2} c_s^2 \frac{\rho_{e1}^2}{\rho_{e0}}. \quad (8.33)$$

Since by definition $\langle \rho_{e1} \rangle = 0$ (where $\langle \rangle$ can be seen as the space average), we need to consider the second-order term for the internal energy. At main order we obtain:

$$\langle E \rangle = \frac{1}{2}\rho_{e0} \langle \mathbf{b}_\perp^2 \rangle + \frac{1}{2}\rho_{e0} \langle b_\|^2 \rangle + \frac{1}{2}\frac{c_s^2}{\rho_{e0}} \langle \rho_{e1}^2 \rangle. \quad (8.34)$$

Using the closure (8.13), the definition of the sound speed $c_s^2 \equiv (\beta_i + \beta_e)b_0^2/2$, and some manipulation, we find eventually:

$$\langle E \rangle = \frac{1}{2}\rho_{e0} \langle \mathbf{b}_\perp^2 \rangle + \frac{1}{2}\mathcal{K}\rho_{e0} \langle b_\|^2 \rangle. \quad (8.35)$$

From the canonical variables, it is easy to prove that in Fourier space:

$$E(\mathbf{k}) = \frac{1}{2}\rho_{e0} \sum_s |A_k^s|^2 = \frac{1}{2}\rho_{e0} k_\perp^2 |\hat{\psi}_k|^2 + \frac{1}{2}\mathcal{K}\rho_{e0} |\hat{b}_k|^2, \quad (8.36)$$

which is the energy per wavevector.

Introducing the canonical variables, we find after some manipulation:

$$\frac{\partial A_k^s}{\partial t} + is\omega_{emhd}A_k^s = \frac{d_e k_\perp}{4\sqrt{\mathcal{K}}} \int_{\mathbb{R}^6} \sum_{s_p s_q} \mathbf{e}_\parallel \qquad (8.37)$$

$$\times (\mathbf{p}_\perp \times \mathbf{q}_\perp) \frac{s_q}{p_\perp} A_p^{s_p} A_q^{s_q} \delta(\mathbf{k} - \mathbf{p} - \mathbf{q}) d\mathbf{p} d\mathbf{q}$$

$$+ \frac{sd_e}{4\sqrt{\mathcal{K}}} \int_{\mathbb{R}^6} \sum_{s_p s_q} \mathbf{e}_\parallel \cdot (\mathbf{p}_\perp \times \mathbf{q}_\perp) \frac{q_\perp}{p_\perp} A_p^{s_p} A_q^{s_q} \delta(\mathbf{k} - \mathbf{p} - \mathbf{q}) d\mathbf{p} d\mathbf{q}.$$

We can put the two integrals together and use the symmetry in $p_\perp$ and $q_\perp$; we find:

$$\frac{\partial A_k^s}{\partial t} + is\omega_{emhd}A_k^s = \frac{sd_e}{8\sqrt{\mathcal{K}}} \int_{\mathbb{R}^6} \sum_{s_p s_q} \frac{\mathbf{e}_\parallel \cdot (\mathbf{p}_\perp \times \mathbf{q}_\perp)}{p_\perp q_\perp} (s_q q_\perp - s_p p_\perp)$$

$$(sk_\perp + s_p p_\perp + s_q q_\perp) A_p^{s_p} A_q^{s_q} \delta(\mathbf{k} - \mathbf{p} - \mathbf{q}) d\mathbf{p} d\mathbf{q}. \qquad (8.38)$$

We introduce a small parameter $\epsilon \ll 1$ such that:

$$A_k^s \equiv \epsilon a_k^s e^{-is\omega_k t}, \qquad (8.39)$$

where, hereafter, ω_k identifies to ω_{emhd}. Then, we obtain:

$$\frac{\partial a_k^s}{\partial t} = \frac{\epsilon d_e}{8\sqrt{\mathcal{K}}} \int_{\mathbb{R}^6} \sum_{s_p s_q} L_{-kpq}^{ss_p s_q} a_p^{s_p} a_q^{s_q} e^{i(s\omega_k - s_p \omega_p - s_q \omega_q)t} \delta(\mathbf{k} - \mathbf{p} - \mathbf{q}) d\mathbf{p} d\mathbf{q}, \qquad (8.40)$$

with:

$$L_{kpq}^{ss_p s_q} = s\frac{\mathbf{e}_\parallel \cdot (\mathbf{p}_\perp \times \mathbf{q}_\perp)}{p_\perp q_\perp} (s_q q_\perp - s_p p_\perp)(sk_\perp + s_p p_\perp + s_q q_\perp). \qquad (8.41)$$

Equation (8.40) describes the slow evolution in time of the kinetic Alfvén wave amplitude. This is a classical form for three-wave interactions, with a term in the right-hand side of weak amplitude (proportional to ϵ), a quadratic nonlinearity, and an exponential which, over long times, will give a nonzero contribution only when its coefficient cancels out. In the framework of incompressible electron MHD ($\mathcal{K} = 1$), Galtier and Bhattacharjee (2003) derived a similar equation where the only notable difference is the presence of a phase term whose origin lies in the use of a complex helicity basis (and whose dependence eventually disappears in the derivation of the kinetic equations).

At this level, two properties can already be deduced. First, we observe that there is no nonlinear coupling when the wavevectors $\mathbf{p}_\perp$ and $\mathbf{q}_\perp$ are collinear. This means in particular that two wave packets travelling in the same direction at the same speed cannot lead to strong nonlinear effects. Second, there is no nonlinear coupling when $p_\perp$ and $q_\perp$ are equal, if at the same time their polarities s_p and s_q are equal. We have already met this property for inertial wave turbulence (see Chapter 6) and, as mentioned, it seems to be quite general for helical waves (Kraichnan, 1973; Waleffe, 1992; Turner, 2000; Galtier, 2003).

8.6.2 Kinetic Equations and Exact Solutions

To derive the kinetic equations we need to use the following properties:

$$L_{kpq}^{ss_ps_q} = L_{kqp}^{ss_qs_p}, \tag{8.42a}$$

$$L_{0pq}^{ss_ps_q} = 0, \tag{8.42b}$$

$$L_{-k-p-q}^{ss_ps_q} = L_{kqp}^{ss_qs_p}, \tag{8.42c}$$

$$L_{kpq}^{-s-s_p-s_q} = -L_{kqp}^{ss_qs_p}, \tag{8.42d}$$

$$L_{pkq}^{s_pss_q} = \frac{s_p p_\perp p_\parallel}{s k_\perp k_\parallel} L_{kpq}^{ss_ps_q}, \tag{8.42e}$$

$$L_{qkp}^{s_qss_p} = \frac{s_q q_\perp q_\parallel}{s k_\perp k_\parallel} L_{kpq}^{ss_ps_q}. \tag{8.42f}$$

We will also use the resonance condition, which can be written after some manipulation as:

$$\frac{s_q q_\perp - s_p p_\perp}{k_\parallel} = \frac{sk_\perp - s_q q_\perp}{p_\parallel} = \frac{s_p p_\perp - sk_\perp}{q_\parallel}. \tag{8.43}$$

Note that this expression is similar to the case of inertial wave turbulence (see Chapter 6). It is interesting to discuss the particular case of strongly local interactions in the perpendicular direction that generally make a dominant contribution to the turbulent dynamics. In this case, we have $k_\perp \simeq p_\perp \simeq q_\perp$, and the previous expression simplifies to:

$$\frac{s_q - s_p}{k_\parallel} \simeq \frac{s - s_q}{p_\parallel} \simeq \frac{s_p - s}{q_\parallel}. \tag{8.44}$$

If $k_\parallel$ is nonzero, the term on the left will give a nonnegligible contribution only when $s_p = -s_q$. We do not consider the case $s_p = s_q$, which is not relevant to the main order in the case of local interactions, as we can see from the expression of the interaction coefficient (8.41). The immediate consequence is that either the middle term or the right-hand term has a numerator that cancels itself out (at main order), which implies that the associated denominator must also cancel (at main order) to satisfy the equality: for example, if $s = s_p$ then $q_\parallel \simeq 0$. This condition means that the transfer in the parallel direction is negligible: indeed, the integration of equation (8.40) in the parallel direction is then reduced to a few modes (since $p_\parallel \simeq k_\parallel$), which strongly limits the transfer between parallel modes. The cascade in the parallel direction is thus possible, but relatively weak compared to that in the perpendicular direction, which justifies a posteriori the assumption made initially, that is, $k_\perp \gg k_\parallel$. Therefore, this situation is comparable to inertial wave turbulence (see Chapter 6).

After a long but standard development, we eventually find:

$$\frac{\partial e^s(\mathbf{k})}{\partial t} = \frac{\epsilon^2 \pi d_e^2}{16\mathcal{K}} \int_{\mathbb{R}^6} \sum_{s_p s_q} |L_{kpq}^{ss_ps_q}|^2 \delta(s\omega_k + s_p\omega_p + s_q\omega_q) \delta(\mathbf{k}+\mathbf{p}+\mathbf{q})$$
$$\frac{e^s(\mathbf{k})e^{s_p}(\mathbf{p})e^{s_q}(\mathbf{q})}{sk_\perp k_\parallel} \left[\frac{sk_\perp k_\parallel}{e^s(\mathbf{k})} + \frac{s_p p_\perp p_\parallel}{e^{s_p}(\mathbf{p})} + \frac{s_q q_\perp q_\parallel}{e^{s_q}(\mathbf{q})} \right] d\mathbf{p} d\mathbf{q}, \tag{8.45}$$

with by definition the energy density tensor $e^s(\mathbf{k})$ for homogeneous turbulence:

$$\langle a_k^s a_{k'}^{s'} \rangle = e^s(\mathbf{k})\delta(\mathbf{k}+\mathbf{k}')\delta_{ss'}, \tag{8.46}$$

where $\delta_{ss'}$ means $s = s'$. Equation (8.45) is the kinetic equation for kinetic Alfvén wave turbulence. The incompressible electron MHD limit can be recovered when the compressibility $\mathcal{K} = 1$. Note that with the current approach, we directly find the anisotropic limit ($k_\perp \gg k_\parallel$) of the incompressible case. From expression (8.45) it is straightforward to show (using the resonance condition) the detailed conservation of the two invariants, the energy and the magnetic helicity.

We can introduce the axisymmetric spectra for the magnetic energy $E_k \equiv E(k_\perp, k_\parallel) = 2\pi k_\perp(e^+(\mathbf{k}) + e^-(\mathbf{k}))$ and the magnetic helicity $H_k \equiv H(k_\perp, k_\parallel) = 2\pi(e^+(\mathbf{k}) - e^-(\mathbf{k}))$. After some calculation where, in particular, the resonance condition in frequency and the symmetry in directional polarity are used, we obtain the following kinetic equations:

$$\partial_t \begin{Bmatrix} E_k \\ H_k \end{Bmatrix} = \frac{\epsilon^2 d_e^2}{64\mathcal{K}} \sum_{ss_ps_q} \int_{\Delta_\perp} sk_\parallel s_p p_\parallel p_\perp \sin\theta_k \tag{8.47}$$
$$\times \left(\frac{s_q q_\perp - s_p p_\perp}{k_\parallel} \right)^2 (sk_\perp + s_p p_\perp + s_q q_\perp)^2$$
$$\times \begin{Bmatrix} \frac{E_q}{q_\perp}(E_k/k_\perp - E_p/p_\perp) + s_q H_q(sH_k - s_p H_p) \\ s\left[\frac{E_q}{q_\perp}(sH_k - s_p H_p) + s_q H_q(E_k/k_\perp - E_p/p_\perp)\right] \end{Bmatrix}$$
$$\times \delta(s\omega_k + s_p\omega_p + s_q\omega_q)\delta(k_\parallel + p_\parallel + q_\parallel)dp_\perp dq_\perp dp_\parallel dq_\parallel,$$

where θ_k is the angle between the wavevectors $p_\perp$ and $q_\perp$ in the triangle $\mathbf{k}_\perp + \mathbf{p}_\perp + \mathbf{q}_\perp = 0$, and $\Delta_\perp$ is the integration domain (infinitely extended band) corresponding to this triangle (see Figure 5.5). These kinetic equations were obtained for the first time by Galtier and Bhattacharjee (2003) in the incompressible limit, then discussed by Galtier and Meyrand (2015) and Passot and Sulem (2019) in the context of kinetic Alfvén waves.

The exact power-law solutions are found by applying the generalized Zakharov transformation to the axisymmetric case (Kuznetsov, 1972): for the nonzero constant energy flux (Kolmogorov–Zakharov spectrum), we have:

$$\boxed{E(k_\perp, k_\parallel) \sim k_\perp^{-5/2} k_\parallel^{-1/2}}, \tag{8.48}$$

and $H(k_\perp, k_\|) \sim k_\perp^{-7/2} k_\|^{-1/2}$. These solutions correspond to a direct cascade of energy. In particular, this means that the spectrum of magnetic helicity is not the consequence of a dynamics specific to helicity, but the trace of a dynamics induced by energy.

8.7 Inertial/Kinetic-Alfvén Wave Turbulence: A Twin Problem

The exact solutions found for the energy (8.48) are similar to the one obtained for inertial wave turbulence (6.23), while the kinetic equations are different. In fact, despite their apparent difference, we can show that these two kinetic equations obey the same nonlinear diffusion equation when only local interactions (in the perpendicular direction) are retained. To prove this claim, we will follow the same method as in Section 6.6.

The kinetic equation takes the following form:

$$\frac{\partial E_k}{\partial t} = \sum_{s s_p s_q} \int T_{\mathbf{kpq}}^{s s_p s_q} dp_\perp dq_\perp dp_\| dq_\| , \qquad (8.49)$$

with the transfer function per mode:

$$T_{\mathbf{kpq}}^{s s_p s_q} = \frac{d_e^2}{64\sqrt{\mathcal{K}}} \frac{s k_\| s_p p_\| p_\perp}{k_\perp p_\perp q_\perp} \sin\theta \left(\frac{s_q q_\perp - s_p p_\perp}{k_\|}\right)^2 (s k_\perp + s_p p_\perp + s_q q_\perp)^2$$
$$E_q(p_\perp E_k - k_\perp E_p) \delta(s\omega_k + s_p \omega_p + s_q \omega_q) \delta(k_\| + p_\| + q_\|). \qquad (8.50)$$

It can be noted that the small parameter ϵ has been absorbed in the time derivative and therefore no longer appears explicitly: this means that we are focusing on the long times of wave turbulence. The transfer function verifies the following symmetry property, which we will be used later:

$$T_{\mathbf{pkq}}^{s_p s s_q} = -T_{\mathbf{kpq}}^{s s_p s_q} . \qquad (8.51)$$

Within the limit of strongly local interactions, we can write:

$$p_\perp = k_\perp(1 + \epsilon_p) \quad \text{and} \quad q_\perp = k_\perp(1 + \epsilon_q), \qquad (8.52)$$

with $0 < \epsilon_p \ll 1$ and $0 < \epsilon_q \ll 1$. We can then introduce an arbitrary function $f(k_\perp, k_\|)$ and integrate the kinetic equation; we get:

$$\frac{\partial}{\partial t}\left(\int f(k_\perp, k_\|) E_k dk_\perp dk_\|\right) = \sum_{s s_p s_q} \int f(k_\perp, k_\|) T_{\mathbf{kpq}}^{s s_p s_q} dk_\perp dk_\| dp_\perp dq_\perp dp_\| dq_\|$$

$$= \frac{1}{2} \sum_{s s_p s_q} \int [f(k_\perp, k_\|) - f(p_\perp, p_\|)] T_{\mathbf{kpq}}^{s s_p s_q} dk_\perp dk_\| dp_\perp dq_\perp dp_\| dq_\| . \qquad (8.53)$$

For local interactions, we have at the main order (we neglect the contribution of the parallel wavenumber):

$$f(p_\perp, p_\|) = f(k_\perp, k_\|) + (p_\perp - k_\perp)\frac{\partial f(k_\perp, k_\|)}{\partial k_\perp}$$
$$= f(k_\perp, k_\|) + \epsilon_p k_\perp \frac{\partial f(k_\perp, k_\|)}{\partial k_\perp}. \tag{8.54}$$

One obtains (as for inertial wave turbulence):

$$\frac{\partial}{\partial t}\left(\int f(k_\perp, k_\|) E_k dk_\perp dk_\|\right) = \tag{8.55}$$
$$-\frac{1}{2}\sum_{s_s s_p s_q}\int \epsilon_p k_\perp \frac{\partial f(k_\perp, k_\|)}{\partial k_\perp} T^{s_s s_p s_q}_{\mathbf{kpq}} dk_\perp dk_\| dp_\perp dq_\perp dp_\| dq_\|.$$

An integration by part of the right-hand term allows us to write:

$$\frac{\partial E_k}{\partial t} = \frac{1}{2}\frac{\partial}{\partial k_\perp}\left(\sum_{s_s s_p s_q}\int \epsilon_p k_\perp T^{s_s s_p s_q}_{\mathbf{kpq}} dp_\perp dq_\perp dp_\| dq_\|\right). \tag{8.56}$$

The local form of the transfer function $T^{s_s s_p s_q}_{\mathbf{kpq}}$ can be deduced by using the locality in the perpendicular direction. In particular, we have:

$$k_\perp^2 p_\perp^2 q_\perp^2 = k_\perp^6, \tag{8.57}$$

$$\left(\frac{s_q q_\perp - s_p p_\perp}{k_\|}\right)^2 = \left(\frac{s_q - s_p + s_q \epsilon_q - s_p \epsilon_p}{k_\|}\right)^2 k_\perp^2, \tag{8.58}$$

$$(sk_\perp + s_p p_\perp + s_q q_\perp)^2 = (s + s_p + s_q)^2 k_\perp^2, \tag{8.59}$$

$$E_q(p_\perp E_k - k_\perp E_p) = -\epsilon_p k_\perp^3 E_k \frac{\partial(E_k/k_\perp)}{\partial k_\perp}, \tag{8.60}$$

$$\sin\theta = \sin(\pi/3) = \frac{\sqrt{3}}{2}, \tag{8.61}$$

$$\delta(s\omega_k + s_p\omega_p + s_q\omega_q) = \frac{\sqrt{\mathcal{K}}}{k_\perp d_e b_0}\delta(sk_\| + s_p p_\| + s_q q_\|). \tag{8.62}$$

After simplification, we arrive at:

$$T^{s_s s_p s_q}_{\mathbf{kpq}} = -\frac{\sqrt{3}}{32}\frac{d_e}{\sqrt{\mathcal{K}} b_0}\frac{s s_p p_\|}{k_\|}\epsilon_p k_\perp^4 E_k \frac{\partial(E_k/k_\perp)}{\partial k_\perp}$$
$$\delta(sk_\| + s_p p_\| + s_q q_\|)\delta(k_\| + p_\| + q_\|), \tag{8.63}$$

where we have considered the dominant transfer for which $s_p s_q = -1$. The resonance condition leads us to two possible combinations for parallel wavenumbers:

$$k_\| + p_\| - q_\| = 0 \quad \text{and} \quad k_\| + p_\| + q_\| = 0, \tag{8.64}$$

or

$$k_\| - p_\| + q_\| = 0 \quad \text{and} \quad k_\| + p_\| + q_\| = 0. \tag{8.65}$$

The solution is either $q_\| = 0$ or $p_\| = 0$, which means that the strong locality assumption does not apply for the parallel direction. The second solution cancels the transfer function, so we will consider only the first solution, for which we obtain:

$$\frac{\partial E_k}{\partial t} = \frac{\sqrt{3}}{64} \frac{d_e}{\sqrt{\mathcal{K}} b_0} \frac{\partial}{\partial k_\perp} \left(k_\perp^7 E_k \frac{\partial (E_k/k_\perp)}{\partial k_\perp} \right) \int_{-\epsilon}^{+\epsilon} \epsilon_p^2 d\epsilon_p \int_{-\epsilon}^{+\epsilon} d\epsilon_q. \tag{8.66}$$

After integration, we finally arrive at the following nonlinear diffusion equation for energy (Passot and Sulem, 2019):

$$\boxed{\frac{\partial E_k}{\partial t} = C \frac{\partial}{\partial k_\perp} \left(k_\perp^7 E_k \frac{\partial (E_k/k_\perp)}{\partial k_\perp} \right)}, \tag{8.67}$$

with the constant $C = \epsilon^4 d_e/(16\sqrt{3}\sqrt{\mathcal{K}} b_0)$. This nonlinear diffusion equation has been rigorously deduced from the kinetic equation. It describes weak kinetic Alfvén wave turbulence in the limit of strongly local interactions in the perpendicular direction. Note that it is also possible to obtain this diffusion equation using phenomenological arguments (David and Galtier, 2019): the calculation is then simpler but it does not allow us to obtain the exact expression of the constant C.

Surprisingly, the diffusion equation (8.67) is the same as for inertial wave turbulence (6.45); only the expression for C differs. This means that the physics of turbulence is essentially the same for both problems. As explained in Chapter 6, the exact solution found previously can be obtained here for a constant flux, and we can prove that this flux is positive, and thus the cascade direct. In Chapter 6, we made an additional comment about this nonlinear diffusion equation, which becomes very important for kinetic Alfvén wave turbulence: the numerical simulation reveals the existence of a steep solution in $k_\perp^{-8/3}$ during the nonstationary phase (see Figure 8.6). This anomalous spectrum is understood as a self-similar solution of the second kind (David and Galtier, 2019), which disappears when viscous scales are reached, with a bounce of the spectrum at small scales and finally the formation of the expected stationary solution in $k_\perp^{-5/2}$. However, unlike water, the solar wind is a collisionless plasma where dissipation involves physical processes completely different from a simple viscous term.[5] Also, the fact that the solar wind spectrum exhibits a scaling often close to $f^{-8/3}$ (see Figure 8.2) can in fact be interpreted as the signature of weak wave turbulence. It is perhaps not

---

[5] In numerical simulations, a viscous term is added to avoid numerical instabilities.

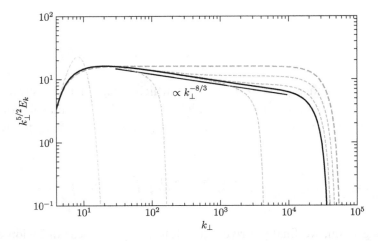

Figure 8.6 Temporal evolution (from bright to dark dashed lines) of the (compensated) magnetic spectrum E_k in KAW turbulence produced with a nonlinear diffusion model. An anomalous $k_\perp^{-8/3}$ spectrum is obtained (solid line) before its bounce at small scales (David and Galtier, 2019). Note that the extension of the inertial range is not physically realistic; however, it is necessary to unambiguously reveal the power-law index. Simulation made by V. David.

so surprising, then, that experimental measurements of intermittency in the fast-rotating case (van Bokhoven et al., 2009) and in the solar wind (Kiyani et al., 2009) give very close self-similar exponents.

8.8 Perspectives

An interesting perspective on compressible plasmas concerns the physics at sub-MHD scales in space plasmas, where kinetic Alfvén and whistler waves are often observed. As shown in Figures 8.2 and 8.3, the turbulent properties of the solar wind vary significantly as we cross the ion Larmor radius that separates MHD scales from sub-MHD scales. In general, the phenomenology of (incompressible) wave turbulence is used to obtain spectral predictions. However, the power-law exponents proposed for the magnetic energy spectrum, namely $-7/3$ and -2.5 for strong and weak wave turbulence respectively, do not match observations, where $-8/3$ is often measured (Alexandrova et al., 2012; Sahraoui et al., 2013; Bale et al., 2019). Additionally, self-similar exponents are found for the structure functions constructed from the magnetic field (Kiyani et al., 2009) or the mass density (Chen et al., 2014; Roberts et al., 2020), giving intermittency a very specific character. Understanding the origin of this spectral difference is a difficult subject because the solar wind plasma is collisionless and, therefore, the dissipation cannot be modeled as a simple viscosity localized at small scales. In particular, this means that the phenomenology of turbulence based implicitly on the equilibrium between a large-scale force and a small-scale dissipation can be questioned. As we have shown numerically in inertial wave turbulence (Chapter 6), the stationary solution is obtained after a bounce of the spectrum at small scales, which is only possible if a viscous term is present. The same behavior is observed for KAW turbulence (David and Galtier, 2019): during the nonstationary phase, a spectrum in $k_\perp^{-8/3}$ is formed, which transforms into $k_\perp^{-2.5}$ only after the rebound of the spectrum at small scales. What would happen without the effect of viscosity? One possible answer is that the initial spectrum in $k_\perp^{-8/3}$ would be maintained, which would explain why standard theories have so far failed to explain the data. This topic is at the heart of current issues related to solar wind turbulence, where the kinetic effects of plasma physics have, undoubtedly, also a role to play (Passot and Sulem, 2015, 2019). In this context, an interesting perspective concerns the direct numerical simulation of the compressible electron MHD equations derived in this chapter to study the differences between the strong and weak KAW turbulence regimes. The reproduction of solar wind intermittency measurements in the weak regime would provide an additional and perhaps definitive argument for a weak wave turbulence interpretation of solar wind turbulence at sub-MHD scales. If we take the comparison further, could we learn more about space plasmas from laboratory experiments using rotating water? In the context of a (compressible) bi-fluid description where there are several types of waves that are related to the ion and electron populations, their nonlinear interactions are not well understood. Theoretical and numerical studies are needed to make progress in this area.

References

Alexandrova, O., Lacombe, C., Mangeney, A., Grappin, R., and Maksimovic, M. 2012. Solar wind turbulent spectrum at plasma kinetic scales. *Astrophys. J.*, **760**(2), 121–127.

Andrés, N., Galtier, S., and Sahraoui, F. 2018. Exact law for homogeneous compressible Hall magnetohydrodynamics turbulence. *Phys. Rev. E*, **97**(1), 013204.

Bale, S. D., Kellogg, P. J., Mozer, F. S., Horbury, T. S., and Reme, H. 2005. Measurement of the electric fluctuation spectrum of magnetohydrodynamic turbulence. *Phys. Rev. Lett.*, **94**(21), 215002.

Bale, S. D., Badman, S. T., Bonnell, J. W. et al. 2019. Highly structured slow solar wind emerging from an equatorial coronal hole. *Nature*, **576**(7786), 237–242.

Bandyopadhyay, R., Sorriso-Valvo, L., Chasapis, A. et al. 2020. In situ observation of Hall magnetohydrodynamic cascade in space plasma. *Phys. Rev. Lett.*, **124**(22), 225101.

Banerjee, S., and Galtier, S. 2013. Exact relation with two-point correlation functions and phenomenological approach for compressible MHD turbulence. *Phys. Rev. E*, **87**(1), 013019.

Banerjee, S., Hadid, L., Sahraoui, F., and Galtier, S. 2016. Scaling of compressible magnetohydrodynamic turbulence in the fast solar wind. *Astrophys. J. Lett.*, **829**(2), L27 (5 pages).

Biskamp, D., Schwarz, E., and Drake, J. F. 1996. Two-dimensional electron magnetohydrodynamic turbulence. *Phys. Rev. Lett.*, **76**(8), 1264–1267.

Bruno, R., and Carbone, V. 2013. The solar wind as a turbulence laboratory. *Living Rev. Solar Physics*, **10**(1), 2.

Carbone, V., Bruno, R., Sorriso-Valvo, L., and Lepreti, F. 2004. Intermittency of magnetic turbulence in slow solar wind. *Plan. Space Sci.*, **52**(10), 953–956.

Chandran, B. D. G. 2008. Weakly turbulent magnetohydrodynamic waves in compressible low-beta plasmas. *Phys. Rev. Lett.*, **101**, 235004.

Chen, C. H. K., Sorriso-Valvo, L., Šafránková, J., and Němeček, Z. 2014. Intermittency of solar wind density fluctuations from ion to electron scales. *Astrophys. J. Lett.*, **789**(1), L8 (5 pages).

Cho, J., and Lazarian, A. 2004. The anisotropy of electron magnetohydrodynamic turbulence. *Astrophys. J. Lett.*, **615**(1), L41–L44.

Cho, J., and Lazarian, A. 2009. Simulations of electron magnetohydrodynamic turbulence. *Astrophys. J.*, **701**(1), 236–252.

David, V., and Galtier, S. 2019. $k_\perp^{-8/3}$ spectrum in kinetic Alfvén wave turbulence: Implications for the solar wind. *Astrophys. J. Lett.*, **880**(1), L10 (5 pages).

Ferrand, R., Galtier, S., and Sahraoui, F. 2021a. A compact exact law for compressible isothermal Hall magnetohydrodynamic turbulence. *J. Plasma Phys.*, **87**(2), 905870220.

Ferrand, R., Sahraoui, F., Laveder, D. et al. 2021b. Fluid energy cascade rate and kinetic damping: New insight from 3D Landau-fluid simulations. *Astrophys. J.*, **923**(1), 122 (8 pages).

Franci, L., Verdini, A., Matteini, L., Landi, S., and Hellinger, P. 2015. Solar wind turbulence from MHD to sub-ion scales: High-resolution hybrid simulations. *Astrophys. J.*, **804**(2), L39 (5 pages).

Galtier, S. 2003. Weak inertial-wave turbulence theory. *Phys. Rev. E*, **68**(1), 015301.

Galtier, S. 2006. Wave turbulence in incompressible Hall magnetohydrodynamics. *J. Plasma Phys.*, **72**, 721–769.

Galtier, S. 2008. Von Kármán–Howarth equations for Hall magnetohydrodynamic flows. *Phys. Rev. E*, **77**(1), 015302.

Galtier, S. 2016. *Introduction to Modern Magnetohydrodynamics*. Cambridge University Press.

Galtier, S., and Bhattacharjee, A. 2003. Anisotropic weak whistler wave turbulence in electron magnetohydrodynamics. *Phys. Plasmas*, **10**, 3065–3076.

Galtier, S., and Meyrand, R. 2015. Entanglement of helicity and energy in kinetic Alfvén wave/whistler turbulence. *J. Plasma Phys.*, **81**(1), 325810106.

Goldstein, M. L., and Roberts, D. A. 1999. Magnetohydrodynamic turbulence in the solar wind. *Phys. Plasmas*, **6**(11), 4154–4160.

Hadid, L. Z., Sahraoui, F., and Galtier, S. 2017. Energy cascade rate in compressible fast and slow solar wind turbulence. *Astrophys. J.*, **838**(1), 9 (11 pages).

Howes, G. G., Dorland, W., Cowley, S. C. et al. 2008. Kinetic simulations of magnetized turbulence in astrophysical plasmas. *Phys. Rev. Lett.*, **100**(6), 065004.

Howes, G. G., Tenbarge, J. M., Dorland, W. et al. 2011. Gyrokinetic simulations of solar wind turbulence from ion to electron scales. *Phys. Rev. Lett.*, **107**(3), 035004.

Kingsep, A. S., Chukbar, K. V., and Yankov, V. V. 1990. Electron magnetohydrodynamics. *Rev. Plasma Phys.*, **16**, 243–291.

Kiyani, K. H., Chapman, S. C., Khotyaintsev, Y. V., Dunlop, M. W., and Sahraoui, F. 2009. Global scale-invariant dissipation in collisionless plasma turbulence. *Phys. Rev. Lett.*, **103**, 075006.

Kraichnan, R. H. 1973. Helical turbulence and absolute equilibrium. *J. Fluid Mech.*, **59**, 745–752.

Kuznetsov, E. A. 1972. Turbulence of ion sound in a plasma located in a magnetic field. *J. Exp. Theor. Phys.*, **35**, 310–314.

Kuznetsov, E. A. 2001. Weak magnetohydrodynamic turbulence of a magnetized plasma. *Sov. J. Exp. Theo. Physics*, **93**(5), 1052–1064.

Marsch, E., Schwenn, R., Rosenbauer, H. et al. 1982. Solar wind protons: Three-dimensional velocity distributions and derived plasma parameters measured between 0.3 and 1 AU. *J. Geophys. Res.*, **87**(A1), 52–72.

Matthaeus, W. H. 2021. Turbulence in space plasmas: Who needs it? *Phys. Plasmas*, **28**(3), 032306.

Matthaeus, W. H., Zank, G. P., Smith, C. W., and Oughton, S. 1999. Turbulence, spatial transport, and heating of the solar wind. *Phys. Rev. Lett.*, **82**, 3444–3447.

Meyrand, R., Kiyani, K. H., Gürcan, O. D., and Galtier, S. 2018. Coexistence of weak and strong wave turbulence in incompressible Hall magnetohydrodynamics. *Phys. Rev. X*, **8**(3), 031066.

Montgomery, M. D., Bame, S. J., and Hundhausen, A. J. 1968. Solar wind electrons: Vela 4 measurements. *J. Geophys. Res.*, **73**(15), 49995003.

Musher, S. L., Rubenchik, A. M., and Zakharov, V. E. 1995. Weak Langmuir turbulence. *Phys. Rep.*, **252**(4), 177–274.

Osman, K. T., Wan, M., Matthaeus, W. H., Weygand, J. M., and Dasso, S. 2011. Anisotropic third-moment estimates of the energy cascade in solar wind turbulence using multispacecraft data. *Phys. Rev. Lett.*, **107**(16), 165001.

Passot, T., and Sulem, P. L. 2015. A model for the non-universal power law of the solar wind sub-ion-scale magnetic spectrum. *Astrophys. J. Lett.*, **812**(2), L37 (5 pages).

Passot, T., and Sulem, P. L. 2019. Imbalanced kinetic Alfvén wave turbulence: From weak turbulence theory to nonlinear diffusion models for the strong regime. *J. Plasma Phys.*, **85**(3), 905850301.

Pilipp, W. G., Miggenrieder, H., Montgomery, M. D. et al. 1987. Characteristics of electron velocity distribution functions in the solar wind derived from the helios plasma experiment. *J. Geophys. Res.*, **92**(A2), 1075–1092.

Politano, H., and Pouquet, A. 1998. Von Kármán–Howarth equation for magnetohydrodynamics and its consequences on third-order longitudinal structure and correlation functions. *Phys. Rev. E*, **57**, R21R24.

Roberts, O. W., Thwaites, J., Sorriso-Valvo, L., Nakamura, R., and Vörös, Z. 2020. Higher-order statistics in compressive solar wind plasma turbulence: High-resolution density observations from the Magnetospheric MultiScale Mission. *Frontiers in Physics*, **8**, 584063.

Sahraoui, F., Belmont, G., and Rezeau, L. 2003. Hamiltonian canonical formulation of Hall-magnetohydrodynamics: Toward an application to weak turbulence theory. *Phys. Plasmas*, **10**(5), 1325–1337.

Sahraoui, F., Huang, S. Y., Belmont, G. et al. 2013. Scaling of the electron dissipation range of solar wind turbulence. *Astrophys. J.*, **777**(1), 15 (11 pages).

Sahraoui, F., Hadid, L., and Huang, S. 2020. Magnetohydrodynamic and kinetic scale turbulence in the near-Earth space plasmas: A (short) biased review. *Rev. Mod. Plasma Phys.*, **4**(1), 4 (33 pages).

Schekochihin, A. A., Cowley, S. C., Dorland, W. et al. 2009. Astrophysical gyrokinetics: Kinetic and fluid turbulent cascades in magnetized weakly collisional plasmas. *Astrophys. J. Supp.*, **182**(1), 310–377.

Sorriso-Valvo, L., Marino, R., Carbone, V. et al. 2007. Observation of inertial energy cascade in interplanetary space plasma. *Phys. Rev. Let.*, **99**(11), 115001.

Turner, L. 2000. Using helicity to characterize homogeneous and inhomogeneous turbulent dynamics. *J. Fluid Mech.*, **408**(1), 205–238.

van Bokhoven, L. J. A., Clercx, H. J. H., van Heijst, G. J. F., and Trieling, R. R. 2009. Experiments on rapidly rotating turbulent flows. *Phys. Fluids*, **21**(9), 096601.

Vasquez, B. J., Smith, C. W., Hamilton, K., MacBride, B. T., and Leamon, R. J. 2007. Evaluation of the turbulent energy cascade rates from the upper inertial range in the solar wind at 1 AU. *J. Geophys. Res.*, **112**, A07101.

Voitenko, Y. M. 1998. Three-wave coupling and weak turbulence of kinetic Alfvén waves. *J. Plasma Phys.*, **60**(3), 515–527.

Waleffe, F. 1992. The nature of triad interactions in homogeneous turbulence. *Phys. Fluids A*, **4**(2), 350–363.

Zakharov, V. E. 1967. Weak-turbulence spectrum in a plasma without a magnetic field. *Sov. J. Exp. Theor. Physics*, **24**, 455459.

Zakharov, V. E. 1972. Collapse of Langmuir waves. *Sov. J. Exp. Theor. Physics*, **35**, 908–914.

9

Gravitational Wave Turbulence

The first direct detection of gravitational waves (GW) by the LIGO–Virgo collaboration (Abbott et al., 2016), a century after their prediction by Albert Einstein (1916), is certainly one of the most important events in astronomy in recent decades. Thanks to this new gravitational astronomy and the fact that GW interact only weakly with matter, the primordial Universe that preceded the first electromagnetic radiation of the Universe, commonly known as the cosmic microwave background (CMB), becomes now potentially accessible. Hence, many efforts are currently made to detect primordial GW. But do we know how the Universe would have behaved if it was made of a sea of GW?

In this final chapter, we shall answer this question and present the regime of weak GW turbulence in the context of primordial cosmology. First, we will give a brief review of cosmology – useful for nonspecialists – with the main steps in the evolution of the Universe, starting from its "birth." Then, we will report the main theoretical properties of weak GW turbulence, which involves four-wave interactions and a dual cascade, with a slow direct energy cascade and an explosive inverse cascade of wave action. This part will be illustrated by direct numerical simulations of Einstein's equations. Finally, we will present an extension to strong GW turbulence with a plausible application: the cosmological inflation produced a fraction of second after the Big Bang.

9.1 Primordial Universe

9.1.1 History of the Universe

Since the observations of Hubble (1929), we have known that the Universe is expanding. This expansion is not uniform, in the sense that the farther away the galaxies are, the faster they move away. Reversing the arrow of time suggests that there was a time (around 13.8 billion years) when the Universe was concentrated in one point. The Big Bang is associated with this theoretical point from which the

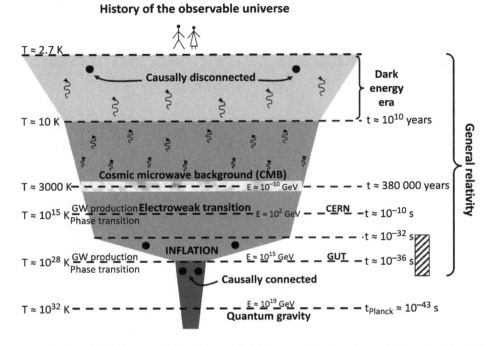

Figure 9.1 A brief history of the observable Universe. Starting from the bottom, the different phases of expansion are symbolized by the widening of the sample of the Universe. Today (top), the observable Universe has a diameter of about 13.8×10^9 light-years, which is only a small fraction of the Universe (whose size could be infinite). By definition, we are at the center of the observable Universe. The timescale on the right starts at the Planck epoch ($t \sim 10^{-43}$ s) to reach the present moment (top). On the left is indicated the temperature of the Universe: the actual temperature corresponds to that of the cosmic microwave background radiation, emitted when the Universe was about 380 000 years old. The primordial Universe, which preceded this moment, can be directly probed only by considering (primordial) GW. These were produced, in particular, during the GUT ($t \sim 10^{-36}$ s) and electroweak ($t \sim 10^{-10}$ s) phase transitions. The inflation phase would have increased the size of the Universe by a factor of at least 10^{28}. A regime of strong GW turbulence is expected close to the birth of the Universe (hatched area), around time $t \sim 10^{-36}$ s $\gg t_{Planck}$.

Universe was created. It is difficult to conceive what the Universe was at the initial instant, especially since the notion of time as we know it today had no meaning. In general, the earliest instant to which we can refer was the first 10^{-43} s, equivalent to Planck time, t_{Planck}, and referred to as the Planck epoch. As proposed by Wheeler (1955), the Universe may have been then in a state of turbulent quantum foam that only quantum gravity could describe, but to date this theory has not yet been established (Weinberg, 2008).

In Figure 9.1 we show schematically what happened next. Globally, the Universe underwent an expansion cooling with several phases. The first phase is usually located around $t \sim 10^{-36}$ s: at this time, the Universe would have undergone a very violent accelerated expansion called inflation – with a speed of expansion faster than the speed of light. This expansion would have occurred at the moment of the GUT (grand unified theory) phase transition, that is, the instant when the strong interaction separated from the electroweak interaction. It is following this phase of inflation that matter would have been created. Let us note that turbulence is a possible mechanism for understanding the post-inflationary transition to thermal equilibrium (Micha and Tkachev, 2003, 2004). Later, around $t \sim 10^{-10}$ s, there was a second phase transition which corresponded to the moment when the weak interaction separated from the electromagnetic interaction. We arrive here at an energy level that CERN instruments can reach, and thus experimentally test the theories. From $t \sim 380\,000$ years, the Universe was no longer opaque and the first electromagnetic radiation could escape: this was the CMB. We enter from this moment into the domain of the observable Universe. Then came the last phase, discovered in 1998 (Perlmutter et al., 1998; Riess et al., 1998): it was a new phase of accelerated expansion, which was due to dark energy. This mysterious denomination mainly reflects our ignorance about the origin of this acceleration.

To explain all these expansion phases, we need to use the equations of general relativity (Einstein, 1915). The decelerated expansion, between the end of inflation and the dark energy era, is well explained by the standard model (Friedmann, 1922). Explaining the accelerated expansion phases it is still an open problem: in general, a nonzero cosmological constant is introduced (or equivalently a negative pressure of quantum origin), but the detailed physical mechanism is unknown. On the other hand, the equations of relativity are not valid near Planck's epoch, where quantum effects are so important that they require a theory of quantum gravity.

9.1.2 Cosmological Inflation

The CMB was measured with unprecedented accuracy by ESA's Planck mission (Planck Collaboration, 2014). These new measurements confirm those made with NASA's COBE and WMAP satellites, that is, that the first electromagnetic in the Universe was of a very high homogeneity, with a temperature around 2.7 K. This property can be explained if a phase of inflation existed in the primordial Universe. To understand this, we must return to Figure 9.1. At the top, the two small black disks symbolize two regions of the sky sufficiently distant from each other to be causally disconnected. If we go back in time, we see that before the inflation phase these two regions were in fact causally connected. This change of state is explained by a supraluminal expansion of the Universe. This expansion at a speed greater than that of light is not in contradiction with the laws of physics, because it is not the propagation of information but the expansion of space-time itself. If a physical process existed (such as turbulence) to homogenize the preinflationary primordial Universe, then inflation would have had the effect of extending the homogenized region over the entire current sky. Note that the role of inflation is not limited to this: for example, it is also a way to explain the flatness of the Universe.

Cosmological inflation was proposed in particular by Guth (1981). This scenario is based on particle physics and the existence of a scalar field called inflaton. However, to date no direct or indirect evidence of its existence has been found. Furthermore, Planck's measures point out the weaknesses of the model, which have led to several criticisms (see, e.g., Ijjas et al., 2013, 2014 and Hollands and Wald, 2002). One of the difficulties is that the classical inflation model requires a very fine tuning of the initial conditions, which is difficult to explain physically. One solution is to evoke the anthropic principle and, for example, the existence of $\sim 10^{100}$ parallel universes. Our universe would therefore have by pure chance the parameters in question.

In Section 9.3 we will see that strong GW turbulence may offer an original mechanism at the origin of the cosmological inflation. To understand the essence of this mechanism, we need to present the main properties of the weak turbulence regime.

9.2 Weak Gravitational Wave Turbulence

9.2.1 Einstein's Equations

The theory of GW turbulence is based on the equations of general relativity, which dynamically describe how space-time is distorted by energy and matter (Einstein, 1915). These equations of relativistic gravitation are written in a general way:

$$R_{\mu\nu} - \frac{1}{2}g_{\mu\nu}R - \Lambda g_{\mu\nu} = \frac{8\pi G}{c^4}T_{\mu\nu}, \tag{9.1}$$

with $R_{\mu\nu}$ the Ricci tensor, R the Ricci scalar, $g_{\mu\nu}$ the metric tensor, Λ the cosmological constant, $T_{\mu\nu}$ the stress-energy tensor, G the gravitational constant, and c the speed of light. For our study, we will neglect the cosmological constant ($\Lambda = 0$). Equations (9.1) relate the energy-matter (right term) to space-time curvature (left term). The fundamental variable to describe the space-time curvature is the metric tensor $g_{\mu\nu}$: it is a 4×4 tensor symmetrical in μ and ν.[1]

It is interesting to make a comparison between Einstein's equations (9.1) and those of Navier–Stokes (2.1). As in hydrodynamics, general relativity is described

---

[1] The metric is what makes it possible to measure the distance between two close points. In classical physics, this distance can be written: $ds^2 = dx^2 + dy^2 + dz^2 = g_{\mu\nu}dx^\mu dx^\nu$. With $x^1 = x$, $x^2 = y$ and $x^3 = z$, we obtain: $g_{\mu\nu} = \delta_{\mu\nu}$.

by nonlinear partial differential equations. The tensors being symmetrical in μ and ν, we have a total of 10 equations. The left-hand terms of (9.1) are for us the core of nonlinearities (time derivatives are also present there), while the right-hand term can be seen as an external force that excites the space-time over a given range of scale (or frequency). For our study, we will place ourselves in the vacuum, which is the same as eliminating the right-hand term of equation (9.1). Under these circumstances, the equations are simplified (for an empty Universe, $R = 0$), and we obtain the equation of general relativity in a vacuum:

$$R_{\mu\nu} = 0. \tag{9.2}$$

By definition:

$$R_{\mu\nu} \equiv \partial_\alpha \Gamma^\alpha_{\mu\nu} - \partial_\mu \Gamma^\alpha_{\alpha\nu} + \Gamma^\beta_{\mu\nu}\Gamma^\alpha_{\alpha\beta} - \Gamma^\alpha_{\mu\beta}\Gamma^\beta_{\alpha\nu}, \tag{9.3a}$$

$$\Gamma^i_{jk} \equiv \frac{1}{2} g^{il}(\partial_j g_{lk} + \partial_k g_{jl} - \partial_l g_{jk}), \tag{9.3b}$$

where Γ^i_{jk} is the Christoffel symbols; Einstein's convention on indices should be used. These expressions are sufficient (we will not give details on the origin of the different Ricci tensor terms) to give us an idea of the type of nonlinearity we are dealing with: they are at least quadratic. Note that the comparison with the Navier–Stokes equations is not complete, since in general relativity there are no pure dissipative terms. However, dissipation of GW by matter is expected through, for example, Landau damping (Baym et al., 2017).

9.2.2 Gravitational Waves

The linear solutions of equation (9.2) are well known: they are gravitational waves (Einstein, 1916, 1918). These solutions are obtained by adding a small perturbation $h_{\mu\nu}$ to the undisturbed (flat) space-time $\eta_{\mu\nu}$, which is simply the Poincaré–Minkowski metric:

$$\eta_{\mu\nu} = \begin{pmatrix} -1 & 0 & 0 & 0 \\ 0 & 1 & 0 & 0 \\ 0 & 0 & 1 & 0 \\ 0 & 0 & 0 & 1 \end{pmatrix}. \tag{9.4}$$

The introduction of the metric:

$$g_{\mu\nu} = \eta_{\mu\nu} + h_{\mu\nu} \tag{9.5}$$

in the linearized equation (9.2) gives the dispersion relation (Maggiore, 2008):

$$\boxed{\omega = kc}. \tag{9.6}$$

Gravitational waves are therefore nondispersive. This could be a problem for the nonlinear treatment because the uniformity of the wave turbulence development is not guaranteed (see Chapter 4). But this potential problem occurs for triadic interactions (as for acoustic waves; see L'vov et al., 1997) while GW turbulence is in fact a problem that is treated at the next order, that is, for quartic interactions (four-wave interactions).

9.2.3 Theory of Weak GW Turbulence

The theory of GW turbulence consists in introducing expression (9.5) into the nonlinear equation (9.2), assuming:

$$|h_{\mu\nu}| \ll 1. \tag{9.7}$$

The objective of this final chapter is not to present the analytical development of the theory but rather to give an idea of the method for arriving at the kinetic equation, whose form is quite classical. In the approximation of waves with weak amplitude (9.7), we see on expressions (9.3) that the dominant nonlinear terms are quadratic. Consequently, the nonlinear treatment will bring out a triadic resonance condition similar to that of acoustic waves, for which the solutions correspond to rays in Fourier space (Newell and Aucoin, 1971; L'vov et al., 1997). From this property concerning the resonant manifold, one can demonstrate that Einstein's equations (for an empty Universe and neglecting the cosmological constant) do not make any contribution to wave turbulence (Galtier and Nazarenko, 2017). Therefore, the theory must be treated at the next order – for quartic interactions – and take into account the cubic nonlinear contributions.

The development of a four-wave theory is more difficult. The equations of general relativity being themselves much more complex than the equations we usually deal with, we must try to simplify the approach. It is known that the search for solutions in general relativity is facilitated by the choice of coordinates. For example, the search for simple solutions associated with spherical objects (black hole type) requires the use of spherical coordinates. We will do the same here and use a metric introduced by Hadad and Zakharov (2014): it is a diagonal metric in Cartesian coordinates, which is written:

$$g_{\mu\nu} = \begin{pmatrix} -(H_0)^2 & 0 & 0 & 0 \\ 0 & (H_1)^2 & 0 & 0 \\ 0 & 0 & (H_2)^2 & 0 \\ 0 & 0 & 0 & (H_3)^2 \end{pmatrix}, \tag{9.8}$$

with H_0, H_1, H_2, and H_3 the Lamé coefficients.[2] These coefficients depend on $x^0 = t$, $x^1 = x$, and $x^2 = y$, but are independent of $x^3 = z$, with by definition the space-time interval: $ds^2 = g_{\mu\nu} dx^\mu dx^\nu$. The Lamé coefficients are defined by the

---

[2] One can verify with this metric that the fourth-order Riemann curvature tensor, as well as the Kretschmann curvature invariant, is nontrivial. This means that the metric used can describe a nontrivial physics that exists independently of the choice of coordinates (Weber and Wheeler, 1957).

following relations:

$$H_0 \equiv e^{-\lambda}(1+\gamma), \quad H_1 \equiv e^{-\lambda}(1+\beta), \quad H_2 \equiv e^{-\lambda}(1+\alpha), \quad H_3 \equiv e^{\lambda}, \quad (9.9)$$

with $\alpha, \beta, \gamma, \lambda \ll 1$. The introduction of these variables into equation (9.2) gives the following system at the main order:

$$\partial_x \dot{\alpha} = -2\dot{\lambda}(\partial_x \lambda), \quad (9.10a)$$
$$\partial_y \dot{\beta} = -2\dot{\lambda}(\partial_y \lambda), \quad (9.10b)$$
$$\partial_{xy} \gamma = -2(\partial_x \lambda)(\partial_y \lambda), \quad (9.10c)$$

with:

$$\partial_t[(1+\alpha+\beta-\gamma)\dot{\lambda}] = \partial_x[(1+\alpha-\beta+\gamma)\partial_x \lambda] + \partial_y[(1-\alpha+\beta+\gamma)\partial_y \lambda], \quad (9.11)$$

where ˙ means the time derivative. As we can see from equation (9.11), GW are indeed linear solutions. We also note that λ is the fundamental variable to describe these waves. The linear solution obtained involves only one wave polarization (the diagonal components whose polarization is noted +), whereas in the general case, there are two polarizations (the + and × polarizations associated with the diagonal and nondiagonal components, respectively (Maggiore, 2008)). This reduction allows a simplification of the problem, while keeping the indispensable element for the theory, the wave.[3]

The analytical development here is based on a relatively classical Hamiltonian approach. The kinetic equation obtained for GW turbulence is finally written (Galtier and Nazarenko, 2017):

$$\partial_t N_{\mathbf{k}} = \epsilon^4 \int_{\mathbb{R}^6} |T_{\mathbf{k}_1 \mathbf{k}_2}^{\mathbf{k} \mathbf{k}_3}|^2 N_{\mathbf{k}} N_{\mathbf{k}_1} N_{\mathbf{k}_2} N_{\mathbf{k}_3} \quad (9.12)$$
$$\times \left[\frac{1}{N_{\mathbf{k}}} + \frac{1}{N_{\mathbf{k}_3}} - \frac{1}{N_{\mathbf{k}_1}} - \frac{1}{N_{\mathbf{k}_2}} \right] \delta_{03,12}^k \delta_{03,12}^{\omega} d\mathbf{k}_1 d\mathbf{k}_2 d\mathbf{k}_3,$$

with by definition $\delta_{03,12}^k \equiv \delta(\mathbf{k} + \mathbf{k}_3 - \mathbf{k}_1 - \mathbf{k}_2)$ and $\delta_{03,12}^{\omega} \equiv \delta(\omega_{\mathbf{k}} + \omega_{\mathbf{k}_3} - \omega_{\mathbf{k}_1} - \omega_{\mathbf{k}_2})$. $N_{\mathbf{k}}$ is the wave action and $T_{\mathbf{k}_1 \mathbf{k}_2}^{\mathbf{k}_3}$ a geometric coefficient verifying certain symmetries that it is not useful to give here. Equation (9.12) has two invariants: energy and wave action. The presence of a second invariant – the wave action – is not automatic for four-wave interactions because it requires the kinetic equation to satisfy additional symmetries (Nazarenko, 2011). In practice, interactions 1 ↔ 3 must be absent, leaving only interactions 2 ↔ 2. Note that the wave action is an invariant that appears in other systems described by quartic interactions, such as nonlinear optics (Dyachenko et al., 1992) or for elastic wave turbulence on

---

[3] The use of a diagonal metric immediately raises the problem of compatibility: the metric has ten elements, and generally four of which can be eliminated by using gauge transformations. This makes the vacuum Einstein equations an overdetermined system of ten equations for six unknowns. This compatibility has been properly checked (without introducing a gauge) by Hadad and Zakharov (2014). In particular, we can see that there are as many equations (9.10)–(9.11) as there are unknowns (9.9).

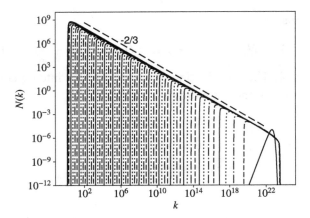

Figure 9.2 Temporal evolution of the (isotropic) wave action spectrum $N(k)$ with an initial spectrum located around $k = 10^{22}$ (decaying simulation of a nonlinear diffusion model). The inverse cascade produces a spectrum close to $k^{-2/3}$ over more than 20 decades (this huge inertial range is probably nonphysical, but necessary here to measure the power-law index unambiguously). Simulation created by É. Buchlin.

a membrane (Hassaini et al., 2019). It is also an invariant that can be found in six-wave problems (Laurie et al., 2012).

Equation (9.12) admits of three exact (isotropic) solutions. First, there is the thermodynamic solution $N_k \sim k^0$. The two others are the Kolmogorov–Zakharov spectra for the wave action:

$$N_k \sim k^{-2/3}, \tag{9.13}$$

and energy (by definition $E_k \equiv \omega_k N_k$):

$$E_k \sim k^0. \tag{9.14}$$

These solutions correspond respectively to an inverse and a direct cascade (see Exercise II.3 and its solutions).[4] The form of the solution for the wave action allows us to affirm that the inverse cascade is explosive (this property is related to the convergence of the integral $\int_0^{k_i} N_k dk$, with k_i the initial wavenumber excitation).

An analysis of the explosiveness of the inverse cascade is made easier if a nonlinear diffusion model is used. As in the case of rotation (see Chapter 6), a nonstationary solution different from the Kolmogorov–Zakharov spectrum can be found. However, the difference in power-law exponent followed by N_k is tiny here ($\simeq -0.6517$ for the nonstationary solution instead of $-2/3$ for the stationary solution). In Figure 9.2, we show the result of a numerical simulation with the formation of a $N(k)$ spectrum over more than 20 decades (Galtier et al., 2019). The expected scenario in weak wave turbulence is therefore the rapid formation of a

---

[4] These spectra are the first exact *statistical* solutions of Einstein's equations ever found.

9.2 Weak Gravitational Wave Turbulence

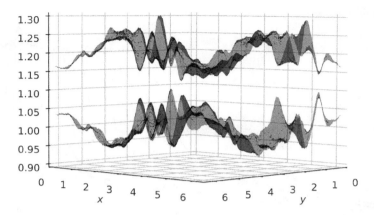

Figure 9.3 Result of a direct numerical simulation of equations (9.10)–(9.11) with a spatial resolution of 512 × 512. Components g_{11} (bottom) and $g_{22} + 0.2$ (top) of the space-time metric produced in weak GW turbulence. There are characterized by small erratic fluctuations around the reference value, the Poincaré–Minkowski flat metric (+1).

large-scale inertial range by an inverse cascade of wave action, and the relatively slow formation of a small-scale inertial range by a direct energy cascade.

9.2.4 Direct Numerical Simulations

The direct numerical simulation of weak GW turbulence is an emerging subject in cosmology: the first simulation in the decaying case was realized by Galtier and Nazarenko (2021) with a pseudospectral code (using FFTW3 and de-aliasing). The main conclusion reached by the authors is that it is possible to produce turbulence in general relativity. Unlike classical hydrodynamic turbulence, it does not consist of randomly interacting vortices but, rather, it takes a form of random interacting waves – the space-time turbulence (see Figure 9.3). Furthermore, the dual cascade was confirmed (see Figure 9.4) with a timescale (transfer time) compatible with four-wave processes, that is, such that $\tau_{tr} \sim \tau_{GW}/\epsilon^4$ (see Chapter 4). In this case, the gravitational time is defined with the initial excitation, that is, $\tau_{GW} = 1/k_i$.

The domain of validity of GW turbulence is limited to a certain range of scales. We have seen that for MHD plasmas (see Chapter 7) the transition between the weak and strong regimes is at a small scale (i.e. at a scale smaller than the initial excitation), whereas in the case of capillary waves the turbulence becomes weaker and weaker at small scales (see Chapter 5). However, in both cases, the cascade of the invariant is direct. In GW turbulence, a transition to strong turbulence is expected at large scales (scales larger than the initial excitation). In other words, the inverse cascade of wave action will necessarily end up (quickly) in the strong regime. In the direct numerical simulation shown in Figure 9.3, the computation

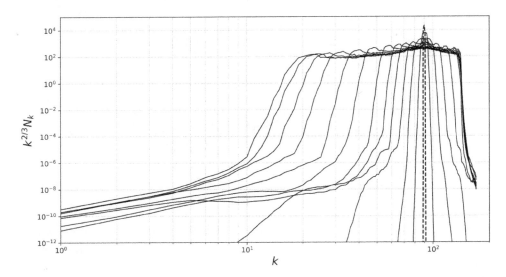

Figure 9.4 Direct evidence of a dual cascade in GW turbulence (same direct numerical simulation as in Figure 9.3). The time evolution of the 1D isotropic compensated wave action spectrum $k^{2/3} N_k$ is shown, with the initial spectrum (located at $k_i = 89$) given in dashed line. The direct cascade is stopped (around $k = 140$) by the additional dissipative term added to the Einstein equations.

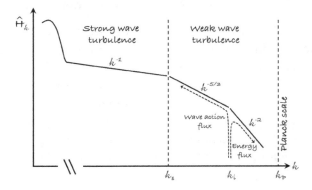

Figure 9.5 Spectrum $\hat{H}_k$ of the space-time metric fluctuations. In these units, the scaling law of the weak inverse cascade is in $k^{-5/3}$ and that of critical balance (from the k_s scale) in k^{-1}. A condensation phenomenon appears at $k = 0$, which leads to an expansion of the Universe (see text). Planck's scale is given for information: close to this scale, the equations of general relativity are no longer valid. Sources of forcing at the k_i scale are assumed sufficiently intense to generate a regime close to that of strong wave turbulence.

stopped when the GW amplitude exceeded 10 percent and the weak turbulence condition (weak wave amplitude) started not to be fulfilled.

9.3 Strong Turbulence and Inflation

Weak GW turbulence is dominated by the inverse cascade of wave action as it is explosive (solution of finite capacity), whereas the direct energy cascade is slow (solution of infinite capacity). This means in principle that modes k smaller than the initial excited mode (k_i) of the space-time can all be excited in a finite time. However, during this process wave turbulence quickly becomes strong, and a phenomenological model is required to pursue the description. As explained in Chapter 7, the critical balance conjecture is a widely used model in turbulence. In the case of GW, the prediction for the large-scale wave action spectrum is $N(k) \sim k^0$, a situation always favorable to an explosive inverse cascade.

What happens next? Although the scenario becomes hypothetical, we can envisage that the inverse cascade finally excites significantly the slow mode $k = 0$ (see Figure 9.5). We have already seen the importance of the slow mode in rotating hydrodynamic turbulence (Chapter 6) and in MHD (Chapter 7). In particular, we know that the slow mode excitation does not violate the principle of causality, because it does not correspond to the propagation of information in the physical space from a given scale to infinity. Note, in passing, that for our analysis we do not need to specify the size of the Universe sample (finite or infinite). In cosmology, the slow mode corresponds to the background level of the metric, which means that if we feed (amplify) this mode, we increase the size of the Universe (phase of expansion). This is where the link is made between turbulence and inflation: due to the explosive nature of the inverse cascade we can expect an equally explosive expansion, that is, a phase of inflation (Galtier et al., 2020).

Is GW turbulence the cause of cosmological inflation? The answer to this question is far from trivial and requires much more work, based, in particular, on direct numerical simulations of strong GW turbulence. However, based on the previous discussion, we can now sketch out a complete scenario. The basic assumption is that the nonlinearities of the equations of general relativity were nonnegligible around $t \sim 10^{-36}$s $\gg t_{Planck}$ (see Figure 9.1). The sharp increase of the condensate described here is necessarily limited in time, because the expansion of the Universe induces a dilution, that is, a decrease in the space-time metric fluctuations, and thus a reduction of nonlinearities, which means the end of the cascade mechanism. Consequently, turbulence offers a natural physical mechanism to stop inflation. Then, we enter a standard expansion phase of the Universe, with a self-similar decay of GW turbulence with a $1/k$ metric spectrum, which corresponds to the critical balance phenomenology. By using a simple dimensional argument, we can deduce the spectrum for the mass density of the primordial Universe following inflation (which is the Harrison–Zeldovich spectrum Harrison, 1970; Zeldovich, 1972). It turns out that this fossil spectrum is in good agreement with Planck/ESA measurements of the CMB (Planck Collaboration, 2016); indeed, the measurements are in correspondence with a $k^{-1.033}$ metric spectrum.

A turbulence-based scenario also has the advantage of offering a natural physical mechanism to homogenize the initial metric fluctuations without fine-tuning (a weakness of current models; Ijjas et al., 2013, 2014). This scenario, which is radically different in nature from the inflation model of particle physics based on a hypothetical scalar field called inflaton, suggests that under extreme conditions (strong nonlinearities) Einstein's gravitation can be a repulsive force.[5]

9.4 Perspectives

Since the equations of general relativity have never failed so far, this cosmological scenario is arguably interesting in helping us to better understand the primordial Universe from first principles. One advantage is that it can be tested directly (a luxury in this field) by direct numerical simulation.[6] This is a promising and exciting prospect in turbulence. In particular, it would be interesting to verify if an anomalous exponent (found in weak GW turbulence and in several other systems – see Chapters 6, 7, and 8) also exists for the inverse cascade in the regime of strong GW turbulence that could lead to a small deviation from the $1/k$ metric spectrum. According to the cosmological observation, a scaling in $k^{-1.033}$ is expected. In this case, that would mean the anomalous exponent of GW turbulence could be at the origin of the formation of the structures in the Universe. The formation of a spectral condensate is not new in turbulence. For example, it is predicted and observed with the nonlinear Schrödinger model (Gross–Pitaevskii equation), which forms a Bose–Einstein condensate (Dyachenko et al., 1992; Zakharov and Nazarenko, 2005; Miller et al., 2013). The numerical study of the formation of such condensate in GW turbulence is an important perspective.

On the other hand, as shown by the theoretical, numerical, and experimental work of Hassaini et al. (2019), there is a strong proximity (type of interactions, symmetry of kinetic equations) between GW turbulence and that of elastic waves in the high tension limit. We can see here another interesting perspective: the development of new laboratory experiments could help us to better understand the primordial Universe.

References

Abbott, B. P., Abbot, R., Abbott, T. D. et al. 2016. Observation of gravitational waves from a binary black hole merger. *Phys. Rev. Lett.*, **116**, 061102.

Baym, G., Patil, S. P., and Pethick, C. J. 2017. Damping of gravitational waves by matter. *Phys. Rev. D*, **96**(8), 084033.

Dyachenko, S., Newell, A. C., Pushkarev, A., and Zakharov, V. E. 1992. Optical turbulence: Weak turbulence, condensates and collapsing filaments in the nonlinear Schrödinger equation. *Physica D*, **57**, 96–160.

---

[5] In the standard model of cosmology (Friedmann, 1922), if we neglect the cosmological constant ($\Lambda = 0$) and if the Universe is empty, the solution is a static Universe.

[6] Other applications of strong GW turbulence, such as the environment of black holes, can also be considered (Green et al., 2014; Yang et al., 2015).

Einstein, A. 1915. Erklarung der Perihelionbewegung der Merkur aus der allgemeinen Relativitatstheorie. *Sitzungsber. preuss. Akad. Wiss*, **47**, 831–839.

Einstein, A. 1916. Näherungsweise Integration der Feldgleichungen der Gravitation. *Sitzungsberichte der Königlich Preusischen Akademie der Wissenschaften (Berlin)*, 688–696.

Einstein, A. 1918. Über Gravitationswellen. *Sitzungsberichte der Königlich Preusischen Akademie der Wissenschaften (Berlin)*, 154–167.

Friedmann, A. 1922. Über die Krümmung des Raumes. *Zeitschrift fur Physik*, **10**, 377–386.

Galtier, S., and Nazarenko, S. V. 2017. Turbulence of weak gravitational waves in the early Universe. *Phys. Rev. Lett.*, **119**, 221101.

Galtier, S., and Nazarenko, S. V. 2021. Direct evidence of a dual cascade in gravitational wave turbulence. *Phys. Rev. Lett.*, **127**, 131101.

Galtier, S., Nazarenko, S. V., Buchlin, É., and Thalabard, S. 2019. Nonlinear diffusion models for gravitational wave turbulence. *Physica D*, **390**, 84–88.

Galtier, S., Laurie, J., and Nazarenko, S. V. 2020. A plausible model of inflation driven by strong gravitational wave turbulence. *Universe*, **6**(7), 98 (16 pages).

Green, S. R., Carrasco, F., and Lehner, L. 2014. Holographic path to the turbulent side of gravity. *Phys. Rev. X*, **4**(1), 011001.

Guth, A. H. 1981. Inflationary universe: A possible solution to the horizon and flatness problems. *Phys. Rev. D*, **23**, 347–356.

Hadad, Y., and Zakharov, V. 2014. Transparency of strong gravitational waves. *J. Geom. Phys.*, **80**, 37–48.

Harrison, E. R. 1970. Fluctuations at the threshold of classical cosmology. *Phys. Rev. D*, **1**(10), 2726–2730.

Hassaini, R., Mordant, N., Miquel, B., Krstulovic, G., and Düring, G. 2019. Elastic weak turbulence: From the vibrating plate to the drum. *Phys. Rev. E*, **99**(3), 033002.

Hollands, S., and Wald, R. M. 2002. Essay: An alternative to inflation. *General Relat. & Grav.*, **34**(12), 2043–2055.

Hubble, E. 1929. A relation between distance and radial velocity among extra-galactic nebulae. *Proc. Natl. Acad. Sci.*, **15**(3), 168–173.

Ijjas, A., Steinhardt, P. J., and Loeb, A. 2013. Inflationary paradigm in trouble after Planck2013. *Phys. Lett. B*, **723**, 261–266.

Ijjas, A., Steinhardt, P. J., and Loeb, A. 2014. Inflationary schism. *Phys. Lett. B*, **736**, 142–146.

Laurie, J., Bortolozzo, U., Nazarenko, S., and Residori, S. 2012. One-dimensional optical wave turbulence: Experiment and theory. *Phys. Rep.*, **514**(4), 121–175.

L'vov, V. S., L'vov, Y., Newell, A. C., and Zakharov, V. E. 1997. Statistical description of acoustic turbulence. *Phys. Rev. E*, **56**(1), 390–405.

Maggiore, M. 2008. *Gravitational Waves*, vol. 1: *Theory and Experiments*. Oxford University Press.

Micha, R., and Tkachev, I. I. 2003. Relativistic turbulence: A long way from preheating to equilibrium. *Phys. Rev. Lett.*, **90**(12), 121301.

Micha, R., and Tkachev, I. I. 2004. Turbulent thermalization. *Phys. Rev. D*, **70**(4), 043538.

Miller, P., Vladimirova, N., and Falkovich, G. 2013. Oscillations in a turbulence-condensate system. *Phys. Rev. E*, **87**(6), 065202.

Nazarenko, S. 2011. *Wave Turbulence*. Lecture Notes in Physics, vol. 825. Springer Verlag.

Newell, A. C., and Aucoin, P. J. 1971. Semi-dispersive wave systems. *J. Fluid Mech.*, **49**, 593–609.

Perlmutter, S., Aldering, G., della Valle, M. et al. 1998. Discovery of a supernova explosion at half the age of the universe. *Nature*, **391**(6662), 51–54.

Planck Collaboration. 2014. Planck 2013 results. I. Overview of products and scientific results. *Astron. Astrophys.*, **571**, A1 (48 pages).

Planck Collaboration. 2016. Planck 2015 results. XX. Constraints on inflation. *Astron. Astrophys.*, **594**, A20 (65 pages).

Riess, A. G., Filippenko, A. V., Challis, P. et al. 1998. Observational evidence from supernovae for an accelerating universe and a cosmological constant. *Astron. J.*, **116**(3), 1009–1038.

Weber, J., and Wheeler, J. A. 1957. Reality of the cylindrical gravitational waves of Einstein and Rosen. *Rev. Modern Phys.*, **29**(3), 509–515.

Weinberg, S. 2008. *Cosmology*. Oxford University Press.

Wheeler, J. A. 1955. Geons. *Phys. Rev.*, **97**, 511–536.

Yang, H., Zimmerman, A., and Lehner, L. 2015. Turbulent black holes. *Phys. Rev. Lett.*, **114**(8), 081101.

Zakharov, V. E., and Nazarenko, S. V. 2005. Dynamics of the Bose–Einstein condensation. *Physica D*, **201**(3–4), 203–211.

Zeldovich, Y. B. 1972. A hypothesis, unifying the structure and the entropy of the Universe. *MNRAS*, **160**, 1P–3P.

Exercises II

II.1 MHD Model of Nonlinear Diffusion

Nonlinear diffusion models are often used in turbulence because their numerical simulations are relatively easy. We have already encountered these models in Chapters 3, 6, and 8. We propose here to develop such a phenomenological model for Alfvén wave turbulence in the balanced case. The particularity of the approach is that it must take into account the anisotropy. In practice, the following equation is considered:

$$\frac{\partial E(\mathbf{k})}{\partial t} = -\nabla \cdot \mathbf{\Pi}(\mathbf{k}),$$

where $E(\mathbf{k})$ is the energy spectrum and $\mathbf{\Pi}(\mathbf{k})$ the associated flux.

(1) What are the conditions of validity of this equation?

(2) Alfvén wave turbulence is strongly anisotropic. Using a coordinate system adapted to this situation, rewrite the diffusion equation.

(3) The transverse component of the energy flux is modeled as follows:

$$\Pi_\perp(\mathbf{k}) = -D(k)\frac{\partial E(\mathbf{k})}{\partial k_\perp},$$

where D is a nonlinear diffusion coefficient which remains to be determined. What is the dimension of D?

(4) We call τ_{tr} the characteristic time of energy transfer. Express τ_{tr} as a function of $E(\mathbf{k})$ and $k_\perp$.

(5) By using the relation $\int E(k_\perp) f(k_\parallel) dk_\perp dk_\parallel = \iiint E(\mathbf{k}) d\mathbf{k}$, show that the nonlinear diffusion equation can be written as:

$$\frac{\partial E(k_\perp)}{\partial t} = C\frac{\partial}{\partial k_\perp}\left(k_\perp^6 E(k_\perp)\frac{\partial (E(k_\perp)/k_\perp)}{\partial k_\perp}\right),$$

where C is a constant.

(6) What is the Kolmogorov–Zakharov spectrum exact solution of the equation? What is the direction of the cascade?

II.2 Four-Wave Interactions

Let us consider the isotropic dispersion relation $\omega = Ak^\alpha$, with A a positive constant and $\alpha \in \mathbb{R}_+^*$. In the two-dimensional case, show graphically that we can always find solutions for four-wave resonant interactions.

II.3 Gravitational Wave Turbulence: Exact Solutions

The objective of this exercise is to obtain analytically the exact solutions of gravitational wave turbulence. We have seen that the kinetic equation is written as follows:

$$\partial_t N_{\mathbf{k}} = \epsilon^6 \int_{\mathbb{R}^6} |T_{\mathbf{k}_1 \mathbf{k}_2}^{\mathbf{k}\mathbf{k}_3}|^2 N_{\mathbf{k}} N_{\mathbf{k}_1} N_{\mathbf{k}_2} N_{\mathbf{k}_3}$$
$$\times \left[\frac{1}{N_{\mathbf{k}}} + \frac{1}{N_{\mathbf{k}_3}} - \frac{1}{N_{\mathbf{k}_1}} - \frac{1}{N_{\mathbf{k}_2}} \right] \delta_{03,12}^k \delta_{03,12}^\omega \, d\mathbf{k}_1 d\mathbf{k}_2 d\mathbf{k}_3,$$

with $\delta_{03,12}^k \equiv \delta(\mathbf{k} + \mathbf{k}_3 - \mathbf{k}_1 - \mathbf{k}_2)$ and $\delta_{03,12}^\omega \equiv \delta(\omega_{\mathbf{k}} + \omega_{\mathbf{k}_3} - \omega_{\mathbf{k}_1} - \omega_{\mathbf{k}_2})$.

We must specify the symmetries verified by the geometric coefficient. We have:

$$T_{\mathbf{k}_1 \mathbf{k}_2}^{\mathbf{k}\mathbf{k}_3} = \frac{1}{4}(W_{\mathbf{k}_1 \mathbf{k}_2}^{\mathbf{k}\mathbf{k}_3} + W_{\mathbf{k}_1 \mathbf{k}_2}^{\mathbf{k}_3 \mathbf{k}} + W_{\mathbf{k}_2 \mathbf{k}_1}^{\mathbf{k}\mathbf{k}_3} + W_{\mathbf{k}_2 \mathbf{k}_1}^{\mathbf{k}_3 \mathbf{k}}),$$

as well as:

$$W_{\mathbf{k}_1 \mathbf{k}_2}^{\mathbf{k}\mathbf{k}_3} = Q_{\mathbf{k}_1 \mathbf{k}_2}^{\mathbf{k}\mathbf{k}_3} + Q_{\mathbf{k} \mathbf{k}_3}^{\mathbf{k}_1 \mathbf{k}_2}.$$

The exact form of $Q_{\mathbf{k}_1 \mathbf{k}_2}^{\mathbf{k}\mathbf{k}_3}$ is not useful (see Galtier and Nazarenko, 2017); only information on its degree of homogeneity is important. We have:

$$T_{a\mathbf{k}_1 \, a\mathbf{k}_2}^{a\mathbf{k} \, a\mathbf{k}_3} = T_{\mathbf{k}_1 \, \mathbf{k}_2}^{\mathbf{k} \, \mathbf{k}_3},$$

which means that this coefficient is dimensionless.

For four-wave interactions, the Zakharov conformal transformation is written:

$$\xi_1 \to \frac{1}{\xi_1}, \quad \xi_2 \to \frac{\xi_3}{\xi_1}, \quad \xi_3 \to \frac{\xi_2}{\xi_1}, \quad \text{(TZa)}$$

$$\xi_1 \to \frac{\xi_3}{\xi_2}, \quad \xi_2 \to \frac{1}{\xi_2}, \quad \xi_3 \to \frac{\xi_1}{\xi_2}, \quad \text{(TZb)}$$

$$\xi_1 \to \frac{\xi_1}{\xi_3}, \quad \xi_2 \to \frac{\xi_2}{\xi_3}, \quad \xi_3 \to \frac{1}{\xi_3}. \quad \text{(TZc)}$$

(1) By applying these transformations to the kinetic equation, find the isotropic exact solutions proposed in Chapter 9.

(2) Deduce the direction of the cascade for each solution.

(3) Give the expression for the Kolmogorov constants.

II.4 Inertial Wave Turbulence: Domain of Locality

Find the domain of locality for inertial wave turbulence.

References

Galtier, S., and Nazarenko, S. V. 2017. Turbulence of weak gravitational waves in the early Universe. *Phys. Rev. Lett.*, **119**, 221101.

Appendix A

Solutions to the Exercises

1.1 1D HD Turbulence: Burgers' Equation

Burgers' equation (Burgers, 1948) is often regarded as a one-dimensional model of compressible hydrodynamic turbulence. This equation is written:

$$\frac{\partial u}{\partial t} + u\frac{\partial u}{\partial x} = \nu\frac{\partial^2 u}{\partial x^2},$$

with u a scalar velocity and ν a viscosity. Burgers' equation is a simple model often used to test new ideas in turbulence. Here we will study the properties of dissipation and then intermittency using the tools introduced for the Navier–Stokes equations.

(1) Demonstrate that:

$$u(x,t) = \frac{1}{t}\left[x - L\tanh\left(\frac{xL}{2\nu t}\right)\right]$$

is an exact solution.

(2) Find the limit of this solution when $\nu \to 0$.

(3) Calculate the mean rate of energy dissipation ε.

(4) Calculate the mean rate of viscous dissipation ε_ν. Conclude.

(5) Find the expression of the smoothed Burgers equation by introducing an anomalous dissipation $\mathcal{D}_I^\ell$.

(6) Calculate the expression of the inertial dissipation $\mathcal{D}_I$ with the exact solution. Conclude.

(7) By using the exact solution in the limit $\nu \to 0$, find the exponents ζ_p by distinguishing the case where $p < 1$ from the case where $p \geq 1$.

Solutions:

(1) To demonstrate that:

$$u(x,t) = \frac{1}{t}\left[x - L\tanh\left(\frac{xL}{2\nu t}\right)\right]$$

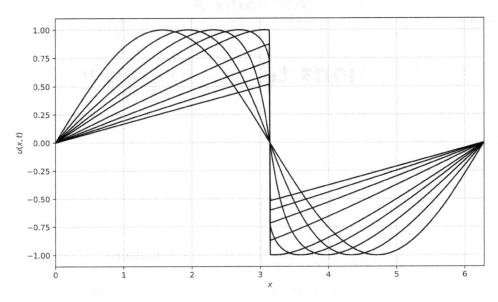

Figure A.1 Temporal evolution of the velocity starting from the initial condition $u(x,0) = \sin(x)$. The formation of the shock is followed by a dissipation phase. Direct numerical simulation of Burgers' equation with a spatial resolution of 32 768 points.

is an exact solution, the following expressions are calculated:

$$\frac{\partial u}{\partial t} = -\frac{x}{t^2} + \frac{L}{t^2}\tanh\left(\frac{xL}{2vt}\right) + \frac{xL^2}{2vt^3}\text{sech}^2\left(\frac{xL}{2vt}\right),$$

$$\frac{\partial u}{\partial x} = \frac{1}{t} - \frac{L^2}{2vt^2}\text{sech}^2\left(\frac{xL}{2vt}\right),$$

$$\frac{\partial^2 u}{\partial x^2} = \frac{L^3}{2v^2 t^3}\sinh\left(\frac{xL}{2vt}\right)\text{sech}^3\left(\frac{xL}{2vt}\right).$$

The weighted sum of these three terms cancels each other out, which proves the result. Note that this solution is unique (Kruzkhov's theorem).

(2) The exact solution is simplified within the limit $v \to 0$ with $x \neq 0$. We use the asymptotic expression of the tanh function and obtain:

$$u(x,t) \xrightarrow[v \to 0^+]{x>0} \frac{1}{t}(x-L),$$

$$u(x,t) \xrightarrow[v \to 0^+]{x<0} \frac{1}{t}(x+L). \tag{I.1.2}$$

This (weak) solution is valid on the $[-L, +L]$ interval around the discontinuity in $x = 0$. This discontinuity is a shock of amplitude $\Delta = 2L/t$. An illustration is given in Figure A.1: these are the results of a direct numerical simulation of Burgers' equation with initially $u(x,0) = \sin(x)$. The numerical grid between 0 and 2π is solved with 32 768 points.

(3) The mean rate of energy dissipation ε within the limit $\nu \to 0$ can be obtained from $\partial \langle u^2/2 \rangle / \partial t$, averaged over the interval $[-L, +L]$. We have:

$$\left\langle \frac{u^2}{2} \right\rangle = \frac{1}{2L} \int_{-L}^{+L} \frac{u^2}{2} dx$$

$$= \frac{1}{4L} \int_{-L}^{0} \left(\frac{x+L}{t} \right)^2 dx + \frac{1}{4L} \int_{0}^{L} \left(\frac{x-L}{t} \right)^2 dx$$

$$= \frac{L^2}{6t^2}.$$

Hence, the mean rate of energy dissipation:

$$\varepsilon = -\frac{\partial}{\partial t} \left\langle \frac{u^2}{2} \right\rangle = \frac{L^2}{3t^3} = \frac{\Delta^2}{12t}.$$

This rate varies over time as the shock dissipates.

(4) The mean rate of viscous dissipation ε_ν is written with an integration by part:

$$\varepsilon_\nu = -\nu \left\langle u \frac{\partial^2 u}{\partial x^2} \right\rangle = \nu \left\langle \left(\frac{\partial u}{\partial x} \right)^2 \right\rangle$$

$$= \nu \left\langle \left(\frac{1}{t} - \frac{L^2}{2\nu t^2} \operatorname{sech}^2 \left(\frac{xL}{2\nu t} \right) \right)^2 \right\rangle.$$

Within the limit $\nu \to 0$, one obtains at the main order:

$$\varepsilon_\nu = \left\langle \frac{L^4}{4\nu t^4} \operatorname{sech}^4 \left(\frac{xL}{2\nu t} \right) \right\rangle = \frac{L^3}{8\nu t^4} \int_{-L}^{+L} \operatorname{sech}^4 \left(\frac{xL}{2\nu t} \right) dx$$

$$= \frac{L^2}{4t^3} \int_{-L^2/2\nu t}^{+L^2/2\nu t} \operatorname{sech}^4(y) dy.$$

By using the relationship:

$$\operatorname{sech}^4(y) = 1 - 2\tanh^2(y) + \tanh^4(y),$$

one can show that $\int_\mathbb{R} \operatorname{sech}^4 y \, dy = 4/3$. Hence, the value:

$$\varepsilon_\nu = \frac{L^2}{3t^3} = \frac{\Delta^2}{12t}.$$

Within the limit $\nu \to 0$, we have therefore $\varepsilon_\nu = \varepsilon$ which is independent of viscosity and positive. Burgers' equation is therefore a model that behaves similarly to hydrodynamics.

(5) We can apply the smoothing function to Burgers' equation. We obtain in the inviscid case ($\nu = 0$):

$$\frac{\partial u^\ell}{\partial t} + \frac{1}{2} \frac{\partial (u^2)^\ell}{\partial x} = 0,$$

hence:
$$\frac{\partial}{\partial t}\left(\frac{uu^\ell}{2}\right) + \frac{u}{4}\frac{\partial(u^2)^\ell}{\partial x} + \frac{u^\ell}{4}\frac{\partial u^2}{\partial x} = 0.$$

The anomalous dissipation, also called inertial dissipation, is defined as:
$$\mathcal{D}_I^\ell \equiv \frac{1}{12}\int \frac{\partial \varphi^\ell}{\partial \xi}(\delta u)^3 d\xi$$
$$= \frac{1}{4}u\partial_x(u^2)^\ell - \frac{1}{4}u^2\partial_x u^\ell - \frac{1}{12}\partial_x(u^3)^\ell.$$

After rearrangement, we get:
$$\frac{\partial}{\partial t}\left(\frac{uu^\ell}{2}\right) + \frac{\partial}{\partial x}\left(\frac{u^2 u^\ell}{4} + \frac{(u^3)^\ell}{12}\right) = -\mathcal{D}_I^\ell.$$

Within the limit $\ell \to 0$, we arrive at the expression:
$$\frac{\partial}{\partial t}\left(\frac{u^2}{2}\right) + \frac{\partial}{\partial x}\left(\frac{u^3}{3}\right) = -\mathcal{D}_I,$$

with the distribution (generalized function) $\mathcal{D}_I \equiv \lim_{\ell \to 0}\mathcal{D}_I^\ell$.

(6) The expression of the anomalous dissipation can be obtained with the exact solution within the limit $\nu \to 0$ and for ℓ small enough. In the integral calculation, care must be taken to separate the contributions of the points to the right and to the left of the reference point x which itself is placed in the vicinity of the shock. We finally get:

$$\mathcal{D}_I^\ell = \frac{1}{12}\int \partial_\xi \varphi^\ell (\delta u)^3 d\xi = \frac{1}{12}\int \partial_\xi \varphi^\ell \Delta^3 d\xi = \frac{\Delta^3}{12}\varphi^\ell(0)$$
$$\xrightarrow[\ell \to 0]{} \mathcal{D}_I = \frac{\Delta^3}{12}\delta(x) = 2L\frac{\Delta^2}{12t}\delta(x) = 2L\varepsilon_\nu \delta(x).$$

In order to make a proper comparison with the mean rate of dissipation, we still have to make an average calculation:

$$\langle \mathcal{D}_I \rangle = \frac{1}{2L}\int_{-L}^{+L} 2L\varepsilon_\nu \delta(x) dx = \varepsilon_\nu.$$

We can therefore see that the averaged anomalous dissipation is identical to ε_ν in the case of Burgers' equation. This one-dimensional compressible model of turbulence is interesting because it allows us to carry out the analytical calculations to the end. Understanding this can help us to interpret (without constituting a proof) the much more difficult case of three-dimensional turbulence. Thus, it would seem that in order to have equality between ε_ν and $\langle \mathcal{D}_I \rangle$, it is necessary to be in an extreme situation in which discontinuities are omnipresent.

(7) We are looking for the expression of the structure function of order p:
$$S_p = \langle |\delta u|^p \rangle,$$

using the relationships (I.1.2). One gets:

$$\delta u \equiv u(x+\ell) - u(x) = \begin{cases} (\ell - 2L)/t & \text{if } 0 \in [x, x+\ell], \\ \ell/t & \text{otherwise}. \end{cases}$$

One obtains:

$$S_p = \frac{1}{2L}\left[\int_{-L}^{-\ell}\left|\frac{\ell}{t}\right|^p dx + \int_{-\ell}^{0}\left|\frac{\ell - 2L}{t}\right|^p dx + \int_{0}^{L}\left|\frac{\ell}{t}\right|^p dx\right]$$

$$= \Delta^p \left[\left(\frac{\ell}{2L}\right)^p \left(1 - \frac{\ell}{2L}\right) + \frac{\ell}{2L}\left(1 - \frac{\ell}{2L}\right)^p\right]$$

$$\simeq \Delta^p \left[\left(\frac{\ell}{2L}\right)^p + \frac{\ell}{2L}\right] \quad \text{with} \quad \ell \ll L.$$

It remains for us to distinguish the case $p < 1$ from the case $p \geq 1$. We finally arrive at the expressions:

$$S_p = \Delta^p \begin{cases} (\ell/2L)^p & \text{if } 0 < p < 1, \\ \ell/2L & \text{if } p \geq 1. \end{cases}$$

Therefore:

$$\zeta_p = \begin{cases} p & \text{if } 0 < p < 1, \\ 1 & \text{if } p \geq 1. \end{cases}$$

This result differs significantly from three-dimensional turbulence: here, ζ_p exponents quickly saturate (because of the shock) and the energy spectrum is in k^{-2} (compatible with $\zeta_2 = 1$).

1.2 Structure Function and Spectrum

We are interested in the relationship between the one-dimensional energy spectrum $E_{1d}(k)$ and the second-order structure function $S_2(r)$ in the case of three-dimensional homogeneous isotropic hydrodynamic turbulence.

(1) Let $R(\mathbf{r})$ be the two-point correlation function of the velocity, that is, $R(\mathbf{r}) = \langle \mathbf{u} \cdot \mathbf{u}' \rangle$. Write down the general relation between this function and the three-dimensional energy spectrum $E_{3d}(\mathbf{k})$.

(2) Focusing on the isotropic case, demonstrate the relationship:

$$S_2(r) = 4\int_0^{+\infty}\left(1 - \frac{\sin kr}{kr}\right)E_{1d}(k)dk,$$

where $S_2(r) = \langle (\delta \mathbf{u})^2 \rangle$.

(3) It is assumed that the one-dimensional energy spectrum is given by the relation $E_{1d}(k) = C_K \varepsilon^{2/3} k^{-5/3}$, where C_K is the Kolmogorov constant. Find the relation:

$$S_2(r) = C_2 \varepsilon^{2/3} r^{2/3},$$

where C_2 is a constant that will be given. Compare your results with the experimental measurements $C_K \simeq 0.5$ and $C_2 \simeq 2.5$ (Sreenivasan, 1995; Welter et al., 2009).

Solutions:

(1) By definition, we have the relations:

$$E_{3d}(\mathbf{k}) = \frac{1}{(2\pi)^3} \int_{\mathbb{R}^3} \frac{1}{2} R(\mathbf{r}) e^{i\mathbf{k} \cdot \mathbf{r}} d\mathbf{r}$$

and

$$R(\mathbf{r}) = 2 \int_{\mathbb{R}^3} E_{3d}(\mathbf{k}) e^{-i\mathbf{k} \cdot \mathbf{r}} d\mathbf{k}.$$

Then, we also have:

$$R(0) = 2 \int_{\mathbb{R}^3} E_{3d}(\mathbf{k}) d\mathbf{k}.$$

(2) For isotropic turbulence:

$$R(\mathbf{r}) = R(r) = 2 \int_0^{+\infty} \int_0^{\pi} \int_0^{2\pi} E_{3d}(k, \emptyset, \emptyset) e^{-i\mathbf{k} \cdot \mathbf{r}} k^2 \sin\theta \, dk \, d\theta \, d\phi$$

$$= 4\pi \int_0^{+\infty} \int_0^{\pi} k^2 E_{3d}(k, \emptyset, \emptyset) e^{-i\mathbf{k} \cdot \mathbf{r}} \sin\theta \, dk \, d\theta.$$

We choose a vector $\mathbf{r}$ along the z-axis; this gives:

$$R(r) = 4\pi \int_0^{+\infty} \int_0^{\pi} k^2 E_{3d}(k, \emptyset, \emptyset) e^{-ikr\cos\theta} \sin\theta \, dk \, d\theta$$

$$= 4\pi \int_0^{+\infty} \int_{-1}^{1} k^2 E_{3d}(k, \emptyset, \emptyset) e^{-ikrX} \, dk \, dX$$

$$= 2 \int_0^{+\infty} 4\pi k^2 E_{3d}(k, \emptyset, \emptyset) \frac{\sin kr}{kr} dk = 2 \int_0^{+\infty} E_{1d}(k) \frac{\sin kr}{kr} dk.$$

In particular:

$$R(0) = 2 \int_0^{+\infty} E_{1d}(k) dk.$$

We also have:

$$S_2(r) \equiv \langle (\delta \mathbf{u})^2 \rangle \equiv \langle (\mathbf{u}' - \mathbf{u})^2 \rangle = \langle (\mathbf{u}'^2 - 2\mathbf{u} \cdot \mathbf{u}' + \mathbf{u}^2) \rangle$$
$$= 2\langle \mathbf{u}^2 \rangle - 2\langle \mathbf{u} \cdot \mathbf{u}' \rangle = 2R(0) - 2R(r).$$

In the isotropic case:
$$S_2(r) = 2R(0) - 2R(r).$$

Hence, this is the relationship we are looking for.

(3) We substitute the proposed spectrum into the relation previously found:

$$\begin{aligned}S_2(r) &= 4C_K\varepsilon^{2/3}\int_0^{+\infty}\left(1-\frac{\sin kr}{kr}\right)k^{-5/3}dk\\&= 4C_K\varepsilon^{2/3}r^{2/3}\int_0^{+\infty}\left(1-\frac{\sin Y}{Y}\right)Y^{-5/3}dY\\&\simeq 4C_K\varepsilon^{2/3}r^{2/3}\times 1.21.\end{aligned}$$

Then, we find the relation $C_2 \simeq 4.84 C_K$ which is compatible with the experimental measurements $C_2 \simeq 2.5$ and $C_K \simeq 0.5$.

I.3 2D HD Turbulence: Detailed Conservation

In the two-dimensional case, the Navier–Stokes equations become simpler and it is possible to demonstrate analytically the existence of a dual cascade of energy and enstrophy (see Chapter 3). For this, we must use the detailed conservation laws for these invariants. The aim of this exercise is to obtain these two laws. We will introduce the stream function ψ such that $\mathbf{u} \equiv \mathbf{e}_z \nabla \psi$. With the use of this function the zero velocity divergence condition is automatically satisfied and the calculations are simplified.

(1) Write the spectral expression of enstrophy conservation.
(2) Demonstrate the detailed conservation of enstrophy using the relationships in the triangle formed by $\mathbf{k}$, $\mathbf{p}$, and $\mathbf{q}$.
(3) Same question for energy.
(4) Generalize the result in the case of statistically isotropic turbulence.

Solutions:
(1) In the two-dimensional case, the vorticity equation is written:
$$\partial_t w_z + \mathbf{u}\cdot\nabla w_z = \nu\Delta w_z,$$

which gives in Fourier space:
$$\partial_t \hat{w}_z(\mathbf{k}) + \nu k^2 \hat{w}_z(\mathbf{k}) = -\int_{\mathbb{R}^4}\hat{\mathbf{u}}(\mathbf{p})\cdot i\mathbf{q}\hat{w}_z(\mathbf{q})\delta(\mathbf{k}-\mathbf{p}-\mathbf{q})d\mathbf{p}d\mathbf{q}.$$

With the introduction of the stream function, we have $\hat{w}_z(\mathbf{k}) = -k^2 \hat{\psi}_\mathbf{k}$, and thus:
$$\partial_t(k^2\hat{\psi}_\mathbf{k}) + \nu k^4 \hat{\psi}_\mathbf{k} = -\int_{\mathbb{R}^4}\hat{\psi}_\mathbf{p}\mathbf{q}\cdot(\mathbf{e}_z\times\mathbf{p})q^2\hat{\psi}_\mathbf{q}\delta(\mathbf{k}-\mathbf{p}-\mathbf{q})d\mathbf{p}d\mathbf{q}.$$

Then, the enstrophy equation is:

$$\partial_t(k^4|\hat{\psi}_\mathbf{k}|^2) + 2\nu k^6|\hat{\psi}_\mathbf{k}|^2 = -\int_{\mathbb{R}^4} k^2 q^2 [\hat{\psi}_\mathbf{k}^* \hat{\psi}_\mathbf{p} \hat{\psi}_\mathbf{q} \\ + \hat{\psi}_\mathbf{k} \hat{\psi}_\mathbf{p}^* \hat{\psi}_\mathbf{q}^*]\mathbf{q}\cdot(\mathbf{e}_z\times\mathbf{p})\delta(\mathbf{k}-\mathbf{p}-\mathbf{q})d\mathbf{p}d\mathbf{q}.$$

By using the relation $\hat{\psi}_\mathbf{k}^* = \hat{\psi}_{-\mathbf{k}}$ and by manipulating the dummy variables $\mathbf{p}$ and $\mathbf{q}$, we arrive at the expression:

$$\partial_t(k^4|\hat{\psi}_\mathbf{k}|^2) + 2\nu k^6|\hat{\psi}_\mathbf{k}|^2 = -\int_{\mathbb{R}^4} k^2 q^2 [\hat{\psi}_\mathbf{k}^* \hat{\psi}_\mathbf{p}^* \hat{\psi}_\mathbf{q}^* \\ + \hat{\psi}_\mathbf{k} \hat{\psi}_\mathbf{p} \hat{\psi}_\mathbf{q}]\mathbf{q}\cdot(\mathbf{e}_z\times\mathbf{p})\delta(\mathbf{k}+\mathbf{p}+\mathbf{q})d\mathbf{p}d\mathbf{q} \\ = \int_{\mathbb{R}^4} k^2 q^2 \hat{\psi}_\mathbf{k} \hat{\psi}_\mathbf{p} \hat{\psi}_\mathbf{q} \mathbf{q}\cdot(\mathbf{e}_z\times\mathbf{p})\delta(\mathbf{k}+\mathbf{p}+\mathbf{q})d\mathbf{p}d\mathbf{q} + c.c.,$$

where $c.c.$ means the complex conjugate. We can make the equation symmetrical in $\mathbf{p}$ and $\mathbf{q}$:

$$\partial_t(k^4|\hat{\psi}_\mathbf{k}|^2) + 2\nu k^6|\hat{\psi}_\mathbf{k}|^2 = \frac{1}{2}\int_{\mathbb{R}^4} k^2 \hat{\psi}_\mathbf{k}\hat{\psi}_\mathbf{p}\hat{\psi}_\mathbf{q}[q^2\mathbf{e}_z\cdot(\mathbf{p}\times\mathbf{q}) \\ + p^2\mathbf{e}_z\cdot(\mathbf{q}\times\mathbf{p})]\delta(\mathbf{k}+\mathbf{p}+\mathbf{q})d\mathbf{p}d\mathbf{q} + c.c. \\ = \int_{\mathbb{R}^4} k^2 S(\mathbf{k},\mathbf{p},\mathbf{q})\delta(\mathbf{k}+\mathbf{p}+\mathbf{q})d\mathbf{p}d\mathbf{q},$$

with by definition:

$$S(\mathbf{k},\mathbf{p},\mathbf{q}) \equiv [(q^2-p^2)(\mathbf{e}_z\cdot(\mathbf{p}\times\mathbf{q}))]\Re[\hat{\psi}_\mathbf{k}\hat{\psi}_\mathbf{p}\hat{\psi}_\mathbf{q}].$$

This gives us expression S from Chapter 3.

(2) The detailed conservation of enstrophy requires the previous expression to be written in the form:

$$\partial_t \int_{\mathbb{R}^2} k^4|\hat{\psi}_\mathbf{k}|^2 d\mathbf{k} + 2\nu \int_{\mathbb{R}^2} k^6|\hat{\psi}_\mathbf{k}|^2 d\mathbf{k} = \frac{1}{3}\int_{\mathbb{R}^6}[k^2 S(\mathbf{k},\mathbf{p},\mathbf{q}) + p^2 S(\mathbf{p},\mathbf{q},\mathbf{k}) \\ + q^2 S(\mathbf{q},\mathbf{k},\mathbf{p})]\delta(\mathbf{k}+\mathbf{p}+\mathbf{q})d\mathbf{k}d\mathbf{p}d\mathbf{q}.$$

We are going to show that the integrand is null. With the definition of S, we have:

$$k^2 S(\mathbf{k},\mathbf{p},\mathbf{q}) + p^2 S(\mathbf{p},\mathbf{q},\mathbf{k}) + q^2 S(\mathbf{q},\mathbf{k},\mathbf{p}) = \Re[\hat{\psi}_\mathbf{k}\hat{\psi}_\mathbf{p}\hat{\psi}_\mathbf{q}] \\ \mathbf{e}_z\cdot[k^2(q^2-p^2)(\mathbf{p}\times\mathbf{q}) + p^2(k^2-q^2)(\mathbf{q}\times\mathbf{k}) + q^2(p^2-k^2)(\mathbf{k}\times\mathbf{p})].$$

Let us introduce the interior angles θ_k, θ_p, and θ_q associated with the triangle $\mathbf{k}+\mathbf{p}+\mathbf{q}=\mathbf{0}$ (see Figure A.2); by using identities in the triangle, we get:

$$k^2 S(\mathbf{k},\mathbf{p},\mathbf{q}) + p^2 S(\mathbf{p},\mathbf{q},\mathbf{k}) + q^2 S(\mathbf{q},\mathbf{k},\mathbf{p}) = \Re[\hat{\psi}_\mathbf{k}\hat{\psi}_\mathbf{p}\hat{\psi}_\mathbf{q}] \\ pq\sin\theta_k[(q^2-p^2)k^2 + (k^2-q^2)p^2 + (p^2-k^2)q^2] = 0,$$

hence the detailed conservation of enstrophy.

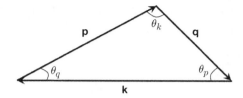

Figure A.2 Angles associated with a triad of interaction $\mathbf{k} + \mathbf{p} + \mathbf{q} = 0$.

(3) To demonstrate the detailed energy conservation, we will use the stream function. We have (see Chapter 3):

$$\partial_t |\hat{\mathbf{u}}(\mathbf{k})|^2 + 2\nu k^2 |\hat{\mathbf{u}}(\mathbf{k})|^2 = \int_{\mathbb{R}^4} S(\mathbf{k}, \mathbf{p}, \mathbf{q}) \delta(\mathbf{k} + \mathbf{p} + \mathbf{q}) d\mathbf{p} d\mathbf{q}.$$

One obtains:

$$\partial_t \int_{\mathbb{R}^2} |\hat{\mathbf{u}}(\mathbf{k})|^2 d\mathbf{k} + 2\nu \int_{\mathbb{R}^2} k^2 |\hat{\mathbf{u}}(\mathbf{k})|^2 d\mathbf{k} = \frac{1}{3} \int_{\mathbb{R}^6} [S(\mathbf{k}, \mathbf{p}, \mathbf{q}) + S(\mathbf{p}, \mathbf{q}, \mathbf{k}) + S(\mathbf{q}, \mathbf{k}, \mathbf{p})] \delta(\mathbf{k} + \mathbf{p} + \mathbf{q}) d\mathbf{k} d\mathbf{p} d\mathbf{q},$$

with:

$$S(\mathbf{k}, \mathbf{p}, \mathbf{q}) + S(\mathbf{p}, \mathbf{q}, \mathbf{k}) + S(\mathbf{q}, \mathbf{k}, \mathbf{p}) = \Re[\hat{\psi}_\mathbf{k} \hat{\psi}_\mathbf{p} \hat{\psi}_\mathbf{q}]$$
$$\mathbf{e_z} \cdot [(q^2 - p^2)(\mathbf{p} \times \mathbf{q}) + (k^2 - q^2)(\mathbf{q} \times \mathbf{k}) + (p^2 - k^2)(\mathbf{k} \times \mathbf{p})].$$

With the relationships in the triangle, we finally get:

$$S(\mathbf{k}, \mathbf{p}, \mathbf{q}) + S(\mathbf{p}, \mathbf{q}, \mathbf{k}) + S(\mathbf{q}, \mathbf{k}, \mathbf{p})$$
$$= \Re[\hat{\psi}_\mathbf{k} \hat{\psi}_\mathbf{p} \hat{\psi}_\mathbf{q}] pq \sin\theta_k [(q^2 - p^2) + (k^2 - q^2) + (p^2 - k^2)] = 0,$$

hence the detailed energy conservation.

(4) With the ensemble average, we find:

$$\partial_t E(\mathbf{k}) + 2\nu k^2 E(\mathbf{k}) = \int_{\mathbb{R}^4} \langle S(\mathbf{k}, \mathbf{p}, \mathbf{q}) \rangle \delta(\mathbf{k} + \mathbf{p} + \mathbf{q}) d\mathbf{p} d\mathbf{q}$$
$$= \int_\Delta (q^2 - p^2) pq\, \mathbf{e_z} \cdot (\mathbf{e_p} \times \mathbf{e_q}) \Re \langle \hat{\psi}_\mathbf{k} \hat{\psi}_\mathbf{p} \hat{\psi}_\mathbf{q} \rangle p d\theta_q dp,$$

where the symbol Δ means that the integral is calculated by satisfying the triangular relationship, and $\mathbf{e_p}$ and $\mathbf{e_q}$ are unit vectors oriented according to $\mathbf{p}$ and $\mathbf{q}$, respectively. Al Kashi's relation in the triangle:

$$q^2 = k^2 + p^2 - 2kp \cos\theta_q,$$

gives us, at k and p fixed, $q dq = kp \sin\theta_q d\theta_q$, hence the expression:

$$\partial_t E(\mathbf{k}) + 2\nu k^2 E(\mathbf{k}) = \int_\Delta (q^2 - p^2) \frac{pq}{\sin\theta_k} \mathbf{e_z} \cdot (\mathbf{e_p} \times \mathbf{e_q}) \Re \langle \hat{\psi}_\mathbf{k} \hat{\psi}_\mathbf{p} \hat{\psi}_\mathbf{q} \rangle dp dq.$$

If the turbulence is statistically isotropic, one can introduce the one-dimensional energy spectrum, $E(k) = 2\pi k E(\mathbf{k})$, hence:

$$\partial_t E(k) + 2\nu k^2 E(k) = T(k),$$

with by definition:

$$T(k) \equiv \int_\Delta 2\pi k(q^2 - p^2)\frac{pq}{\sin\theta_k} \mathbf{e_z} \cdot (\mathbf{e_p} \times \mathbf{e_q}) \Re\langle \hat{\psi}_\mathbf{k} \hat{\psi}_\mathbf{p} \hat{\psi}_\mathbf{q}\rangle dpdq$$

$$\equiv \int_\Delta T(k,p,q) dpdq.$$

Detailed energy conservation is easily demonstrated from the relationship:

$$\int_\mathbb{R} T(k)dk = \frac{1}{3}\int_{\mathbb{R}^3} [T(k,p,q) + T(p,q,k) + T(q,k,p)] dkdpdq.$$

We have for a given triad:

$$T(k,p,q) + T(p,q,k) + T(q,k,p) = 2\pi kpq[(q^2 - p^2) + (k^2 - q^2)$$
$$+ (p^2 - k^2)]\Re\langle\hat{\psi}_\mathbf{k}\hat{\psi}_\mathbf{p}\hat{\psi}_\mathbf{q}\rangle = 0,$$

hence the detailed statistical energy conservation.

For enstrophy, we introduce the spectrum $\Omega(\mathbf{k}) \equiv k^4 \langle |\hat{\psi}_\mathbf{k}|^2\rangle$ and we obtain in a similar way:

$$\partial_t \Omega(\mathbf{k}) + 2\nu k^2 \Omega(\mathbf{k}) = \int_{\mathbb{R}^4} k^2 \langle S(\mathbf{k},\mathbf{p},\mathbf{q})\rangle \delta(\mathbf{k}+\mathbf{p}+\mathbf{q}) dpdq$$
$$= \int_\Delta k^2 (q^2 - p^2)\frac{pq}{\sin\theta_k} \mathbf{e_z} \cdot (\mathbf{e_p} \times \mathbf{e_q}) \Re\langle\hat{\psi}_\mathbf{k}\hat{\psi}_\mathbf{p}\hat{\psi}_\mathbf{q}\rangle dpdq.$$

For the one-dimensional enstrophy spectrum this gives:

$$\partial_t \Omega(k) + 2\nu k^2 \Omega(k) = k^2 T(k) = \int_\Delta k^2 T(k,p,q) dpdq.$$

We have then:

$$\int_\mathbb{R} k^2 T(k) dk = \frac{1}{3}\int_{\mathbb{R}^3} [k^2 T(k,p,q) + p^2 T(p,q,k) + q^2 T(q,k,p)] dkdpdq,$$

with per triad:

$$k^2 T(k,p,q) + p^2 T(p,q,k) + q^2 T(q,k,p) = 2\pi kpq[k^2(q^2 - p^2) + p^2(k^2 - q^2)$$
$$+ q^2(p^2 - k^2)]\Re\langle\hat{\psi}_\mathbf{k}\hat{\psi}_\mathbf{p}\hat{\psi}_\mathbf{q}\rangle = 0,$$

which demonstrates the detailed statistical conservation of enstrophy.

II.1 MHD Model of Nonlinear Diffusion

Non-linear diffusion models are often used in turbulence because their numerical simulations are relatively easy. We have already encountered these models in Chapters 3 and 6. We propose here to develop such a phenomenological model for Alfvén wave turbulence in the balanced case. The particularity of the approach is that it must take into account the anisotropy. In practice, the following equation is considered:

$$\frac{\partial E(\mathbf{k})}{\partial t} = -\nabla \cdot \mathbf{\Pi}(\mathbf{k}),$$

where $E(\mathbf{k})$ is the energy spectrum and $\mathbf{\Pi}(\mathbf{k})$ the associated flux.

(1) What are the conditions of validity of this equation?

(2) Alfvén wave turbulence is strongly anisotropic. Using a coordinate system adapted to this situation, rewrite the diffusion equation.

(3) The transverse component of the energy flux is modeled as follows:

$$\Pi_\perp(\mathbf{k}) = -D(k)\frac{\partial E(\mathbf{k})}{\partial k_\perp},$$

where D is a non-linear diffusion coefficient which remains to be determined. What is the dimension of D?

(4) We call τ_{tr} the characteristic time of energy transfer. Express τ_{tr} as a function of $E(\mathbf{k})$ and $k_\perp$.

(5) By using the relation $\int E(k_\perp) f(k_\parallel) dk_\perp dk_\parallel = \iiint E(\mathbf{k}) d\mathbf{k}$, show that the non-linear diffusion equation can be written as:

$$\frac{\partial E(k_\perp)}{\partial t} = C \frac{\partial}{\partial k_\perp} \left(k_\perp^6 E(k_\perp) \frac{\partial (E(k_\perp)/k_\perp)}{\partial k_\perp} \right),$$

where C is a constant.

(6) What is the Kolmogorov–Zakharov spectrum exact solution of the equation? What is the direction of the cascade?

Solutions:

(1) The equation describes the physics of the inertial range in wave turbulence. This means that the external force acts on a larger scale than the scales of the inertial range and that dissipation acts on a smaller scale. At the same time, we need to check the wave turbulence conditions. We know that a transition to strong turbulence is possible at small scales. Therefore, an additional condition exists; this one is written $\tau_A \ll \tau_{NL}$, that is:

$$k_\perp z \ll k_\parallel b_0,$$

with b_0 the uniform magnetic field and z the Elsässer field.

(2) Cylindrical coordinates are used, which are particularly well adapted to this problem. We obtain in the case of axisymmetric turbulence:
$$\frac{\partial E(\mathbf{k})}{\partial t} = -\frac{1}{k_\perp}\frac{\partial (k_\perp \Pi_\perp(\mathbf{k}))}{\partial k_\perp}.$$

(3) Dimensionaly we find:
$$D \sim \frac{k_\perp^2}{\tau_{tr}},$$
with τ_{tr} the characteristic transfer time in Alfvén wave turbulence.

(4) The transfer time is:
$$\tau_{tr} \sim \omega \tau_{NL}^2 \sim \frac{k_\| b_0}{k_\perp^2 z^2} \sim \frac{b_0}{k_\perp^4 E(\mathbf{k})}.$$

(5) Using the previous results, we can write:
$$\frac{\partial E(\mathbf{k})}{\partial t} \sim \frac{1}{k_\perp}\frac{\partial}{\partial k_\perp}\left(k_\perp^3 \tau_{tr}^{-1}\frac{\partial E(\mathbf{k})}{\partial k_\perp}\right) \sim \frac{1}{b_0 k_\perp}\frac{\partial}{\partial k_\perp}\left(k_\perp^7 E(\mathbf{k})\frac{\partial E(\mathbf{k})}{\partial k_\perp}\right).$$

The axisymmetric spectrum is then introduced:
$$\frac{\partial (k_\perp k_\| E(\mathbf{k}))}{\partial t} \sim \frac{\partial (k_\| f(k_\|) E(k_\perp))}{\partial t} \sim \frac{k_\| f^2(k_\|)}{b_0}\frac{\partial}{\partial k_\perp}\left(k_\perp^6 E(k_\perp)\frac{\partial (E(k_\perp)/k_\perp)}{\partial k_\perp}\right).$$

After simplification, we obtain the following non-linear diffusion equation:
$$\boxed{\frac{\partial E(k_\perp)}{\partial t} = C\frac{\partial}{\partial k_\perp}\left(k_\perp^6 E(k_\perp)\frac{\partial (E(k_\perp)/k_\perp)}{\partial k_\perp}\right)},$$

with C a constant depending, among others, on b_0 and $f(k_\|)$. Note that this diffusion equation was deduced directly from the MHD kinetic equations by considering the approximation of local interactions (Galtier and Buchlin, 2010).

(6) To find the exact solutions, we introduce the spectrum $E(k_\perp) = A k_\perp^x$ (with $A > 0$) and the flux $\Pi_\perp(k_\perp)$. For this analysis, we take $C = 1$, which means that we place ourselves on the characteristic time of the Alfvén wave turbulence cascade. We get:
$$\Pi_\perp = A^2(1-x)k_\perp^{2x+4}.$$

The first solution $x = 1$ is a spectrum with zero energy flux: this is the thermodynamics solution. The second solution $x = -2$ corresponds to the Kolmogorov–Zakharov spectrum; in this case,
$$\boxed{\Pi_\perp = 3A^2 > 0}.$$

The flux is positive and therefore the cascade direct.

II.2 Four-Wave Interactions

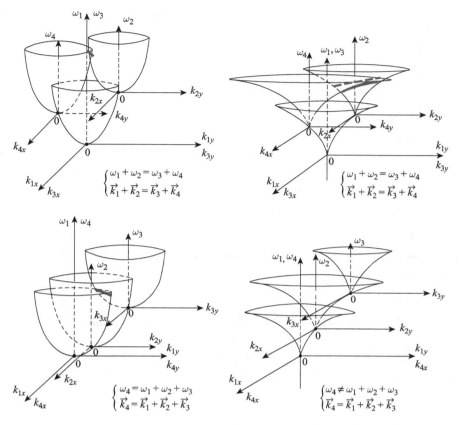

Figure A.3 Solutions to the resonance conditions (left: $\alpha > 1$; right: $\alpha < 1$; top: $2 \to 2$ interactions; bottom: $3 \to 1$ interactions) are always possible except for the last case (bottom right).

II.2 Four-Wave Interactions

Let us consider the isotropic dispersion relation $\omega = Ak^\alpha$, with A a positive constant and $\alpha \in \mathbb{R}_+^*$. In the two-dimensional case, show graphically that we can always find solutions for four-wave resonant interactions.

Solutions:
We shall consider three cases: $\alpha > 1$, $\alpha = 1$, and $\alpha < 1$. As shown graphically in Figures A.3 and A.4, solutions are always possible for interactions of the type $2 \to 2$:
$$\omega_k + \omega_{k_1} = \omega_{k_2} + \omega_{k_3},$$
$$\mathbf{k} + \mathbf{k}_1 = \mathbf{k}_2 + \mathbf{k}_3,$$
while for $3 \to 1$ one cannot find solutions when $\alpha < 1$.

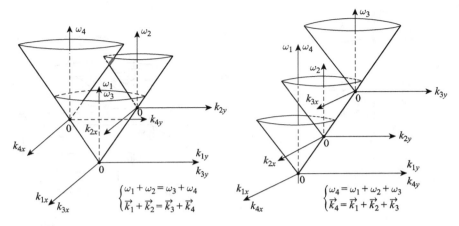

Figure A.4 Solutions to the resonance conditions for $\alpha = 1$ (left: $2 \to 2$ interactions; right: $3 \to 1$ interactions). For $3 \to 1$ interactions, solutions correspond to collinear wavevector.

II.3 Gravitational Wave Turbulence: Exact Solutions

The objective of this exercise is to obtain analytically the exact solutions of gravitational wave turbulence. We have seen that the kinetic equation is written as follows:

$$\partial_t N_{\mathbf{k}} = \epsilon^6 \int_{\mathbb{R}^6} |T^{\mathbf{k}\mathbf{k}_3}_{\mathbf{k}_1 \mathbf{k}_2}|^2 N_{\mathbf{k}} N_{\mathbf{k}_1} N_{\mathbf{k}_2} N_{\mathbf{k}_3}$$
$$\times \left[\frac{1}{N_{\mathbf{k}}} + \frac{1}{N_{\mathbf{k}_3}} - \frac{1}{N_{\mathbf{k}_1}} - \frac{1}{N_{\mathbf{k}_2}} \right] \delta^k_{03,12} \delta^\omega_{03,12} \, d\mathbf{k}_1 d\mathbf{k}_2 d\mathbf{k}_3 \, ,$$

with $\delta^k_{03,12} \equiv \delta(\mathbf{k} + \mathbf{k}_3 - \mathbf{k}_1 - \mathbf{k}_2)$ and $\delta^\omega_{03,12} \equiv \delta(\omega_{\mathbf{k}} + \omega_{\mathbf{k}_3} - \omega_{\mathbf{k}_1} - \omega_{\mathbf{k}_2})$.

We must specify the symmetries verified by the geometric coefficient. We have:

$$T^{\mathbf{k}\mathbf{k}_3}_{\mathbf{k}_1 \mathbf{k}_2} = \frac{1}{4}(W^{\mathbf{k}\mathbf{k}_3}_{\mathbf{k}_1 \mathbf{k}_2} + W^{\mathbf{k}_3 \mathbf{k}}_{\mathbf{k}_1 \mathbf{k}_2} + W^{\mathbf{k}\mathbf{k}_3}_{\mathbf{k}_2 \mathbf{k}_1} + W^{\mathbf{k}_3 \mathbf{k}}_{\mathbf{k}_2 \mathbf{k}_1}),$$

as well as:

$$W^{\mathbf{k}\mathbf{k}_3}_{\mathbf{k}_1 \mathbf{k}_2} = Q^{\mathbf{k}\mathbf{k}_3}_{\mathbf{k}_1 \mathbf{k}_2} + Q^{\mathbf{k}_1 \mathbf{k}_2}_{\mathbf{k}\,\mathbf{k}_3} .$$

The exact form of $Q^{\mathbf{k}\mathbf{k}_3}_{\mathbf{k}_1 \mathbf{k}_2}$ is not useful (see Galtier and Nazarenko, 2017); only information on its degree of homogeneity is important. We have:

$$T^{a\mathbf{k}\, a\mathbf{k}_3}_{a\mathbf{k}_1\, a\mathbf{k}_2} = T^{\mathbf{k}\, \mathbf{k}_3}_{\mathbf{k}_1\, \mathbf{k}_2},$$

which means that this coefficient is dimensionless.

For four-wave interactions, the Zakharov conformal transformation is written:

$$\xi_1 \to \frac{1}{\xi_1}, \quad \xi_2 \to \frac{\xi_3}{\xi_1}, \quad \xi_3 \to \frac{\xi_2}{\xi_1}, \quad \text{(TZa)}$$

$$\xi_1 \to \frac{\xi_3}{\xi_2}, \quad \xi_2 \to \frac{1}{\xi_2}, \quad \xi_3 \to \frac{\xi_1}{\xi_2}, \quad \text{(TZb)}$$

$$\xi_1 \to \frac{\xi_1}{\xi_3}, \quad \xi_2 \to \frac{\xi_2}{\xi_3}, \quad \xi_3 \to \frac{1}{\xi_3}. \quad \text{(TZc)}$$

(1) By applying these transformations to the kinetic equation, find the isotropic exact solutions proposed in Chapter 9.
(2) Deduce the direction of the cascade for each solution.
(3) Give the expression for the Kolmogorov constants.

Solutions:
(1) We introduce the isotropic wave action spectrum:

$$N_i \equiv 2\pi k_i N_{\mathbf{k}_i},$$

with in particular $k_0 \equiv k$. We obtain the kinetic equation:

$$\partial_t N_0 = \int C_{k_1 k_2}^{k k_3} N_0 N_1 N_2 N_3 \left[\frac{k}{N_0} + \frac{k_3}{N_3} - \frac{k_1}{N_1} - \frac{k_2}{N_2} \right] \delta_{03,12}^{\omega} \, dk_1 dk_2 dk_3,$$

with:

$$C_{k_1 k_2}^{k k_3} \equiv \frac{\epsilon^4}{4\pi^2} \int |T_{k_1 k_2}^{k k_3}|^2 \delta_{03,12}^{k} \, dC_0 dC_1 dC_2 dC_3,$$

where C_i are circles of radius k_i. We will look for nontrivial solutions in power law. We introduce $N_i = A k_i^x$ (with $A > 0$), and the dimensionless wavenumber $\xi_i \equiv k_i/k$; we get:

$$\partial_t N_0 = A^3 k^{3x+1} \int C_{\xi_1 \xi_2}^{1 \xi_3} (\xi_1 \xi_2 \xi_3)^x (1 + \xi_3^{1-x} - \xi_1^{1-x} - \xi_2^{1-x}) \delta_{03,12}^{\xi} \, d\xi_1 d\xi_2 d\xi_3.$$

Two trivial solutions appear:

$$\boxed{x = 0 \quad \text{and} \quad x = 1}.$$

These are zero-flux (thermodynamic) solutions associated with wave action and energy, respectively.
To find the nontrivial solutions, we decompose the integral into four identical parts and apply the three Zakharov transformations on three integrals. In particular, we use the following properties:

$$C^{1\xi_3}_{\xi_1\xi_2} \xrightarrow{TZa} \xi_1^2 C^{\xi_1\xi_2}_{1\xi_3} = \xi_1^2 C^{1\xi_3}_{\xi_1\xi_2},$$

$$C^{1\xi_3}_{\xi_1\xi_2} \xrightarrow{TZb} \xi_2^2 C^{\xi_2\xi_1}_{\xi_3 1} = \xi_2^2 C^{1\xi_3}_{\xi_1\xi_2},$$

$$C^{1\xi_3}_{\xi_1\xi_2} \xrightarrow{TZc} \xi_3^2 C^{\xi_3 1}_{\xi_1\xi_2} = \xi_3^2 C^{1\xi_3}_{\xi_1\xi_2},$$

obtained using the symmetries of the geometric coefficient. We obtain the expression:

$$\partial_t N_0 = \frac{1}{4} A^3 k^{3x+1} \left\{ \int C^{1\xi_3}_{\xi_1\xi_2} (\xi_1\xi_2\xi_3)^x (1 + \xi_3^{1-x} - \xi_1^{1-x} - \xi_2^{1-x}) \delta^\xi_{03,12} \, d\xi_1 d\xi_2 d\xi_3 \right.$$

$$+ \int \xi_1^2 C^{1\xi_3}_{\xi_1\xi_2} \frac{(\xi_2\xi_3)^x}{\xi_1^{3x}} \frac{(\xi_1^{1-x} + \xi_2^{1-x} - 1 - \xi_3^{1-x})}{\xi_1^{1-x}} \xi_1 \delta^\xi_{12,03} \frac{d\xi_1 d\xi_2 d\xi_3}{\xi_1^4}$$

$$+ \int \xi_2^2 C^{1\xi_3}_{\xi_1\xi_2} \frac{(\xi_1\xi_3)^x}{\xi_2^{3x}} \frac{(\xi_2^{1-x} + \xi_1^{1-x} - \xi_3^{1-x} - 1)}{\xi_2^{1-x}} \xi_2 \delta^\xi_{21,30} \frac{d\xi_1 d\xi_2 d\xi_3}{\xi_2^4}$$

$$+ \left. \int \xi_3^2 C^{1\xi_3}_{\xi_1\xi_2} \frac{(\xi_1\xi_2)^x}{\xi_3^{3x}} \frac{(\xi_3^{1-x} + 1 - \xi_1^{1-x} - \xi_2^{1-x})}{\xi_3^{1-x}} \xi_3 \delta^\xi_{30,12} \frac{d\xi_1 d\xi_2 d\xi_3}{\xi_3^4} \right\}.$$

By rearranging the terms, we find:

$$\partial_t N_0 = \frac{1}{4} A^3 k^{3x+1} \int C^{1\xi_3}_{\xi_1\xi_2} (\xi_1\xi_2\xi_3)^x (1 + \xi_3^{1-x} - \xi_1^{1-x} - \xi_2^{1-x})$$

$$(1 - \xi_1^{-3x-2} - \xi_2^{-3x-2} + \xi_3^{-3x-2}) \delta^\xi_{03,12} d\xi_1 d\xi_2 d\xi_3$$

$$= k^{3x+1} I(x).$$

Therefore, the non-zero constant flux solutions (Kolmogorov–Zakharov spectrum) are:

$$\boxed{x = -\frac{2}{3} \quad \text{and} \quad x = -1}.$$

(2) The first exact solution corresponds to an inverse cascade of wave action and the second to a direct cascade of energy. To demonstrate this association, we must introduce the wave action and energy fluxes, Ξ and Π, respectively.

In the first case, we have:

$$\partial_t N_0 = k^{3x+1} I(x) = -\frac{\partial \Xi}{\partial k},$$

hence:

$$\Xi(k) = -\frac{k^{3x+2}}{3x+2} I(x).$$

The constant (nonzero) wave action flux solution is obtained for $x = -2/3$, a value that cancels the integral I, hence (with L'Hôpital's rule):

$$\lim_{x \to -2/3} \Xi(k) \equiv \zeta = -\lim_{x \to -2/3} \frac{I(x)}{3x+2} = -\frac{\partial I(z)}{\partial z}\bigg|_{z=0},$$

with $z = 3x + 2$. We finally obtain:

$$\zeta = -\frac{A^3}{4} \int C_{\xi_1\xi_2}^{1\xi_3}(\xi_1\xi_2\xi_3)^{-2/3}(1 + \xi_3^{5/3} - \xi_1^{5/3} - \xi_2^{5/3}) \ln\left(\frac{\xi_1\xi_2}{\xi_3}\right) \delta_{03,12}^\xi d\xi_1 d\xi_2 d\xi_3$$
$$= -A^3 J_1.$$

An evaluation of the sign of the integrand shows that $\boxed{\zeta < 0}$: the cascade is thus inward.

In the second case, we have $E_0 = kN_0$ and:

$$\partial_t E_0 = k^{3x+2} I(x) = -\frac{\partial \Pi}{\partial k}.$$

Hence the relation:

$$\Pi(k) = -\frac{k^{3x+3}}{3x+3} I(x).$$

The solution with constant (nonzero) energy flux is obtained for $x = -1$, a value that cancels the integral I, hence:

$$\lim_{x \to -1} \Pi(k) \equiv \varepsilon = -\lim_{x \to -1} \frac{I(x)}{3x+3} = -\frac{\partial I(\tilde{z})}{\partial \tilde{z}}\Big|_{\tilde{z}=0},$$

with $\tilde{z} = 3x + 3$. We finally get the expression:

$$\varepsilon = -\frac{A^3}{4} \int C_{\xi_1\xi_2}^{1\xi_3}(\xi_1\xi_2\xi_3)^{-1}(1 + \xi_3^2 - \xi_1^2 - \xi_2^2)$$
$$(\xi_1 \ln \xi_1 + \xi_2 \ln \xi_2 - \xi_3 \ln \xi_3) \delta_{03,12}^\xi d\xi_1 d\xi_2 d\xi_3$$
$$= A^3 J_2.$$

An evaluation of the sign of the integrand shows that $\boxed{\varepsilon > 0}$: the cascade is therefore foreward.

(3) To find the expressions of the Kolmogorov constants, we just need to express A according to the flux. In the case of the wave action, we have:

$$N_0 = C_N (-\zeta)^{1/3} k^{-2/3},$$

with the Kolmogorov constant, $\boxed{C_N = J_1^{-1/3}}$.

For energy, we obtain:

$$N_0 = C_E \varepsilon^{1/3} k^{-1},$$

with the Kolmogorov constant, $\boxed{C_E = J_2^{-1/3}}$.

II.4 Inertial Wave Turbulence: Domain of Locality

Find the domain of locality for inertial wave turbulence.

Solutions:
We start with the kinetic equation describing the evolution of the kinetic energy (see Chapter 6):

$$\partial_t E_k \propto \sum_{ss_ps_q} \int_{\Delta_\perp} \frac{sk_\| s_p p_\|}{k_\perp^2 p_\perp^2 q_\perp^2} \left(\frac{s_q q_\perp - s_p p_\perp}{\omega_k}\right)^2 (sk_\perp + s_p p_\perp + s_q q_\perp)^2 \sin\theta$$
$$E_q(p_\perp E_k - k_\perp E_p)\delta(s\omega_k + s_p\omega_p + s_q\omega_q)\, \delta(k_\| + p_\| + q_\|) dp_\perp dq_\perp dp_\| dq_\|\,,$$

where $E_k \equiv E(k_\perp, k_\|) \sim k_\perp^x |k_\||^y$ is the axisymmetric spectrum. By introducing the dimensionless wavenumbers $\tilde{p}_\perp \equiv p_\perp/k_\perp$, $\tilde{q}_\perp \equiv q_\perp/k_\perp$, $\tilde{p}_\| \equiv p_\|/k_\|$ and $\tilde{q}_\| \equiv q_\|/k_\|$, we find the expression:

$$\partial_t E_k \propto \sum_{ss_ps_q} \int_{\Delta_\perp} \frac{ss_p \tilde{p}_\|}{\tilde{p}_\perp^2 \tilde{q}_\perp^2} (s_q\tilde{q}_\perp - s_p\tilde{p}_\perp)^2 (s + s_p\tilde{p}_\perp + s_q\tilde{q}_\perp)^2$$
$$\sqrt{1 - \left(\frac{1+\tilde{p}_\perp^2 - \tilde{q}_\perp^2}{2\tilde{p}_\perp}\right)^2} \tilde{q}_\perp^x \tilde{q}_\|^y (\tilde{p}_\perp - \tilde{p}_\perp^x \tilde{p}_\|^y)$$
$$\delta\left(s + s_p \frac{\tilde{p}_\|}{\tilde{p}_\perp} + s_q \frac{\tilde{q}_\|}{\tilde{q}_\perp}\right) \delta(1 + \tilde{p}_\| + \tilde{q}_\|) d\tilde{p}_\perp d\tilde{q}_\perp d\tilde{p}_\| d\tilde{q}_\|\,,$$

for which we will evaluate the convergence in three regions of nonlocality.

Region A: $\tilde{p}_\perp = 1 + r\cos\beta$, $\tilde{q}_\perp = r\sin\beta$, with $r \ll 1$ and $\beta \in [\pi/4, 3\pi/4]$. At leading order, we have (we use the other delta function):

$\delta(s + s_p\frac{\tilde{p}_\|}{\tilde{p}_\perp} + s_q\frac{\tilde{q}_\|}{\tilde{q}_\perp}) = \delta(sr\cos\beta + s_q\tilde{q}_\|/(r\sin\beta))$ when $s = s_p$,
$\delta(s + s_p\frac{\tilde{p}_\|}{\tilde{p}_\perp} + s_q\frac{\tilde{q}_\|}{\tilde{q}_\perp}) = \delta(2 + s_q\tilde{q}_\|/(r\sin\beta))$ when $s = -s_p$,
$(s_q\tilde{q}_\perp - s_p\tilde{p}_\perp)^2 = 1$,
$(s + s_p\tilde{p}_\perp + s_q\tilde{q}_\perp)^2 = 4$ when $s = s_p$,
$(s + s_p\tilde{p}_\perp + s_q\tilde{q}_\perp)^2 = r^2(s_q\sin\beta - s\cos\beta)^2$ when $s = -s_p$,
$\sqrt{1 - \left(\frac{1+\tilde{p}_\perp^2 - \tilde{q}_\perp^2}{2\tilde{p}_\perp}\right)^2} = r\sqrt{\sin^2\beta - \cos^2\beta}$.

These estimates lead to $\tilde{q}_\| = -ss_q r^2 \cos\beta \sin\beta$, $\tilde{p}_\| = -1$ when $s = s_p$, and to $\tilde{q}_\| = -2s_q r\sin\beta$, $\tilde{p}_\| = -1$ when $s = -s_p$. Note that the use of the integrations over $\tilde{p}_\|$ and $\tilde{q}_\|$ leads to the emergence of the factor $r|\sin\beta|$ in both cases.

For $s = s_p$, we find the following contribution to the kinetic equation:

$$\int_0^R \int_{\pi/4}^{3\pi/4} \frac{1}{r^2 \sin^2\beta} r\sqrt{\sin^2\beta - \cos^2\beta}\, r|\sin\beta|(r\sin\beta)^x$$
$$(r^2|\sin\beta\cos\beta|)^y (1 + r\cos\beta - (1 + r\cos\beta)^x) dr d\beta\,,$$

which gives:

$$\int_0^R \int_{\pi/4}^{3\pi/4} r^{x+2y+2} \sqrt{\sin^2\beta - \cos^2\beta}(\sin\beta)^{x-2} |\cos\beta|^y |\sin\beta|^{y+1} \cos\beta\, dr d\beta\,.$$

The condition for convergence is $x + 2y > -3$, however, because of the symmetry in β the integral is zero. The next order correction leads to the condition $x + 2y > -4$.

For $s = -s_p$, we find the following contribution:

$$\int_0^R \int_{\pi/4}^{3\pi/4} \frac{1}{r^2 \sin^2 \beta} r^2 (s_q \sin \beta - s \cos \beta)^2 r \sqrt{\sin^2 \beta - \cos^2 \beta} \, r |\sin \beta|$$
$$(r \sin \beta)^x (r|\sin \beta|)^y (1 + r \cos \beta - (1 + r \cos \beta)^x) dr d\beta,$$

which gives (no cancellation from β is expected):

$$\int_0^R r^{x+y+4} dr.$$

We find the condition for convergence $x + y > -5$.

Region B: $\tilde{p}_\perp = r \cos \beta$, $\tilde{q}_\perp = 1 + r \sin \beta$, with $r \ll 1$ and $\beta \in [-\pi/4, +\pi/4]$.
At leading order, we have (we use the other delta function):

$$\delta(s + s_p \frac{\tilde{p}_\parallel}{\tilde{p}_\perp} + s_q \frac{\tilde{q}_\parallel}{\tilde{q}_\perp}) = \delta(s_q r \sin \beta + s_p \tilde{p}_\parallel/(r \cos \beta)) \text{ when } s = s_q,$$
$$\delta(s + s_p \frac{\tilde{p}_\parallel}{\tilde{p}_\perp} + s_q \frac{\tilde{q}_\parallel}{\tilde{q}_\perp}) = \delta(2 + s_p \tilde{p}_\parallel/(r \cos \beta)) \text{ when } s = -s_q,$$
$$(s_q \tilde{q}_\perp - s_p \tilde{p}_\perp)^2 = 1,$$
$$(s + s_p \tilde{p}_\perp + s_q \tilde{q}_\perp)^2 = 4 \text{ when } s = s_q,$$
$$(s + s_p \tilde{p}_\perp + s_q \tilde{q}_\perp)^2 = r^2 (s_p \cos \beta - s \sin \beta)^2 \text{ when } s = -s_q,$$
$$\sqrt{1 - \left(\frac{1+\tilde{p}_\perp^2 - \tilde{q}_\perp^2}{2\tilde{p}_\perp}\right)^2} = \sqrt{1 - \sin^2 \beta / \cos^2 \beta}.$$

These estimates lead to $\tilde{p}_\parallel = -ss_p r^2 \cos \beta \sin \beta$, $\tilde{q}_\parallel = -1$ when $s = s_q$, and to $\tilde{p}_\parallel = -2s_p r \cos \beta$, $\tilde{q}_\parallel = -1$ when $s = -s_q$. Note that the use of the integrations over $\tilde{p}_\parallel$ and $\tilde{q}_\parallel$ leads to the emergence of the factor $r|\cos \beta|$ in both cases.

For $s = s_q$, we find the following contribution to the kinetic equation:

$$\int_0^R \int_{-\pi/4}^{+\pi/4} \frac{r^2 \cos \beta \sin \beta}{r^2 \cos^2 \beta} \sqrt{1 - \sin^2 \beta / \cos^2 \beta} \, r |\cos \beta|$$
$$(r \cos \beta - (r \cos \beta)^x)(r^2|\cos \beta \sin \beta|^y) dr d\beta,$$

which gives for the problematic integral:

$$\int_0^R \int_{-\pi/4}^{+\pi/4} r^{x+2y+2} \sin \beta (\cos \beta)^{x+y} \sqrt{1 - \sin^2 \beta / \cos^2 \beta} |\sin \beta|^y dr d\beta.$$

The condition for convergence is $x + 2y > -3$, however, because of the symmetry in β the integral is zero. The next order correction leads to the condition $x + 2y > -4$.

For $s = -s_q$, we find the following contribution:

$$\int_0^R \int_{-\pi/4}^{+\pi/4} \frac{r \cos \beta}{r^2 \cos^2 \beta} \sqrt{1 - \sin^2 \beta / \cos^2 \beta} \, r |\cos \beta|$$

$$r^2(s_p\cos\beta - s\sin\beta)^2(r\cos\beta - (r\cos\beta)^x)(r|\cos\beta|^y)rdrd\beta,$$

which gives for the problematic integral (no cancellation from β is expected):

$$\int_0^R r^{x+y+3}dr.$$

We find the condition for convergence $x + y > -4$ which is stronger than for region A.

Region C: $\tilde{p}_\perp = (\tau_1 + \tau_2)/2$, $\tilde{q}_\perp = (\tau_1 - \tau_2)/2$, with $\tau_1 \gg 1$ and $\tau_2 \in [-1, +1]$. At leading order, we have (we use the other delta function):

$$\delta(s + s_p\frac{\tilde{p}_\parallel}{\tilde{p}_\perp} + s_q\frac{\tilde{q}_\parallel}{\tilde{q}_\perp}) = \delta(s - 4\tilde{p}_\parallel\tau_2/\tau_1^2) \text{ when } s_p = s_q,$$

$$\delta(s + s_p\frac{\tilde{p}_\parallel}{\tilde{p}_\perp} + s_q\frac{\tilde{q}_\parallel}{\tilde{q}_\perp}) = \delta(s + 4\tilde{p}_\parallel/\tau_1) \text{ when } s_p = -s_q,$$

$$(s_q\tilde{q}_\perp - s_p\tilde{p}_\perp)^2 = \tau_2^2 \text{ when } s_p = s_q,$$

$$(s_q\tilde{q}_\perp - s_p\tilde{p}_\perp)^2 = \tau_1^2 \text{ when } s_p = -s_q,$$

$$(s + s_p\tilde{p}_\perp + s_q\tilde{q}_\perp)^2 = \tau_1^2 \text{ when } s_p = s_q,$$

$$(s + s_p\tilde{p}_\perp + s_q\tilde{q}_\perp)^2 = \tau_2^2 \text{ when } s_p = -s_q,$$

$$\sqrt{1 - \left(\frac{1+\tilde{p}_\perp^2 - \tilde{q}_\perp^2}{2\tilde{p}_\perp}\right)^2} = \sqrt{1 - \tau_2^2 - 2\tau_2/\tau_1 + 2\tau_2^3/\tau_1} = \sqrt{1 - \tau_2^2}.$$

These estimates lead to $\tilde{p}_\parallel = s\tau_1^2/(4\tau_2)$, $\tilde{q}_\parallel = -\tilde{p}_\parallel$ when $s_p = s_q$, and to $\tilde{p}_\parallel = -s\tau_1/4$, $\tilde{q}_\parallel = \tilde{p}_\parallel$ when $s_p = -s_q$. Note that the use of the integrations over $\tilde{p}_\parallel$ and $\tilde{q}_\parallel$ leads to the emergence of a factor $|\tau_1^2/\tau_2|$ and $|\tau_1|$ for $s_p = s_q$ and $s_p = -s_q$, respectively.

For $s_p = s_q$, we find the following contribution to the kinetic equation:

$$\int_\tau^{+\infty}\int_{-1}^{+1} \frac{\tau_1^2}{\tau_2}\frac{1}{\tau_1^4}\tau_2^2\tau_1^2\sqrt{1-\tau_2^2}\tau_1^x\left|\frac{\tau_1^2}{\tau_2}\right|^y(\tau_1 - \tau_1^x\left|\frac{\tau_1^2}{\tau_2}\right|^y)\frac{\tau_1^2}{|\tau_2|}d\tau_1 d\tau_2,$$

which gives for the problematic integral:

$$\int_\tau^{+\infty}\int_{-1}^{+1} \tau_1^{x+2y+3}\sqrt{1-\tau_2^2}\frac{\tau_2}{|\tau_2|^{y+1}}d\tau_1 d\tau_2.$$

The condition for convergence is $x + 2y < -4$, however, because of the symmetry in τ_2 the integral is zero. The next order correction leads to the condition $x + 2y < -3$.

For $s_p = -s_q$, we find the following contribution:

$$\int_\tau^{+\infty}\int_{-1}^{+1} \tau_1\frac{1}{\tau_1^4}\tau_1^2\tau_2^2\sqrt{1-\tau_2^2}\tau_1^x|\tau_1|^y(\tau_1 - \tau_1^x|\tau_1|^y)|\tau_1|d\tau_1 d\tau_2,$$

which gives for the problematic integral (no cancellation from τ_2 is expected):

$$\int_\tau^{+\infty}\int_{-1}^{+1} \tau_1^{x+y+1}\tau_2^2\sqrt{1-\tau_2^2}d\tau_1 d\tau_2.$$

We find the condition for convergence $x + y < -2$.

In conclusion, the conditions of locality (convergence of integrals) are:

$$\boxed{-4 < x + 2y < -3},$$

$$\boxed{-4 < x + y < -2}.$$

The Kolmogorov–Zakharov spectrum corresponds to $x = -5/2$ and $y = -1/2$ which places this solution exactly in the middle of the intervals of convergence.

References

Burgers, J. M. 1948. A mathematical model illustrating the theory of turbulence. *Adv. Appl. Mech.*, **1**, 171–199.

Galtier, S., and Buchlin, E. 2010. Nonlinear diffusion equations for anisotropic magnetohydrodynamic turbulence with cross-helicity. *Astrophys. J.*, **722**(2), 1977–1983.

Galtier, S., and Nazarenko, S. V. 2017. Turbulence of weak gravitational waves in the early universe. *Phys. Rev. Lett.*, **119**, 221101.

Sreenivasan, K. R. 1995. On the universality of the Kolmogorov constant. *Phys. Fluids*, **7**(11), 2778–2784.

Welter, G. S., Wittwer, A. R., Degrazia, G. A. et al. 2009. Measurements of the Kolmogorov constant from laboratory and geophysical wind data. *Phys. A: Stat. Mech. Appl.*, **388**(18), 3745–3751.

Appendix B

Formulary

Vector Identities

- Multiple products:

$$(\mathbf{A} \times \mathbf{B}) \cdot (\mathbf{C} \times \mathbf{D}) = (\mathbf{A} \cdot \mathbf{C})(\mathbf{B} \cdot \mathbf{D}) - (\mathbf{A} \cdot \mathbf{D})(\mathbf{B} \cdot \mathbf{C})$$
$$\mathbf{A} \cdot (\mathbf{B} \times \mathbf{C}) = \mathbf{B} \cdot (\mathbf{C} \times \mathbf{A}) = \mathbf{C} \cdot (\mathbf{A} \times \mathbf{B})$$
$$\mathbf{A} \times (\mathbf{B} \times \mathbf{C}) = \mathbf{B}(\mathbf{A} \cdot \mathbf{C}) - \mathbf{C}(\mathbf{A} \cdot \mathbf{B})$$

- Rules for products with derivatives:

$$\nabla(fg) = f(\nabla g) + g(\nabla f)$$
$$\nabla(\mathbf{A} \cdot \mathbf{B}) = \mathbf{A} \times (\nabla \times \mathbf{B}) + \mathbf{B} \times (\nabla \times \mathbf{A}) + (\mathbf{A} \cdot \nabla)\mathbf{B} + (\mathbf{B} \cdot \nabla)\mathbf{A}$$
$$\nabla \cdot (f\mathbf{A}) = f(\nabla \cdot \mathbf{A}) + \mathbf{A} \cdot (\nabla f)$$
$$\nabla \cdot (\mathbf{A} \times \mathbf{B}) = \mathbf{B} \cdot (\nabla \times \mathbf{A}) - \mathbf{A} \cdot (\nabla \times \mathbf{B})$$
$$\nabla \times (f\mathbf{A}) = f(\nabla \times \mathbf{A}) - \mathbf{A} \times (\nabla f)$$
$$\nabla \times (\mathbf{A} \times \mathbf{B}) = (\mathbf{B} \cdot \nabla)\mathbf{A} - (\mathbf{A} \cdot \nabla)\mathbf{B} + \mathbf{A}(\nabla \cdot \mathbf{B}) - \mathbf{B}(\nabla \cdot \mathbf{A})$$

- Rules for second-order derivatives:

$$\nabla \cdot (\nabla \times \mathbf{A}) = 0$$
$$\nabla \times (\nabla f) = \mathbf{0}$$
$$\nabla \times (\nabla \times \mathbf{A}) = \nabla(\nabla \cdot \mathbf{A}) - \Delta \mathbf{A}$$

Vectorial Derivatives

We list here the vectorial derivatives in the three usual coordinate systems (see Figure B.1).

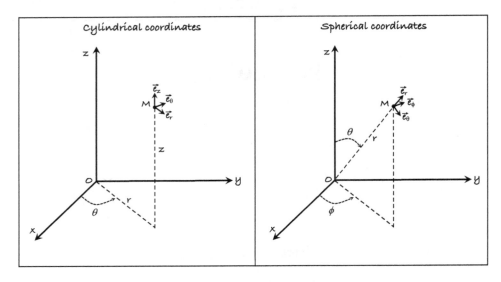

Figure B.1 Systems of cylindrical polar and spherical coordinates.

- Cartesian coordinates:
  $d\ell = dx\,\mathbf{e_x} + dy\,\mathbf{e_y} + dz\,\mathbf{e_z}$
  $dV = dx\,dy\,dz$
  Gradient: $\nabla \Psi = \frac{\partial \Psi}{\partial x}\mathbf{e_x} + \frac{\partial \Psi}{\partial y}\mathbf{e_y} + \frac{\partial \Psi}{\partial z}\mathbf{e_z}$
  Divergence: $\nabla \cdot \mathbf{u} = \frac{\partial u_x}{\partial x} + \frac{\partial u_y}{\partial y} + \frac{\partial u_z}{\partial z}$
  Rotational: $\nabla \times \mathbf{u} = \left(\frac{\partial u_z}{\partial y} - \frac{\partial u_y}{\partial z}\right)\mathbf{e_x} + \left(\frac{\partial u_x}{\partial z} - \frac{\partial u_z}{\partial x}\right)\mathbf{e_y} + \left(\frac{\partial u_y}{\partial x} - \frac{\partial u_x}{\partial y}\right)\mathbf{e_z}$
  Scalar Laplacian: $\Delta \Psi = \frac{\partial^2 \Psi}{\partial x^2} + \frac{\partial^2 \Psi}{\partial y^2} + \frac{\partial^2 \Psi}{\partial z^2}$
  Vectorial case: $\Delta \mathbf{u} = \Delta u_x \mathbf{e_x} + \Delta u_y \mathbf{e_y} + \Delta u_z \mathbf{e_z}$

- Cylindrical coordinates:
  $d\ell = dr\,\mathbf{e_r} + rd\theta\,\mathbf{e_\theta} + dz\,\mathbf{e_z}$
  $dV = r\,dr\,d\theta\,dz$
  Gradient: $\nabla \Psi = \frac{\partial \Psi}{\partial r}\mathbf{e_r} + \frac{1}{r}\frac{\partial \Psi}{\partial \theta}\mathbf{e_\theta} + \frac{\partial \Psi}{\partial z}\mathbf{e_z}$
  Divergence: $\nabla \cdot \mathbf{u} = \frac{1}{r}\frac{\partial}{\partial r}(ru_r) + \frac{1}{r}\frac{\partial u_\theta}{\partial \theta} + \frac{\partial u_z}{\partial z}$
  Rotational: $\nabla \times \mathbf{u} = \left(\frac{1}{r}\frac{\partial u_z}{\partial \theta} - \frac{\partial u_\theta}{\partial z}\right)\mathbf{e_r} + \left(\frac{\partial u_r}{\partial z} - \frac{\partial u_z}{\partial r}\right)\mathbf{e_\theta} + \frac{1}{r}\left(\frac{\partial}{\partial r}(ru_\theta) - \frac{\partial u_r}{\partial \theta}\right)\mathbf{e_z}$
  Scalar Laplacian: $\Delta \Psi = \frac{1}{r}\frac{\partial}{\partial r}\left(r\frac{\partial \Psi}{\partial r}\right) + \frac{1}{r^2}\frac{\partial^2 \Psi}{\partial \theta^2} + \frac{\partial^2 \Psi}{\partial z^2}$
  Vectorial case: $\Delta \mathbf{u} = \left(\Delta u_r - \frac{u_r}{r^2} - \frac{2}{r^2}\frac{\partial u_\theta}{\partial \theta}\right)\mathbf{e_r} + \left(\Delta u_\theta - \frac{u_\theta}{r^2} + \frac{2}{r^2}\frac{\partial u_r}{\partial \theta}\right)\mathbf{e_\theta} + \Delta u_z \mathbf{e_z}$
  $\mathbf{u} \cdot \nabla \mathbf{b} = \left(\mathbf{u} \cdot \nabla b_r - \frac{u_\theta b_\theta}{r}\right)\mathbf{e_r} + \left(\mathbf{u} \cdot \nabla b_\theta + \frac{u_\theta b_r}{r}\right)\mathbf{e_\theta} + (\mathbf{u} \cdot \nabla b_z)\mathbf{e_z}$

- Spherical coordinates:
  $d\ell = dr\,\mathbf{e_r} + rd\theta\,\mathbf{e_\theta} + r\sin\theta\,d\phi\,\mathbf{e_\phi}$
  $dV = r^2 \sin\theta\,dr\,d\theta\,d\phi$

Gradient: $\nabla \Psi = \frac{\partial \Psi}{\partial r} \mathbf{e_r} + \frac{1}{r} \frac{\partial \Psi}{\partial \theta} \mathbf{e_\theta} + \frac{1}{r \sin \theta} \frac{\partial \Psi}{\partial \phi} \mathbf{e_\phi}$

Divergence: $\nabla \cdot \mathbf{u} = \frac{1}{r^2} \frac{\partial (r^2 u_r)}{\partial r} + \frac{1}{r \sin \theta} \frac{\partial (\sin \theta\, u_\theta)}{\partial \theta} + \frac{1}{r \sin \theta} \frac{\partial u_\phi}{\partial \phi}$

Rotational: $\nabla \times \mathbf{u} = \frac{1}{r \sin \theta} \left(\frac{\partial (\sin \theta\, u_\phi)}{\partial \theta} - \frac{\partial u_\theta}{\partial \phi} \right) \mathbf{e_r} + \frac{1}{r} \left(\frac{1}{\sin \theta} \frac{\partial u_r}{\partial \phi} - \frac{\partial (r u_\phi)}{\partial r} \right) \mathbf{e_\theta} + \frac{1}{r} \left(\frac{\partial (r u_\theta)}{\partial r} - \frac{\partial u_r}{\partial \theta} \right) \mathbf{e_\phi}$

Scalar Laplacian: $\Delta \Psi = \frac{1}{r^2} \frac{\partial}{\partial r} \left(r^2 \frac{\partial \Psi}{\partial r} \right) + \frac{1}{r^2 \sin \theta} \frac{\partial}{\partial \theta} \left(\sin \theta \frac{\partial \Psi}{\partial \theta} \right) + \frac{1}{r^2 \sin^2 \theta} \frac{\partial^2 \Psi}{\partial \phi^2}$

Vectorial case: $\Delta \mathbf{u} = \left(\Delta u_r - \frac{2}{r^2} \left(\frac{1}{\sin \theta} \frac{\partial (\sin \theta\, u_\theta)}{\partial \theta} + \frac{1}{\sin \theta} \frac{\partial u_\phi}{\partial \phi} + u_r \right) \right) \mathbf{e_r} +$
$\left(\Delta u_\theta + \frac{1}{r^2} \left(\frac{\partial 2 u_r}{\partial \theta} - \frac{u_\theta}{\sin^2 \theta} - \frac{2 \cos \theta}{\sin^2 \theta} \frac{\partial u_\phi}{\partial \phi} \right) \right) \mathbf{e_\theta} +$
$\left(\Delta u_\phi + \frac{1}{r^2} \left(\frac{2}{\sin \theta} \frac{\partial u_r}{\partial \phi} + \frac{2 \cos \theta}{\sin^2 \theta} \frac{\partial u_\theta}{\partial \phi} - \frac{u_\phi}{\sin^2 \theta} \right) \right)$

$\mathbf{e_\phi} \mathbf{u} \cdot \nabla \mathbf{b} = \left(\mathbf{u} \cdot \nabla b_r - \frac{u_\theta b_\theta + u_\phi b_\phi}{r} \right) \mathbf{e_r} + \left(\mathbf{u} \cdot \nabla b_\theta + \frac{u_\theta b_r - u_\phi b_\phi \cot \theta}{r} \right) \mathbf{e_\theta} + \left(\mathbf{u} \cdot \nabla b_\phi + \frac{u_\phi b_r + u_\phi b_\theta \cot \theta}{r} \right) \mathbf{e_\phi}$

Fundamental Theorems

- Gradient theorem:
$$\int_a^b (\nabla f) \cdot d\boldsymbol{\ell} = f(b) - f(a)$$

- Divergence (or Ostrogradsky) theorem:
$$\iiint (\nabla \cdot \mathbf{A}) d\mathcal{V} = \oiint \mathbf{A} \cdot d\mathcal{S}$$

- Rotational (or Stokes) theorem:
$$\iint (\nabla \times \mathbf{A}) \cdot d\mathcal{S} = \oint \mathbf{A} \cdot d\boldsymbol{\ell}$$

Index

β–model, 49
2D turbulence, 82

acoustic wave, 109
acoustic wave turbulence, 60, 63, 116, 237
Alfvén wave, 179, 182, 209
anisotropy, 157, 165, 182, 188, 212, 217, 246
anomalous dissipation, 33, 36, 40, 75
anomalous exponent, 47, 253
anomalous spectrum, 226
anticyclone, 159
astrophysics, 57, 179
asymptotic closure, 81, 121
autocorrelation, 17

Batchelor, 7, 74
Benney, 106
Bernoulli equation, 127
Big Bang, 232
Burgers' equation, 45, 99, 249
Burgers' spectrum, 59
Burgers' turbulence, 53
butterfly effect, 14

canonical variable, 132, 190, 220
capillary wave, 104, 107, 108, 127, 129
capillary wave turbulence, 130, 139
cascade, 4, 77
chaos, 12
closure, 74, 78
complex helicity decomposition, 161
compressible phenomenology, 63
compressible turbulence, 57, 58, 60, 206
compression, 62

Coriolis force, 155
correlation, 4
cosmological inflation, 232, 242
critical balance phenomenology, 182, 183, 195, 209, 218, 242
cumulant, 19, 78, 119, 138
cyclone, 159

damping rate, 79
detailed conservation, 73, 82, 86, 100, 106, 140, 223, 255
deterministic, 14
DIA, 7, 73, 81
dilatation, 62
direct cascade, 30, 36, 49, 147, 260, 265
direct numerical simulation, 10, 149, 197
discretization effect, 121
dual cascade, 8, 84, 92, 93, 240
Duchon & Robert's law, 44
Duffing equation, 111

eddy, 15
eddy turbulence, 12, 69, 78
eddy turnover time, 39
EDQNM, 7, 79, 161
Einstein equations, 232
elastic wave, 110, 243
electron MHD, 213
Elsässer variable, 190
energy cascade, 4
energy conservation, 31
energy flux, 74, 89, 145, 169
energy spectrum, 5, 70, 74
energy transfer, 30, 121
ensemble average, 17, 31, 70, 119, 136
enstrophy, 8, 31, 84, 89
enstrophy flux, 90

ergotic, 17
Euler equation, 32
exact law, 61, 212

Fjørtoft phenomenology, 84
four-fifths law, 5, 38, 46
four-thirds law, 38
four-wave interaction, 105, 207, 232, 240, 247, 261
fractal, 6, 46, 49
fractal law, 52

general relativity, 234
gravitational wave, 111, 116, 232, 237
gravitational wave turbulence, 247, 262
gravito-capillary wave, 147
gravity wave, 104, 108, 127, 129

Hölder's condition, 45
Hall effect, 207
Harrison–Zeldovich spectrum, 242
Hasselmann, 105
heating, 27
homogeneity, 19
homogeneous turbulence, 7, 33, 34, 70, 119, 138, 223

inertial dissipation, 43, 99, 249, 252
inertial range, 35, 62, 75, 77, 186
inertial wave, 108, 156, 162
inertial wave turbulence, 121, 155, 248, 265
intermittency, 6, 18, 46, 96, 157, 175, 197, 209, 227
internal energy, 60
internal gravity wave, 108
interstellar turbulence, 57, 58
inverse cascade, 8, 10, 84, 85, 239, 265
ion inertial length, 214
Iroshnikov–Kraichnan spectrum, 187
irreversibility, 139
isothermal, 60
isotropic turbulence, 38, 71, 74
isotropy, 20

Joule, 27

Kármán–Howarth (weak formulation), 44
Kármán–Howarth equation, 35
Kelvin wave, 110
kinetic Alfvén wave, 207, 217
kinetic equation, 107, 121, 139, 165, 191, 223, 238

Kolmogorov, 5, 9, 53
Kolmogorov constant, 76, 132, 147, 194, 254
Kolmogorov phenomenology, 39, 76
Kolmogorov spectrum, 76, 94
Kolmogorov–Zakharov spectrum, 9, 144, 166, 170, 223, 239, 246
Kraichnan, 7, 73

Leray, 41
local interaction, 73, 145, 167, 226
locality, 35
log-normal model, 53, 54
log-Poisson model, 55, 56
Lorenz, 12

magnetic Reynolds number, 181
magnetohydrodynamics, 8, 179, 209
magnetostrophic wave, 109
mean rate of energy dissipation, 31, 45, 76, 99, 249
mean rate of energy transfer, 39, 130, 163, 186, 210
mirror symmetry, 20
moment, 19, 78, 119
multifractal law, 53
multiple scale method, 106, 113, 117

natural closure, 106, 137
Navier–Stokes equations, 27, 104, 156
non-Gaussian wings, 48
nonlinear diffusion, 93, 169, 197, 224, 240, 246, 259
nonlinear time, 39
nonlocal interaction, 73
nonstationary solution, 95, 170

Obukhov, 5, 53
Onsager's conjecture, 42
optical wave, 110

phenomenology, 39, 40, 63, 130, 162, 185, 217
plasma physics, 206
plasma wave, 109
primordial Universe, 232
probability density function, 18, 157
projection operator, 72

QN, 78
quantum turbulence, 110
quasi-resonance, 121
quasi-resonant interaction, 175

Index

random phase approximation, 107
resonance condition, 105, 107, 116, 140, 164, 169, 190, 222
resonant interaction, 105
Reynolds, 2
Reynolds number, 2, 16, 27, 45, 181
Richardson, 4
Riemann–Lebesgue lemma, 119, 139
Rossby number, 156, 157, 202
Rossby wave, 109
rotation, 155

scale invariance, 20
sea, 127
secular term, 106, 112, 135
self-similar hypothesis, 46
self-similar solution, 170
self-similarity, 175
shell model, 96
shock, 59
singularity, 40
six-wave interaction, 110
slow mode, 159, 163, 175, 190, 195, 242
solar wind, 185, 188, 207, 213, 217
sonic scale, 63
space-time, 235
spatiotemporal memory, 18, 48
spectral analysis, 4
spectral bounce, 228
spectral correlator, 136
spectral theory, 7
spontaneous symmetry breaking, 33
stationarity, 20
stationary turbulence, 37, 75
stream function, 86
structure function, 19, 46, 61, 99, 157, 197, 209, 253, 254
Sun, 180, 198

supersonic, 57, 63
surface wave, 104, 105, 108
symmetry, 19

Taylor, 30, 36
Taylor hypothesis, 208
temporal memory, 81
thermodynamic spectrum, 144, 170, 193, 239
three-wave interaction, 140, 219
tokamak, 190
transfer function, 74, 167
transfer time, 122, 188, 240
triadic interaction, 72, 87, 107, 116, 185
turbulence, 2
turbulent diffusion, 4
turbulent Mach number, 60

uniform expansion, 112, 114, 119
universality, 35
unpredictability, 14

viscous dissipation, 44
vorticity filament, 157
vorticity tube, 36

wave action, 9, 107, 239
wave breaking, 132
wave packet, 122
wave turbulence, 12
weak solution, 41
whistler wave, 217, 218
whitecaps, 132
wind tunnel, 4, 207

Zakharov, 9, 107
Zakharov transformation, 12, 87, 107, 141, 166, 223, 247, 263
zeroth law, 32